Minority Report

Collective Studies in Knowledge and Society

Series Editor: James H. Collier, Associate Professor of Science and Technology in Society, Virginia Tech

This is an interdisciplinary series published in collaboration with the Social Epistemology Review and Reply Collective. It addresses questions arising from understanding knowledge as constituted by, and constitutive of, existing, dynamic, and governable social relations.

The Future of Social Epistemology: A Collective Vision
Edited by James H. Collier

Social Epistemology and Technology: Toward Public Self-Awareness Regarding Technological Mediation
Edited by Frank Scalambrino

Socrates Tenured: The Institutions of 21st-Century Philosophy
Adam Briggle and Robert Frodeman

Social Epistemology and Epistemic Agency
Edited by Patrick J. Reider

Democratic Problem-Solving: Dialogues in Social Epistemology
Justin Cruickshank and Raphael Sassower

The Kuhnian Image of Science: Time for a Decisive Transformation?
Edited by Moti Mizrahi

Taking Conspiracy Theories Seriously
Edited by M. R. X. Dentith

Overcoming Epistemic Injustice: Social and Psychological Perspectives
Edited by Benjamin R. Sherman and Stacey Goguen

Heraclitus Redux: Technological Infrastructures and Scientific Change
Joseph C. Pitt

Epistemic Paternalism: Conceptions, Justifications and Implications
Edited by Amiel Bernal and Guy Axtell

Philosophy in the Age of Science? Inquiries into Philosophical Progress, Method, and Societal Relevance
Edited by Wouter Floria Kalf, Michael Klenk, Jeroen Hopster, and Julia Hermann

Minority Report: Dissent and Diversity in Science
William T. Lynch

Minority Report

Dissent and Diversity in Science

William T. Lynch

ROWMAN & LITTLEFIELD
Lanham • Boulder • New York • London

Published by Rowman & Littlefield
An imprint of The Rowman & Littlefield Publishing Group, Inc.
4501 Forbes Boulevard, Suite 200, Lanham, Maryland 20706
www.rowman.com

6 Tinworth Street, London SE11 5AL, United Kingdom

British Library Cataloguing in Publication Data

Library of Congress Cataloging-in-Publication Data

Names: Lynch, William, 1965- author.
Title: Minority report : dissent and diversity in science / William T. Lynch.
Description: London ; New York : Rowman & Littlefield Publishing Group, [2020] | Series: Collective studies in knowledge and society | Includes bibliographical references and index.
Identifiers: LCCN 2020025685 (print) | LCCN 2020025686 (ebook) | ISBN 9781786612373 (cloth) | ISBN 9781786612380 (epub)
Subjects: LCSH: Science—Philosophy. | Science—Social aspects.
Classification: LCC Q175 .L96 2020 (print) | LCC Q175 (ebook) | DDC 500—dc23
LC record available at https://lccn.loc.gov/2020025685
LC ebook record available at https://lccn.loc.gov/2020025686

In loving memory of Elfriede Lynch

Contents

Acknowledgments

I am very grateful for my education in Science and Technology Studies and the willingness of my teachers and fellow students to allow and encourage interdisciplinary and integrative exploration. I began with a minor in Science and Technology Studies at Rensselaer Polytechnic Institute before moving on to graduate study at Virginia Polytechnic Institute and State University and at Cornell University. A postdoc at the Max Planck Institute for the History of Science offered further insights and opportunities.

My argument in this book brings together revisionist history of the philosophy of science, explicitly interdisciplinary approaches in Science and Technology Studies, old school history and philosophy of science, social epistemology, evolutionary epistemology, and social theory, among other fields. Because I have tried to address the relationship between approaches that rarely communicate (or systematically miscommunicate) within the broader fields that study science, I have included footnotes, rather than endnotes, to document and contextualize (or recontextualize) my claims and to offer readers a place to follow up on them to see if they agree with my interpretation. The short title version of the citations should help the reader see what sources I am relying upon without thumbing to the back of the book or switching to the endnotes of an e-book every time I make a controversial or unfamiliar claim. In many cases, the footnotes offer different points of view and more complexity and nuance, in good Lakatosian form.

In 1998, I accepted a position in the Interdisciplinary Studies program at Wayne State University, as part of a unit on science and technology that carried forward something of the James Conant program for educating non-science majors, in this case predominantly nontraditional students, through case studies in the history of science. I am very grateful to my colleagues in this unit, who provided models for integrating scientific and historical interests:

Marsha Richmond, Cliff Maier, Moti Nissani, and David Bowen. My understanding of specific themes developed in this book has also benefited from Interdisciplinary Studies colleagues Julie Klein, Stuart Henry, Lisa Maruca, Fran Shor, and Ron Aronson.

In the infinite wisdom of the powers that be, the Interdisciplinary Studies program was dissolved; and Marsha, Fran, Ron, and I joined the History Department. I am very grateful for the collegial environment and support I have experienced. My growth as a historian has been made possible by my new colleagues, among whom mention should be made of two scholars with expertise in history of science and technology, Eric Ash and Jennifer Hart. I am grateful to the department chairs during my time in the department—Marc Kruman and Liz Faue—who have supported me and never once questioned if what I was doing was really history.

Early versions of the argument in this book were presented at Virginia Tech, Cornell University, Wayne State University, and at the Society for Social Studies of Science annual meeting in Paris. Chapter 4 is a significantly revised version of a paper published in *Philosophy of the Social Sciences*, modified to include discussion of Paul Feyerabend's criticism of Wittgenstein and a more developed discussion of implications of the argument for the field of Science and Technology Studies.

Chapter 6 is a revised version of two articles, one in *Science and Engineering Ethics* and one in *Technology's Stories*, the Society for the History of Technology's online magazine. The impetus for my work on engineering ethics came from a fortuitous collaboration with Ron Kline during the two years following the completion of my dissertation before finding academic employment. Early versions of the argument in chapter 6 were presented at Wayne State University, Cornell University, and the Cleveland meeting of the Society for Social Studies of Science. I am grateful for comments by Ron Kline, Park Doing, Gary Downey, Tracy Spaight, Jack Brown, and Suzanne Moon.

I was involved in a short exchange about Lakatos's Hungarian background in the vibrant science studies journal associated with the Russian Academy of Scholars, *Epistemology & Philosophy of Science*. I am grateful for permission to republish portions of my article and my reply to the peer exchange in chapters 2 and 7. I am grateful to the respondents, Vladimir N. Porus, Ilya T. Kasavin, Lada V. Shipovalova, and Svetlana V. Shibarshina. I am also very grateful for many discussions with Eddie Benjamin about Marxism, politics, and history. I miss you and I salute you.

I am grateful for the support of Jim Collier in bringing this book to fruition and for his vision of social epistemology and publicly engaged scholarship. I have benefited from comments by three reviewers who have offered distinct but complementary advice. I have done my best to take it. I am grateful for the editorial help of Isobel Cowper-Coles, Scarlet Furness, Rebecca Anastasi, Dhara Snowden, and Janice Braunstein.

Finally, I would like to thank my family for their support, encouragement, and patience. Stacy and Liam have been there supporting me throughout the process of researching and writing this book. They have brought me joy and love, and I love them with all my heart. I am grateful for the support of my parents, Bill and Elfriede, my sister Elfriede, my brothers Dan and Ed, and their families. While I was in the process of completing this book, my mom passed away unexpectedly in her sleep. She was a remarkable person, and we all miss her greatly. This book is dedicated to her.

Chapter One

Minority Report

Research Programs and the Debate about Climate Change

In the 1956 Philip K. Dick short story "The Minority Report," adapted into a movie starring Tom Cruise in 2002, mutant human beings capable of seeing future crimes are relied upon by a "precrime" enforcement agency. Three "precogs" are strapped down, with their prophetic ramblings recorded and analyzed by computer to identify and arrest individuals about to commit murders before they have done so. Typically, the predictions of each precog differ and the computer synthesizes a majority report based on the predictions of the two precogs whose predictions have the greatest amount of overlap. When the commissioner of precrime, John A. Anderton, is indicted for a future murder, he suspects a conspiracy and seeks out the "minority report," detailing the suppressed testimony of the third precog.

Science works a lot like this science fiction story. Contrary to the view that scientists in a field all share the same "paradigm," as Thomas Kuhn famously argued in *The Structure of Scientific Revolutions* in 1962, scientific experts support different, and competing, research programs and come to different assessments of how things work. Moreover, statements of scientific consensus need to be actively synthesized out of the work of different scientists, and not all work by scientists will be equally credited by science as a whole. While this system works well enough for most purposes, it is possible for minority views to fail to get the hearing that they deserve, especially in retrospect, as scientific paradigms or research programs rise and fall over time, depending upon their changing credibility and support within science. It is even possible that statements of consensus can result from the interests of a few gaming the system rather than reflecting the considered judgment of the social institution of science.

In the normal course of events, most people are not the victims of an elaborate conspiracy designed to mislead, despite the paranoid fantasies fueled by the omnipresence of minority reports online and the cultivation of a politics of suspicion that targets ominous authorities with thwarting access to the truth. It is probably too much to believe that elaborate scientific and technological conspiracies are responsible for hoodwinking the public and stealing their money because majority reports state that we landed on the moon, we evolved by natural selection, humans have caused climate change, pharmaceutical medicines treat disease, or the collapse of the Twin Towers resulted from hijackers crashing planes. Stories like "The Minority Report" entertain us because they show how far afield knowledge claims can go and how corruption can erode even the most carefully cultivated system for producing knowledge. It is an interesting sociological question (to which we will return) to consider why so many people today feel like John A. Anderton, heroically uncovering conspiracies designed to support the denial of the plain truth available to those courageous enough to look for themselves or who find alternative expert informants.

While trying to avoid wild conspiracy theories, is it enough just to accept that the statements of mainstream science are authoritative? Would this apply to all disciplines and how exactly does one identify the state of consensus within science in the first place? When should minority views be given a hearing and, perhaps, the means to investigate further? Should we look for ways to encourage greater diversity in science and technology, or should we take it for granted that the social system of science has evolved to best winnow truth from error? How should an understanding of the necessary role of social support and patronage for science and technology inform a critical assessment of our ability to uncover truth and to promote invention? In short, how much should we tinker with the social institutions of science and technology to better advance our knowledge and understanding of the world?

In this book, I look to answer some of these questions by reconsidering the theories and narratives put forward by scholars writing about the history, philosophy, and social study of science and technology. I take seriously the findings of this literature, while allowing myself the freedom to develop alternative interpretations and conclusions about how knowledge works. In the end, I suggest that it is possible to develop an approach to evaluating science and technology that allows for the careful, rather than indiscriminate, cultivation of past and present minority views, bringing them into dialogue with dominant perspectives or giving them space to develop an alternative approach.

SCIENCE AND TECHNOLOGY STUDIES AND THE CRITIQUE OF EXPERTS

When I began, in the late 1980s, to study the interdisciplinary field of Science and Technology Studies (STS or sometimes S&TS), which subjects science itself to critical scrutiny based on the history of its social organization, it seemed to me that the field offered a critical counterweight to the excessive trust in experts in modern society. Study of the complex historical and sociological processes leading to the construction of allegedly pure knowledge seemed a worthy correction to blind faith in a human institution subject to some of the same foibles as any other human institution. Moreover, examining the relationship of science to power, economic and political, and the role of orthodoxy and prejudice among its practitioners allowed us to consider how our own sense of the options available to us to arrange our lives is shaped by experts.

The concern that the shaping of inquiry within science and technology had been impacted negatively (or, at least, less democratically) than it might have been by capitalism or militaristic governments shaped generations of critics, from the left-wing British scientists borrowing from the cryptically unorthodox externalist Marxism of Boris Hessen, victim of Stalin's purges, to early proponents of "Science for the People," to programs in "Science, Technology, and Society," all in one way or another calling science to account for too often reinforcing power rather than being a beacon for truth.[1]

If the concern with capitalism or the military-industrial complex did not grab everyone's attention, a concern that the image of science as an open and objective institution did not fit the practice of science garnered the attention of others. The attention to groupthink begun by Ludwig Fleck and Thomas Kuhn led to a steady dismantling of the idea that science earned its authority through an automatic and objective process of winnowing truth from error. Whether the critique was political or epistemological, the urgent message that many of us wished to convey to the world is that skepticism of expert knowledge was needed.

Meanwhile, the world changed beneath our feet. While we were teaching our students that experts did not always have to be believed without question, the explosion of the World Wide Web and the proliferation of political and institutional skepticism about experts of all sorts took place in a way that seemed to encourage a general climate of suspicion about motives, a move that revived and intensified earlier movements that helped motivate the initial

1. Werskey, *The Visible College*; Graham, "The Socio-Political Roots of Boris Hessen"; Moore, *Disrupting Science*, ch. 6.

study of the sociology of knowledge itself.[2] Now the concern was that unregulated skepticism eroded all intellectual authority and leveled all distinctions. The interrogation of motives replaced the search for truth and consensus formation in science was seen as a conspiracy by one's political enemies.

As a result of this generalized skepticism and the technological means for its dissemination, we are confronted with numerous challenges to established intellectual authority in today's world. What most scientists say is a very well-supported theory, the view that attributes climate warming to human activities, is rejected in numerous quarters as nothing more than a make-work project for scientists and a liberal or socialist conspiracy to crash the economy. Others reject the theory of evolution by natural selection, seen by biologists as the linchpin of all scientific explanations of living forms, for a literal interpretation of Genesis or a less-specific account of an intelligent designer.

While both of these debates will be seen by many readers as an unfortunate case where ideology trumps science, where politics, financial interests, or religious beliefs interfere with rational evaluation of the evidence, they are only the tip of the iceberg when it comes to questioning scientific consensus. Criticisms of mainstream science are by no means limited to corporate polluters, right-wing politicians, and religious zealots. Many people challenge or criticize established views about medical and dietary issues, from the proponents of low-carb diets to parents who blame vaccines for their child's autism. Critics blame Big Pharma for distorting scientific research to hook us all on medicines to treat our physical and mental ailments.

Moreover, it is clear from a close examination of scientific controversies that science by no means operates under conditions of complete agreement about the current state of evidence in any field. While Thomas Kuhn's famous theory of scientific paradigms rightly called attention to the persistence of anomalies in mature science (things that could not be explained by the accepted theory), he tried to argue that dominant paradigms ended up displacing alternative approaches to scientific questions and that this suppression of minority opinion was actually a sign of scientific maturity. Yet every field is torn by methodological and theoretical controversies for those that look closely, as scholars in the field of Science and Technology Studies have found as they have sought to apply the tools of modern sociological and anthropological criticism to science itself. So how should today's citizen, floating freely among competing claims on the internet, judge scientific controversies when they encounter them? How should we judge what is true and what is false, what is well supported and what is spun out of someone's paranoid imagination?

---

2. Meja and Stehr, *Knowledge and Politics*. The connection between the origins of the sociology of knowledge and a generalized suspicion of motivated reasoning underlying claims to truth is the (polemical) theme of Remmling, *Road to Suspicion*.

The situation is only made more complicated by the fact that many defenders of scientific orthodoxy rely upon the appeal to authority to justify their claims about what we ought rationally to believe. For example, climate change denialists are repudiated by pointing out that the peer-reviewed scientific literature is almost completely devoid of scientific articles that reject the mainstream view that humans are contributing to a global greenhouse effect by their fossil fuel use.[3] Anti-evolutionists are similarly ridiculed by calling attention to the pervasiveness of evolutionary explanation in modern biology, and, increasingly, science across the board.

While pointing out that very intelligent people in an institution geared toward systematic criticism find no cause to question certain well-established views might be a clue that there is something to be said for these theories, a simple appeal to consensus is not very satisfying. At one time, the idea that the earth moves was seen as an absurd view, decisively refuted by Aristotle and opposed by the consensus of the relevant scientific community, scholastic philosophers versed in his arguments. Even though virtually every crank since Galileo has appealed to his example to justify his or her own unorthodox views in a self-serving way, dissident views are, in fact, often rehabilitated in the course of the history of science or implicated in discoveries neglected by dominant approaches.[4]

For critics of mainstream science, such consensus is the result of systematic bias on the part of the individuals who make up the scientific enterprise. Again, while such elaborate conspiracies against the truth might well strike us as absurd, it does, in some ways, just continue the appeal to authority made by defenders of institutional science. Where defenders of orthodoxy count scientists who support the "consensus" view as evidence that the view should be believed, critics of orthodoxy look to examples of scientists who do not share the dominant view. Such appeals are usually accompanied by an account of how the "voice in the wilderness" was suppressed by the institutional power of science. Rarely do critics take up the arguments on the opposing side, so much as they make their own appeal to authority, albeit the authority of those in a position to dissent from the mainstream view and testify to the mechanisms that block their view from a fair hearing.

Why would some have personal immunity from the coercive social pressure within science, and why would that make their own view correct? Should science as an institution avoid social mechanisms that enforce consensus entirely? Could it? Most sociologists that have looked into the social structure of science in the past have done so on the assumption that the institution of

3. Oreskes, "The Scientific Consensus on Climate Change."
4. A start at cataloguing the variety of roles for dissenting scientists can be found in Harman and Dietrich, *Rebels, Mavericks, and Heretics in Biology*.

science is well designed to produce knowledge and that the social organization of science needs to be coercive of individual behavior to be effective at producing knowledge. Kuhn spoke of the function of "dogma" in science, and Polanyi argued that the scientific community "must often refuse publication merely because the conclusion of a paper appears to be unsound in the light of current scientific knowledge" and disregarded if they are published.[5]

To be sure, some of the earlier rules and norms that have been proposed to underlie scientific institutions by structural-functionalist sociologists turn out to be, at the least, incomplete if not distorted in their picture of the operation of science. Robert Merton famously argued that science as a knowledge-producing enterprise was underwritten by social norms that enforced behavior effective for truth-seeking. The "ethos of science" was understood as "that affectively toned complex of values and norms which is held to be binding on the man of science."[6]

Still, even critics of Robert Merton's norms of science—universalism, communism, disinterestedness, and organized skepticism—have not substituted a picture of science in which the behavior of scientists is unregulated by social rules. Still working within the structural-functionalist tradition within sociology, incorporating Merton and Elinor Barber's theory of sociological ambivalence whereby institutions evidence conflicting normative impulses, Ian Mitroff pointed to the important role of "counternorms" in science. Counternorms, in the right context, are fully legitimate parts of the behavior of science: particularism, solitariness or secretiveness, the role of particular interests, and organized dogmatism.[7]

Crucial to our current discussion is the observation that scientists often make important contributions to knowledge not by acting impartially and remaining above the fray, but by forcefully defending a particular view with a particular theoretical agenda in mind. The role of "bad behavior" in advancing good science—albeit often at the cost of failing to recognize the contributions of others—is made clear in James Watson's autobiographical account of the discovery of DNA, where the contributions of Rosalind Franklin's X-ray diffraction data are exploited, but her scientific judgment as a woman is questioned, in the race to discovery and publication.[8]

---

5. Kuhn, "The Function of Dogma in Science"; Polanyi, "The Republic of Science," p. 57. Tellingly, Polanyi uses the example of research into ESP as an example, suggesting that there *is* something unscientific going on in paranormal research. Compare Collins and Pinch, "The Construction of the Paranormal."

6. Merton, "The Normative Structure of Science," pp. 268–69.

7. Mitroff, "Norms and Counter-Norms in a Select Group of the Apollo Moon Scientists," 587–93; Merton, *Sociological Ambivalence, and Other Essays.*

8. Watson, *The Double Helix*, pp. 56, 68–70, 148, 164–67, 209–12; Watson, *Avoid Boring People*, pp. 104–7; Maddox, *Rosalind Franklin*, chs. 10–13. On the erasure of serendipity in the original

An even more radical approach to the social norms underlying the institution of science was introduced by Michael Mulkay, who helped set the stage for new, social constructivist approaches to the sociology of science that have dominated the field in recent decades. Even here, however, no grounds are given for presuming that science operates—or ought to operate—free from social constraint. While Mulkay rejects the idea that the concept of norms makes a good explanation of the behavior of scientists, since both norms and their opposite counternorms can be invoked selectively to explain any kind of behavior whatever, he does argue that the rhetorical form of the scientific journal article is socially enforced and does account for the *appearance* that science follows Merton's norms, an appearance that Mulkay calls an *ideology*.[9]

In short, the view that science has an effective social organization for enforcing individual behavior in the laboratory is rejected for the view that it leaves such behavior alone but instead regulates the production of written journal articles. Science is not unregulated socially—it is just regulated at a different part of the process of science than previously believed. Social organization is presumed to be inadequate to regulate how scientists conduct research, but sufficient to enforce a particular ideological account of science in official publications. The role of publication is to weed out the imperfect traces of the historical process leading up to a knowledge claim, while leaving behind a cleaned up picture of scientific discovery that is distinctly ideological, providing social justification for supporting science and deflecting potential criticisms of the processes and products of science.

This emphasis on the mechanisms regulating the production of texts would be an important component of the new self-styled "radical" sociology of scientific knowledge emerging in the 1970s and 1980s, most notably in Bruno Latour and Steve Woolgar's *Laboratory Life: The Social Construction of Scientific Facts*.[10] In practice, this new approach shifted the emphasis of the sociology of science away from identifying key, stable institutional supports of modern science to focusing on highly contingent processes underlying the active construction of knowledge claims about the world. The new focus became controversies that are settled through social processes of negotiation and conflict. In style, this new "sociology of scientific knowledge" (or SSK, one of a flurry of new acronyms adopted to describe the field) sought to take science down a peg, by showing it not to be the automatic and transpersonal method for discovering uncontested facts and uncontestable theories, but a messy, human process not different in kind from any other human activity.

---

scientific journal article reporting the discovery of DNA and its restoration in later popular accounts by Watson and Crick, see Merton and Barber, *Travels and Adventures of Serendipity*, pp. 278–84.

9. Mulkay, "Norms and Ideology in Science."

10. Latour and Woolgar, *Laboratory Life*.

In some ways, SSK developed as the theoretical superstucture for the growing distrust of experts in modern society, now pervasive in society, but having one origin among scientists and humanists who taught in programs in Science, Technology, and Society and questioned whether scientists always knew quite what they were doing when they developed weapons for the military in Vietnam, when they prescribed methods from physics as suitable for the social sciences, or when they developed new technologies that led to social dislocation and environmental devastation. While this so-called low-church STS had a different constellation of interests and history from later programs in "high-church" Science and Technology Studies that were more explicitly based upon the sociology of scientific knowledge (SSK), they both shared an interest in challenging the unquestioned claims of experts.[11]

To some extent, this odd conjunction of "STSers" set the stage for later critical interrogations of scientific authority, whether mild or wild. While low-church STSers might tend to come from a left-wing background, and high-church STSers would constantly reiterate their support for the institution of science even while they challenged what they believed to be mythological accounts of how science operated, nothing prevented other interests from using the impartial and symmetrical presentation of both sides to a controversy to other purposes than scholarly myth-busting. This focus on even-handed treatment of scientific controversies was a key methodological tenet of the new sociology and an ironic displacement of Merton's norm of disinterestedness from a norm for science to a norm for Science and Technology Studies itself. Whether borrowing directly from STS research itself, or developing cognate approaches independently, a host of corporate and religious criticisms of orthodox science developed. In corporate America, industries that felt they had suffered economic harm from particular findings of science sought to sow doubt and to treat seemingly settled scientific controversies as deserving to be reopened. The pattern was set by tobacco companies challenging the scientific consensus that tobacco damaged the health of smokers and was extended to promote climate denialism and the like.[12]

Sociologists had already warned that "neutral" and "impartial" dissection of a scientific controversy where one side had become the acknowledged winner would appear to scientists as if it was designed to support the refuted view.[13] Merton's norms were finding no more solid ground in the study of sci-

11. The conceptual distinction between "high-church" and "low-church" STS emerged from a 1991 National Endowment for the Humanities–sponsored Summer Institute titled "Science as Cultural Practice," which took place at Wesleyan University, Connecticut, directed by Steve Fuller and Joseph Rouse. See Fuller and Raman, *Teaching Science and Technology Studies*. The distinction is discussed in Fuller and Collier, *Philosophy, Rhetoric, and the End of Knowledge*, pp. xii–xiii.

12. Oreskes and Conway, *Merchants of Doubt*, ch. 1.

13. Scott, Richards, and Martin, "Captives of Controversy," pp. 476–77, 489–91.

ence than in science itself. The field's penchant for rehabilitating underdogs led STS into a cultural war about the authority of science itself. In effect, the homeopathic dose of skepticism that SSK introduced to discussions about the intellectual authority of science gave way to a cultural free-for-all where all claims by scientists are potentially suspect not only as to their truth value but in terms of their basic honesty of motive.

Senator James Inhofe took this paranoid view to the Senate floor when he proclaimed, "With all of the hysteria, all of the fear, all of the phony science, could it be that man-made global warming is the greatest hoax ever perpetrated on the American people? It sure sounds like it."[14] The "growth of suspicion" that haunted the twentieth century and fueled the initial development of the sociology of knowledge—which, for the most part, initially exempted science from criticism—exploded into a situation where the interrogation of motives seems to be the starting point for analyzing controversial scientific claims.

At the same time, the awareness of the existence of counternorms seemed to fuel *perspectivism* among critics of science themselves. Perspectivism is the view that critics have the right to adopt alternative points of view to established paradigms or hard cores of research programs, which can be used to frame and interpret scientific evidence in a different way. The idea is that we all have our biases and perspectives, even within science, so perhaps any perspective, by "right," ought to be able to get into the game of generating and evaluating scientific hypotheses. Again, a Pandora's box was opened, as STS scholars had argued that androcentric hypotheses had dominated science and ought to be checked by the explicit introduction of feminist alternatives.[15] Fundamentalist Christians or oil companies could surely play this game, too.

I first became aware of the possibility for perspectivist appropriations of work in STS at the 1993 Society for the Social Studies of Science annual meeting when another conference participant introduced himself to me as working for the "STS wing of the Hare Krishnas." His name was Michael Cremo and he had written a book, *Forbidden Archaeology*, that had sought to apply the assumptions of Sanskrit religious texts, the Vedas, to archaeological questions, in effect developing the opposite critique of mainstream archaeology than that put forward by young earth creationists, though curiously praised by them on the sleeve of the book. Rather than assuming that human history (and the rest of creation) was much younger than believed, on the order of 6,000 years, Cremo followed the Vedic belief that a single "day" was 4.3 billion years and that our species had been around for much longer than argued by modern science, for hundreds of millions of years rather than the accepted view that Homo sapiens date to about 100,000 years or so, with

---

14. Inhofe, "Science of Climate Change."
15. Longino, *Science as Social Knowledge*, pp. 3–13.

the evolutionary path leading to us diverging from the ape line only about seven million years ago. He tells us that his

> research into humanity's hidden history was inspired by [his] study of the ancient Sanskrit writings of India, collectively known as the Vedas. Among these Vedic writings are the Puranas, or histories, which tell of human civilizations existing on this planet for tens of millions, even hundreds of millions of years.[16]

How does one begin to establish credibility for such a viewpoint that is so out of line with what scientists believe?

First, Cremo points out that modern Darwinians were once in the same position as Cremo, endorsing a view rejected by virtually all the experts, showing that consensus does not entail truth and that science has changed its position before and may do so again. After this bit of space-clearing, he then suggests that someone examining the evidence beginning from a standpoint as a member of the Krishna movement would see and interpret the existing evidence differently, seeing anomalous evidence as suggesting an entirely different conclusion, just as Thomas Kuhn had suggested occurs during scientific revolutions such as the shift to Copernican astronomy or to quantum physics.

What could be seen as a liability, Cremo's religious "bias" in conducting research on archaeological topics, is converted into a reason to believe he could see what others could not, precisely because his prior interpretive perspective allowed him to see anomalies for the mainstream view as the basis for a new interpretation of the evidence. For Cremo, a "Vedic archaeologist" would predict that "scientists digging into the earth should find signs of a human presence going back hundreds of millions of years," which he attempts to document in the book, and that "this evidence will largely be ignored because it radically contradicts the ideas of human origins currently held by the scientific community."[17]

Crucially, this rhetorical move—what we can call a *perspectivist* appeal—could only be carried out if one already accepts that interpretation is indispensable to observation, that is, if one has already rejected a positivist or empiricist view of scientific observation for one that endorses the "theory-laden" character of observation. Cremo employs much the same idea in referring to a "knowledge filter" in the operation of science that lets dominant ideas pass through the scientific literature with less scrutiny than ideas that challenge accepted views. Ideas radically incompatible with the

---

16. Cremo, "Forbidden Archaeology." The argument is laid out in Cremo and Thompson, *The Hidden History of the Human Race*, and in a longer, 952-page version titled *Forbidden Archaeology*. For a review, see Wodak and Oldroyd, "'Vedic Creationism.'"

17. Cremo, "Forbidden Archaeology," p. 391.

dominant theory are rejected, to be "forgotten, set aside, or, in some cases, actively suppressed."

In addition to highlighting the role of perspective in interpreting or filtering observation, Cremo also mines the history of archaeology for observations that are anomalous from the standpoint of the dominant perspective. Either the alleged fact was ignored, explained away, or rejected by the dominant perspective and Cremo's contribution is understood to be looking at the anomaly from a different perspective that is not biased toward rejecting the uncomfortable facts. For instance, historical reports of human remains found in geological strata as far older than the accepted age of Homo sapiens are "uncovered" by Cremo, reports of which are ignored or dismissed by modern scientists as the result of "ad hoc" explanations of how recent remains ended up in older strata or dismissal of unreliable observations made by amateurs before the advent of sophisticated archaeological excavation procedures. In effect, Cremo facilitates his critique by a historical excavation of his own, unearthing a variety of anomalies that can be recast as the basis for a new dominant theory, what we might call *anomaly mongering*.

Anomaly mongering can be defined as a strategy of focusing upon what a dominant research program *fails* to explain in order to provide plausibility for an alternative research program. As a strategy for introducing or rehabilitating a minority approach to knowledge, it suffers from the fact that *all* research programs are riddled with anomalies, with much of the routine work of science defined by using the resources of the dominant approach to develop successful explanations of some of the existing anomalies, while ignoring more intractable anomalies on the assumption that they will be explained sometime in the future. While in one sense, the mere existence of anomalies cannot be the basis for rejecting a scientific theory, since a later refinement of that theory within a larger trajectory of a research program may produce a solution, critics of dominant research programs are able to appeal to anomalies in the hope of effecting a Kuhnian scientific revolution, introducing a new research program where previously neglected anomalies can be explained more elegantly by a new approach that starts from scratch.[18]

What historians of science observe as a factual occurrence gets reframed as a right. Cremo exploits the recognition of the existence of anomalies in all sciences (as argued by Kuhn and other scholars of science) with a suggestion

18. For Lakatos's argument that negative heuristics protect the hard core of research programs against falsification, see Lakatos, "Falsification and the Methodology of Scientific Research Programs." At the University of Vienna in the 1940s, before turning to philosophy, Feyerabend studied with the physicist Felix Ehrenhaft, who lost the dispute with Robert Millikan over the interpretation of the oil drop experiments leading to knowledge of the unit charge of an electron. Oberheim, *Feyerabend's Philosophy*, p. 121, concludes that this experience shaped his later philosophy of science because it showed that "an unexplainable experimental effect by itself lacks the epistemological power to challenge a well-established fundamental theory."

that fairness requires that competing perspectives, with their different filters and attitudes toward existing anomalies that help convert facts to evidence for specific theories (in line with the Duhem-Quine thesis), receive a hearing that the dogmatic social organization of modern science disallows. Thus, this anomaly mongering is coupled with the recognition of socially enforced perspectivism to convert an empiricist liability (distorting evidence by reference to a preferred outside "bias") to a post-positivist virtue (using a counterhegemonic standpoint to call attention to neglected facts that the conceptually conservative institution of science ignores or actively suppresses). One might even speak of the need for "affirmative action" programs for neglected ideas in science, as has been suggested in the case of intelligent design-based challenges to evolutionary science.[19] New approaches to philosophy and sociology of science had inadvertently provided rhetorical ammunition to proponents of extreme conspiracy theories about science.

That being said, the new sociology of science can hardly be blamed for every misuse of its findings that uses them to paint a distorted picture of science crushing minority views based upon inadequate evidence. At the same time, it can be asked what the sociology of science suggests about how we might evaluate the legitimacy of scientific theories, mainstream and contrarian. Are they all on the same plane or is there good reason to accept that mainstream science has most of it right compared to its critics? What remains unclear for our purposes is what exactly this new approach to the sociology of science should tell us about how we judge and evaluate science from the outside (and we all judge most science from the outside since even scientists are not specialists in all areas). After all, there are a lot of questions we could ask about the shift to the study of scientific controversies.

Two contrary, but not quite contradictory, tendencies underlay the new focus on scientific controversies: *skepticism* and *cognitive authoritarianism*. Scientific conclusions were to be seen as provisional, established and shaped by culture, rather than given once and for all by reason or nature itself. At the same time, the question of whom to believe about scientific affairs was to be settled by deferring to the judgment of the expert scientists themselves. If the institution of science was satisfied that an issue was settled, no amount of critical history of the questionable steps leading to that consensus by the sociologist was considered sufficient to challenge this belief. Simultaneously, science was less solid in its conclusions than previously believed, but this seemed to leave no other basis for scientific authority than simple deference to those with the appropriate training and expertise.

This combination of a "radical" interpretation of science with a deferential approach to scientific expertise has its roots in the work of Michael Polanyi

19. Kitzmiller v. Dover 2005, pp. 70–71; Fuller, *Dissent over Descent*, pp. 144–45.

and Thomas Kuhn and was immediately criticized by philosophers coming out of the school of Karl Popper, most notably Imre Lakatos and Paul Feyerabend, who emphasized that the progress of science depends upon the willingness to support criticism of well-established points of view. I will consider these criticisms in chapter 2. At that point, building on their views, I will develop an approach to evaluating the diversity of approaches within any scientific field in chapter 3. For now, let us examine skepticism and cognitive authoritarianism in more detail.

Skepticism reflected the observation that controversies were rarely settled based upon decisive rational criteria that could be agreed upon by all observers. Instead, "social" dimensions of the controversy were held to explain why scientific consensus developed in the first place. This skepticism, based upon what came to be known as the Duhem-Quine thesis, provided a rationale to extend "social" explanation into the core of scientific work itself, even if this tacitly preserved a distinction between "social" causes and "rational" causes that no longer made sense when everything became social.[20] Crucial to the methodological punch of this move was the belief that social causes were contingent and capricious, hence changeable, given a different way of organizing science or a different past history of scientific culture, rather than given by the nature of reality or by the rules of proper reasoning.

The corollary of this skeptical approach was that scientific conclusions could be quite different if the social conditions fostering scientific work were different. It is this emphasis on the idea that things could have been different, even in science itself, that made the new sociology of scientific knowledge seem "radical" rather than any specific political valence of the new approach. Indeed, many proponents of the old "low-church" STS found that there was a move *away* from explicit political criticism, as political criticism was held to be in conflict with the methodological imperative to analyze scientific controversies impartially and symmetrically.[21] That being said, the actual counterfactual implications of this claim that science could have been different were remarkably underexplored, even obscured, for reasons we will explore in chapter 6.

Once a rationalist philosophy of science was rejected as an adequate explanation of the closure of a controversy, sociologists were left to put back together an account of why science ought to be believed, at least provisionally, in providing accounts of how the natural world worked. Here is where cognitive authoritarianism came back into the story, in one way or another. While anti-authoritarians might have looked to "open up" seemingly settled scientific controversies, sociologists were much more likely to see their task as describing

20. Fuller, *Social Epistemology*; Longino, *The Fate of Knowledge*, ch. 4.
21. Winner, "Upon Opening the Black Box," pp. 371–76.

how expertise works without dismantling the authority experts were to be accorded. If the best expertise that was possible was highly contingent on historical and cultural circumstance, then that was all the expertise that could be demanded. It's as if we did, indeed, pay attention to the man behind the curtain but insisted that he ought to be treated as a wizard nonetheless.

This was even more the case since the model was explicitly sociological, and it was admitted that rational models of the growth of scientific knowledge could not decide which view was truly correct. As a corollary, closer examination by the sociologist of the intellectual reasons that supported one side over the other would not work. In short, the sociologist could show how controversies were settled sociologically but had to defer to the scientist for judgment about the truth or rationality of the theories. To see this coupling of skepticism with cognitive authoritarianism, let's examine the development of Harry Collins's sociology of science, from his "empirical program of relativism" (EMPOR) to his recent models of expertise.

FROM METHODOLOGICAL RELATIVISM TO DEFERENCE TO EXPERTS

Accompanying a series of case studies, Collins laid out a programmatic approach to the sociology of scientific knowledge in a series of publications in the 1970s that built on concepts like Polanyi's "tacit knowledge" and Wittgenstein's "form of life" to introduce what amounted to a sociological displacement of the traditional topics of epistemology. The skepticism of Collins's approach reflected his dismantling of epistemological accounts of how scientists settle their disputes about how to represent the natural world and substitution of a purely sociological account based upon the conflict between different groups of scientists endorsing different interpretations of the experimental evidence.

Key to this disciplinary colonialism—replacing philosophy of science with sociology—is the observation, or interpretation thereof, that scientific evidence did not decide the conflict between different theories because each side interpreted the evidence differently, and in ways that favored the continuance of their own theoretical perspectives. If the theory-ladenness of observation was taken to its limit, and the explanation for why a given scientist endorsed a particular theory was related to social dimensions including social position and historical background, then experimental evidence *by itself* was powerless to settle debate over experiments and enforce consensus in science. Taken to its extreme, Collins can be read as suggesting that experiment did not matter.

Belief that experiment settles controversies in science arises because of an illusion created by distance—in time or space—from the point of controversy, as well as from the retelling of why closure occurred after the fact relying upon the mythical explanations of traditional epistemology. Traditional philosophy and the post hoc statements of scientists themselves miss what scientists in the thick of a controversy implicitly understand that institutional and interpersonal conflicts and the mechanisms to shore up one side over the other decide outcomes by exploiting the inherent openness in the scientific interpretation of experimental evidence to support one view over another, equally workable view.

Ironically, the words of scientists themselves in the heat of battle can be used as evidence against their own epistemological accounts of the success of science. The conclusion that social reasons decide scientific controversies is seen by Collins as a straightforward empirical finding of the sociology of science (here interpretation is downplayed to the extent that sociologists are expected to see things the same way if they simply look closely enough at science in practice). As Collins put it at the time, "recent work in the sociology of scientific knowledge suggests that proximity to the seat of creation of scientific facts usually has the effect of creating *uncertainty*," not certainty.[22]

The main empirical finding of SSK for Collins is that "the natural world plays a small or non-existent role in the construction of scientific knowledge."[23] Considered as a methodological injunction, he argues that "[t]he appropriate attitude for conducting this inquiry is to 'assume that the natural world in no way constrains what is believed to be.'"[24] But if science itself does not establish its interpretations of how nature works by use of empirical evidence that definitively establishes one interpretation over another on rational, rather than social grounds, then does that not mean that Collins's interpretation of empirical evidence in the sociology of science is subject to the same considerations?

Collins's answer is to reject the "reflexivity" proposed in the competing strong program in the sociology of scientific knowledge. Instead of calling for the implications of SSK to be applied to itself, to understand its conclusions to be subject to the same kind of interpretive flexibility and social reasons behind closure as found in the natural sciences, Collins argues that such an approach would be disabling and counterproductive, amounting to abandonment of the distinct disciplinary perspective of the sociologist. Here, I am less concerned with whether such a rejection of reflexivity is justified or not than

22. Collins, "Public Experiments and Displays of Virtuosity," p. 725.
23. Collins, "Stages in the Empirical Programme of Relativism," p. 3.
24. Collins, "Son of Seven Sexes," p. 54.

to tease out a corollary of this view as it relates to the cognitive authority of intellectual communities.

For Collins, the appropriate disciplinary attitude of a natural scientist is to proceed *as if* scientific work would lead to definitive conclusions about science and once scientists had decided as a collective which interpretation of the experimental evidence they were going to accept, that was sufficient for settling the issue. Those not engaged in the scientific work itself, including sociologists who study how the conclusions were arrived at, were not to be allowed grounds for challenging those scientific conclusions. Scientists were the experts on their own work and were to be credited with the authority to represent nature following the establishment of scientific consensus.

Sociologists, on the other hand, were to be credited as experts of a different sort, who could demonstrate that the actual reasons for the united front presented by mature science were different from what scientists or philosophers might claim after the fact. Different interpretations of how scientific controversies were settled were not to be credited after sociologists had settled on their explanation, *any more than the scientists' views were to be questioned once they resolved their disputes*. Sociological explanation is not to be the prelude to criticism of existing scientific explanations—sociological explanation "leaves pure science very nearly as it is; in the main it simply redescribes it."[25] Sociologists are just the type of community that gives social explanations of scientific controversies and are to be credited as experts once their own disputes have reached closure.

This emphasis on the acceptance of disciplinary authority on what might be called internal and institutional grounds provides a clue to Collins's seeming prevarication about whether we discover or assume that nature plays no role in scientific reasoning when we act as sociologists.[26] In line with his account of an experimenter's regress, whereby scientific evidence can only have the force it is credited with if social reasons endorse one interpretation of that evidence over competing interpretations, the social interpretation of scientific reasoning is simultaneously an unproblematic empirical finding of the sociologist and a disciplinary convention enabling sociological work to proceed in the first place.

Sociologists, especially sociologists who have read Polanyi and Wittgenstein, are able to engage in an explanatory enterprise where social reasons are invoked to explain scientific consensus. Lacking that background and training, or lacking the nerve to apply it to the core of science, leaves the sociology of science undone and incomplete, as new wave SSKers argued

25. Collins, "An Empirical Relativist Programme," p. 98.

26. See Collin, *Science Studies as Naturalized Philosophy*, pp. 99–100; Pinch, "Relativism: Is It Worth the Candle?," pp. 41–44.

about Mertonian sociology of science and about the sociology of knowledge up to that point.

The key point here is that it is a social convention that sociologists seek out certain types of explanations and that this convention ensures that a sociological explanation of scientific consensus will be produced when sociologists are allowed to carry out their disciplinary work *without the interference of other disciplines, such as philosophy or that of the scientists under study*. While eschewing reflexivity, Collins in fact gives a thoroughgoing and consistent reflexive account, albeit one that is dependent upon *disciplinary autonomy* and implicitly crediting expert communities with the right to their own conclusions about their object of study.

When the science wars exploded, and scientists themselves became critical of the explanations proffered of their actions by social scientists, Collins had a ready-made basis for a response and defense of the field, what amounted to a kind of détente for science studies. Social scientists were to be granted autonomy in giving explanations of the behavior of scientists, but scientists in turn were to be given autonomy in representing nature itself. Scientists were not experts on their own expertise, but experts on the natural objects of their study. Sociologists were experts on their objects of study, the scientists, though not, in turn, on their own actions.

The task of a sociology of the sociology of scientific knowledge should not be carried out by the sociologists of science (or perhaps, by different sociologists of science). Each expert community was to proceed with their usual practice of trying to turn contestable evidence about their objects of study into the unquestionable conclusions of a scientific community. If sociological explanation of either community's actions can be expected to show the sleight of hand involved in this use of evidence, this was not taken to be grounds for criticizing those conclusions.

In other words, *sociological explanation of any intellectual community is not grounds for criticism of its conclusions*. And, as a corollary, *experts are not to be challenged when they are limiting their claims to the proper scope of their expertise*. To be sure, experts often claimed to speak on matters that went beyond their expertise and here it was appropriate to question their claims. For instance, public demonstrations might seek to directly appeal to the wider citizenry rather than letting fellow scientists settle the issue based on real familiarity with the phenomenon at hand.[27]

Scientists might also seek to use "unfair" means of excluding those with a legitimate claim to speak as experts on the same topic, so outsiders could use the findings of the sociology of science to spot when this exclusion would occur. In

27. Collins, "Public Experiments and Displays of Virtuosity"; Collins and Pinch, *Golem at Large*, ch. 3.

short, the "core set" of scientists who are legitimately debating the interpretation of scientific evidence in a field should not be either artificially restricted by excluding experts with competing interpretations or artificially expanded by including those without actual expertise. But "actual" experts are not to be challenged by outsiders, whether this involves non-scientists challenging scientific conclusions, or non-sociologists challenging sociological interpretations of the practice of science. Experts are experts and are only answerable to other experts studying the same thing, not to non-researchers or outside experts.

Collins and Evans have extended this approach to intellectual authority to a full-blown theory of expertise, which sets the stage for a defense of expert knowledge rather than a basis for its critique. They leveled their sights on sociological work that argued for the existence of lay expertise, knowledge held by non-credentialed parties with a stake in the knowledge produced, such as Cumbrian farmers observing the effects of radiation on their flock of sheep or AIDS activists seeking out alternative approaches to the understanding and treatment of a new, devastating disease. They argued that expertise was a non-negotiable concomitant of their experience and training. Whereas a positivist or empiricist could criticize a scientific finding if the evidence did not support the claim, Collins and Evans cut off any basis for outsiders to challenge scientific claims.[28]

A sociological understanding of knowledge communities might have been taken to imply that all knowledge claims are "relative" to the communities who produce them. This is the basis of relativistic theories of knowledge, which are held by critics to take skepticism too far, leaving no basis for explaining why any knowledge claim should be more widely believed. Aren't experts just people who believe whatever they have been trained or brainwashed into believing? Are sociological theories of knowledge too corrosive of intellectual authority? Collins and Evans argue that their program provides the basis for a theory of expertise that distinguishes expert from non-expert judgment by coupling their sociological skepticism with the cognitive authoritarianism we've seen in the work of Polanyi and Kuhn.

Collins and Evans argue that experts are members of *self-warranting communities*, people who by dint of their social position and training are able to judge who is to be believed and who is not. This knowledge of who the fellow experts are is, like knowledge itself, tacit rather than explicit. For Collins, "the acquisition of nearly every expertise, if not all of them, depends on the acquisition of the tacit knowledge pertaining to the expert domain in question. Tacit knowledge can be acquired only by immersion in the society of those who already possess it."[29]

---

28. Collins and Evans, *Rethinking Expertise*. In discussing lay expertise, they are responding to arguments in Wynne, "Sheep Farming after Chernobyl" and Epstein, *Impure Science*.
29. Collins, "Three Dimensions of Expertise," p. 254.

Experts cannot tell you why their list of approved experts are to be believed, just that they are as a result of their deep familiarity with the field. If knowledge is primarily tacit, cognitive authoritarianism follows, as only those who are trained by other experts possess knowledge. Although they possess it, they cannot "show" it publicly—it is just part of their embodied engagement with the phenomena in question. Thus, Collins and Evans's theory of expertise follows from Collins's appropriation of Polanyi's concept of tacit knowledge. Summarizing this aspect of Collins and Evans's account, we can say that whoever knows more than the public cannot prove it, while those who know less than experts cannot judge them. The public (or even experts in different fields) must take experts at their word, since they "possess" knowledge quite literally in their bones.[30] This knowledge can be represented verbally to the public by those in the know, but it cannot be shown to fit some independent basis for judging the validity of the knowledge. There is no independent check on the validity of expert knowledge.

How does one even know who these experts are? Collins and Evans give two answers to this question. First, they identify expertise with experience so outsiders would be well advised to look for scientists with experience in their field. Since Collins and Evans state that the knowledge that scientists possess, being tacit knowledge, is built up from experience over time, we could determine if they have the requisite experience. How much time and exposure to the phenomena under investigation do the experts have? (It is here, that Collins and Evans concede that the Cumbrian sheep farmers and AIDS activists, by dint of their experience, may in fact have expertise, albeit uncredentialed expertise that may lead them to be unfairly excluded from the core group.)

The second answer they give to the question of how to identify experts is to ask the experts. While we might want to try and figure out how much experience putative experts have, in truth, knowing which experts to defer to is itself something only the core experts know. They possess tacit knowledge of who has tacit knowledge, or "unspoken understandings of who is to be trusted among those who work in the esoteric core of the science."[31] The public (and journalists) who bought into the mistaken view that vaccines cause autism were duped by a charismatic scientist who did not deserve credit. Similarly, patients who dig up health information on the internet, or

30. In Collins and Evans, *Rethinking Expertise*, ch. 3, they consider just how much embodiment is necessary to "possess" knowledge, arguing against Dreyfus's position that full embodiment is necessary. Other constructivist sociologists deny that physical embodiment is the primary way that knowledge exists in the material world, seeing knowledge as materially distributed across brains, databases, books, journals, institutional routines, interactional practices, and the like. In this sense, knowledge is not something that is "possessed" by a body but an alienated product of human activity. Such an approach follows Hegel (or Marx) rather than Polanyi. See the discussion in chapter 5 below.

31. Ibid., 20.

even consult primary source medical journals, cannot know who deserves to be read and who ought to be ignored.[32] One of the key things that experts "know" is who to pay attention to in the first place and who to ignore. This is not only a sociological interpretation of science but one where scientific communities are seen as *self-warranting*. Their own activity within a particular community in which tacit knowledge of a certain kind is cultivated is all the justification that they need.

This view meshes very well with the kind of cognitive authoritarianism that Lakatos and Feyerabend saw in the approach of Polanyi and Kuhn. In their view, this perspective put distinct limits on the tradition of critical inquiry allegedly informing science itself. Where experts ignore the work of some in the core group because they tacitly "know" the work is no good, Collins and Evans endorse the validity of this kind of judgment. However, while core group "gossip" of this kind may imply that scientists know who to believe, who is doing good work, it may also be that they are prejudiced by their training in a particular way of viewing the world. Collins and Evans, like Thomas Kuhn before, do not propose that their work "unmasks" the authority of science. Where many readers of Kuhn may have read that paradigms channel inquiry in selected ways as prima facie grounds for questioning whether their conclusions are authoritative, Kuhn saw the limitations of paradigms as self-imposed by scientists, who would also know when to change paradigms.[33] Likewise, Collins and Evans see outsiders as in no position to critique the core knowledge of experts since the form of life that experts inhabit is only truly understood by other participants.

As we will explore more in chapter 5, there is a long tradition of associating the ability of the public to verify that scientific claims are supported by evidence with science's "democratic" or "emancipatory" character. Some of the early versions of this idea had been heavily criticized, such as the verifiability criterion within logical positivism, whereby a statement was only meaningful when the terms for confirming it through observation were specified.

Karl Popper developed a thorough criticism of the inadequacy of this approach for capturing how science developed, substituting a falsificationist methodology that likewise was intended to provide an independent logical check on authority claims within science. The challenge that Thomas Kuhn presented to the Popperian school stemmed from his identification of the presence in good science—not just bad science or pseudoscience—of dogma and resistance to falsifications, now dubbed anomalies. Anomalies could be taken either as puzzles to be solved in the course of time or as threats to the existence of the dominant paradigm.

---

32. Collins, *Are We All Scientific Experts Now?*, ch. 3.
33. Fuller, *Thomas Kuhn*.

There was nothing inherent in the existence of any anomaly that defined how one should handle it—that was left to the judgment of the scientist. Proponents of the dominant paradigm (which would be most scientists most of the time) would, in a sense, avoid seeing anomalies as potential falsifications and instead see them as puzzle pieces that would need to be made to fit with the dominant approach by force, as it were. In Popper's view, this kind of approach reflected the intellectual sins of the Freudian or Marxist, rather than the critical thinking inherent to science that also served as a model for an open society.

To be sure, Popper himself granted that empirical statements in science were "conventions," allowing that social processes of inquiry would guide what came to be seen as a fact. Once a fact was established, however, it was imperative that the facts were used to attempt to disprove one's assumptions rather than reinforce them. Kuhn's significance as a threat to the Popperian project came from extending the "conventional" component of science to the logic of scientific justification and not just the logic of discovery. Such an approach seemed to leave little room for challenging entrenched positions within science except for rare maverick scientists who would remain within what Collins called the core set. Outside critics who felt knowledge claims were questionable would have no recourse to object, at least if the body of knowledge in question met the parameters of Kuhn's mature science.

Crucially for Kuhn, mature science emerged when the existence of multiple perspectives and the debate about fundamentals was abandoned for a dogmatic agreement on fundamentals. Groupthink replaced critical thinking as the basis of science. Those in the Popperian school thought they could see the pernicious effects of Kuhn's perspective when self-styled Kuhnians in sociology or other "immature" sciences sought to impose uniformity on their field based on the new view of science.

Though Lakatos and Feyerabend were both critical of this limitation on outside, critical scrutiny of expertise, they saw much of value in emphasizing the conventional elements of science straight through to include standards of justification as well as discovery. Lakatos's approach can be seen as a way to incorporate as much of the messiness of the history of science as possible within a view that still allowed judgments of the broad trends in scientific history to be made on a rational basis.

Feyerabend became less concerned with this project of specifying the residual basis for rationality in history, but he, too, was critical of the idea that criticism should be excluded from normal science. Both pointed to the continuing existence of multiple, competing perspectives even within mature science, an observation that remains the most compelling counterexample (anomaly?) to Kuhnian history. More important, Lakatos and Feyerabend both feared that acceptance of Kuhn's view would kill the spirit of critical thinking and proliferation of alternatives vital not only to science, but to a functioning democratic society as well.

In chapters 2 and 3, we will explore in greater depth the dialectically linked projects of Lakatos and Feyerabend, and their significance for developing an approach that can build upon the insights of STS while facilitating outside criticism of the processes and products of science. For now, it is enough to show how Lakatos's idea that we can evaluate a research program's *track record* over time at generating novel, confirmed predictions can be used as a way to begin to evaluate the relative strength of scientists' claims about the world, while also using that as a basis for suggesting cases where minority perspectives may deserve more support for promising, but unproven, ideas.

Whereas the approaches of Polanyi, Kuhn, and Collins tell us to leave science alone as an integral, self-regulating enterprise, a view of the significance of STS that borrows from the perspectives of Lakatos and Feyerabend leaves much more room for using the history of science as a launching pad for critical intervention into scientific theories, methods, and means of organizing inquiry. In contrast to cognitive authoritarianism, I argue that scientific communities are not self-warranting but are part of larger historical projects that are always inadequate and incomplete, whether understood in their own terms or in comparison with competing projects. In Lakatos's words, the history of science is "a complex war of attrition between . . . research programmes," one where it is possible to "establish what the relative strengths (i.e., imaginative resources and empirical luck)" of the competing programs were over time.[34]

Understood in these terms, the approach I advocate here can be seen as part of the project of what has been called social epistemology. Evidence is not related to warrant by logic but neither is it given at once by community membership. The justification behind claims to knowledge is an evolving relationship within and between traditions of inquiry and scientific groups, not one that can be analyzed either by logical relationships alone or by appeal to the competence and possession of tacit knowledge by experts. An understanding that scientific knowledge is social and historical in its essence should be seen as grounds for criticism and intervention in science, where insiders and outsiders can by right engage the theories, methods, and practices behind knowledge claims and suggest changes in how science is carried out or supported. Science as it exists is not the best of all possible knowledges, and an understanding of its history and social organization should be used to suggest ways that we can do better.

---

34. Lakatos, "Anomalies versus 'Crucial Experiments,'" p. 212.

The approach I advocate differs from similar proposals that pay less attention to "track records," however. Since inquiry is organized and channeled in ways that develop over time, a first step to evaluating science is to ask where it has been and where it is headed. Steve Fuller's influential approach to social epistemology, as well as some of the approaches to the feminist and political critique of science, while key influences on the approach developed here, takes a different approach to encouraging minority views. I do not develop an approach rehabilitating minority perspectives primarily based upon a conception of their "political" rights to representation, a "multicultural" approach that demands "affirmative action" for any and all contenders who feel alienated from modern, materialistic approaches to science.[35] What matters more crucially is whether or not a case can be made that the *early track record* of an alternative approach has enough promise to demand further support at launching and carrying out the program. While being aware of anomalies inherent to research programs is important, we should reject approaches to social epistemology that use perspectivism to engage in anomaly mongering unlikely to lead to new insights.

ANOMALY MONGERING VERSUS MINORITY PROGRESS

Social epistemology calls for a truly normative approach to science, building upon constructivist sociology and, as such, finds itself opposed to the cognitive authoritarianism inherent to Polanyi, Kuhn, and Collins.[36] This approach is shared by a wide variety of critical perspectives that seek to bridge the gap between descriptive sociology and normative philosophy, including feminist, postcolonial, socialist, reformist, and philosophical writers of various stripes. However, social epistemologists are split on how far to take the challenge to the status quo and on what basis. To see the difference between these two camps in social epistemology, the one spearheaded by Fuller and a diffuse

35. Constructivist STS, as well as philosophy of science, is split on whether it shares the "materialism" or "physicalism" of modern science or sees that as an imposed and dispensable assumption. See for instance, the discussion about metaphysical realism dividing the strong program from Collins's empirical program of relativism, discussed in Pinch, "Relativism," 40–41. The approach I seek to develop here can be seen in some ways to be an attempt to develop the suppressed alternative to Collins's empirical and relativist approach found in elements of the strong program, which sought to establish SSK as a reflexive, scientific approach to the study of science that takes for granted an underlying shared metaphysical reality, albeit not one represented as such in any particular scientific (or metascientific) theory.

36. See Fuller, *Social Epistemology*, for the founding call to arms, as well as the approaches found in the journal of the same name he established. In addition to this constructivist-friendly social epistemology, there is a "veristic" variety of social epistemology that engages analytical philosophy of science. See Goldman, "Foundations of Social Epistemics"; Goldman and Blanchard, "Social Epistemology."

group operating with a different approach, consider how each would handle challenges to dominant perspectives within science.

To a first approximation, we can divide these two camps depending upon whether they ground critique in a call for *political representation* or *dialectical engagement*. Political representation grounds critique in traditional, political demands for inclusion and democratic representation, seeing claims to truth as instances of power *tout court*.[37] Dialectical engagement grounds critique in a historical assessment of the prospects and problems of ongoing attempts to understand the world, realizing that critique never constructs alternatives de novo but instead transforms what exists. Considered in terms of minority research programs, a dialectical approach looks to see whether the program is capable of generating novel insights from its core insights rather than demand support due to a static conception of empirical support or political representation. (We will see in chapter 2 how Lakatos and Feyerabend drew creatively on Hegelian and Marxist dialectics to provide the basis for this fusion of historical and normative approaches.)

Fuller's view advocates perspectivist affirmative action and anomaly mongering, in effect calling for opening up the institution of science to a wider range of inputs from the larger society, exploiting anomalies in dominant theories as a wedge for critique. For example, in his writings in defense of intelligent design, he argues that the secular orientation of modern science wastes human potential by unfairly stigmatizing and excluding highly religious students from pursuing a career in science.[38] In the process, ideas that offer potential solutions to anomalies in the dominant perspective, such as explaining the emergence of complex, improbable biological entities by appeal to intelligent design, fail to get a hearing. The job of the social epistemologist is to force science to be more responsive to excluded ideas held by constituent groups within society.

Fuller's view can be seen as a form of populism within science, call it *cognitive populism*, whereby greater social inclusion within the process of generating the metaphysical content of potential research programs is called for on the grounds of political fairness and inclusion. In the context of U.S. religious politics, this particular proposal on behalf of intelligent design might be seen as largely a form of right-wing populism, as is the critique of anthropogenic climate change Fuller has also at times endorsed.[39] However, Fuller and other social epistemologists have also proposed Left versions of the same idea, suggesting that the exclusion of women or other subordinated groups should lead to changes in the methods and content of knowledge, as excluded groups

37. Rouse, *Knowledge and Power*.
38. Fuller, "What Has Atheism Ever Done for Science?"
39. Fuller, *Knowledge*, p. 279; Fuller and Lipińska, *The Proactionary Imperative*.

are brought into science in greater numbers.[40] It is political exclusion and the authority structure of science that testify to the fact that science has not lived up to its universalist claims. This idea that the political marginalization of competing ideas is inherent to science is taken to require intervention to achieve a more democratic outcome at the level of knowledge itself.

Approaches emphasizing *dialectical engagement* do not deny that excluded groups should be recruited to science or even that such inclusion may (or should) transform the content and methods of science. However, authors advocating this approach require more than lack of political inclusion as a basis for altering the metaphysical content of research programs in science. They require early promise that a dynamic, fruitful alternative approach is in the offing—that alternative research trajectories promise to be able to deliver their own thing rather than focus primarily on the criticism of androcentric perspectives in science, for instance. How different, initially relatively autonomous research traditions eventually interact, altering and incorporating elements of competing approaches over time, is also a component of a more dialectical approach—one particularly evident in Feyerabend's simultaneous endorsement of a "breathing space" for new perspectives and the use of incommensurable observation languages for qualitative transformation of dominant approaches (discussed in chapter 2).

Notice that this kind of distinction cuts across other distinctions familiar within social epistemology broadly speaking. For example, the distinction between standpoint theory and feminist empiricism within feminist approaches to science might at first seem to correspond to this distinction, with standpoint theorists emphasizing including women to transform androcentric science as a political project and feminist empiricism focusing on the epistemological case that can be made for alternatives to androcentric science. However, it is possible for standpoint theorists to focus on how the participation of women in science might propel the development of new approaches that deserve a wider hearing based on what the work uncovers. Evelyn Fox Keller does this in her examination of the neglected research program of Barbara McClintock, an approach out of step with mainstream molecular biology of her time. Keller shows that McClintock's approach, facilitated by her role as an outsider and a woman, anticipated later criticisms of a strict interpretation of the central dogma of genetics.[41]

Similarly, feminist epistemologists might focus primarily on the anomalies generated by androcentric science as part of an argument for political inclusion as a remedy. Feminist empiricist approaches that go further than this

---

40. See the classic discussions of standpoint theory in Harding, *The Science Question in Feminism*; and Haraway, "Situated Knowledges."

41. Keller, *A Feeling for the Organism*.

and include examination of how pioneering scientists developed alternative research trajectories would be closer to the approach of dialectical engagement advocated here. An example of the latter might be Allison Wylie's work on new approaches pioneered by women archaeologists, focused not just on gender-driven anomalies within archaeology, but alternative patterns of inference from artifacts to historical reconstruction that show promise of generating their own novel insights.[42]

Lakatos's idea of focusing on a track record of generating novel predictions, some of which are corroborated, links well with this kind of approach. Such a congruence between critical political critiques of science and Lakatosian philosophy of science has been hard to see since Lakatos's reputation has come to be seen as a defender of internalism, a philosopher's bulwark against sociological approaches. Such an interpretation is mistaken, as we will see in the next chapter. The congruence between Lakatos's approach and feminist critics of science is also hard to see since Lakatos's own political philosophy, not to mention patterns of behavior relevant to gender dynamics, would not win him any converts among feminists. Lakatos supported Arthur Jensen's argument that racial differences in intelligence were real, urging his graduate student, Peter Urbach, to defend this minority view using his Methodology of Scientific Research Programs.[43] Publication of the correspondence between Lakatos and Feyerabend has also exposed sexist attitudes on his part. What distinguishes Lakatos's own suppressed approach to science studies is the way it captures well how judgments about the development of research trajectories over time can be compared. Criticism and evaluation of research programs in science is not left to the scientists alone, nor is criticism made simply on the basis of including excluded views if they don't seem a viable basis for further work.

If we take this approach seriously, we can see how some minority positions have little to offer, while others deserve preliminary support based on early fruitful work. An example of a minority position lacking much hope of challenging a scientific consensus is the challenge to anthropogenic climate change by climate skeptics, discussed in the remainder of this chapter. Here the track record—going back over a century—is strong for the hypothesis of anthropogenic climate change and particularly weak for most scientific critics of the consensus.

Examples of the minority positions developing fruitful alternatives that deserve a hearing include new claims of "Lamarckian" inheritance associated with the science of epigenetics (discussed in chapter 2), the current advocacy

42. Wylie, "The Engendering of Archaeology."

43. Lakatos and Feyerabend, *For and Against Method*, pp. 348–49, n. 200; Urbach, "Progress and Degeneration in the 'IQ Debate.'"

of an extended evolutionary synthesis (ESS) challenging genetic reductionism in evolutionary biology (discussed in chapter 3 and contrasted with intelligent design theory), and low-carb critics of the diet heart hypothesis connecting saturated fat consumption to heart disease (discussed in chapter 4). Later chapters will consider more complex social and political dimensions affecting minority positions in science. Chapters 5 and 6 will examine the place of social movements in shaping scientific debates, including movements of patients opposed to standard medical treatment of thyroid and adrenal disease (chapter 5) and environmental critics of the toxic impact of petrochemicals and radioactive by-products on human health (chapter 6). Chapters 5 and 6 raise the issue of when critics of dominant approaches deserve special consideration, with critics reviving neglected but not refuted or unfruitful perspectives. These chapters show how political and historical critique can connect to real epistemological issues within science, as well as the obstacles placed in their way by larger social forces. Chapter 6 also looks at how the economic and political patronage of science, exemplified by Cold War government and corporate sponsorship of science, can advance some scientific research programs to a dominant status, while inhibiting other, potentially worthy programs at the same time. The conclusion looks at the current erosion of the scientific ethos as neoliberal commodification of science proceeds apace, particularly in pharmaceutical research, which has challenged elements of the institution of science that are central to its effective functioning.

In the remainder of this chapter, I need to first establish that an emphasis on the potential rationality of dissenting and minority positions does *not* mean that *any* criticism of existing scientific work, however loud, deserves much attention. To be sure, Lakatos and Feyerabend were defenders of *liberty* in science, an approach I wholeheartedly endorse. Robust versions of academic freedom are needed that allow scientists and scholars the right to pursue unpopular positions. Lakatos and Feyerabend differed on how to carry out a defense of liberty in science. Lakatos was much more likely to see academic freedom as requiring that faculty be protected from political interference by student radicals, for instance, while Feyerabend was more concerned to bring in student radicals to challenge the orthodoxies of their professors.[44]

Both were adamantly opposed to using some kind of standard of empirical justification to exclude minority views from the academy or block their right to be heard. We can recognize that some degenerating perspectives are not worth investing further time and resources without denying them the right to speak, to be heard, and to keep their jobs. Some minority positions, however,

44. Lakatos, "A Letter to the Director of the London School of Economics" in *Philosophical Papers*, II, 247–53; Feyerabend, "Consolations for the Specialist"; Lakatos and Feyerabend, *For and Against Method*. For discussion of the role of student protests in shaping the philosophies of Lakatos and Feyerabend, see Martin, "'The Battle Is On.'"

can make a claim that their approach shows promise that deserves support and further consideration by other core group scientists. In some cases, the dogmatism and inertia of scientific institutions is something that we ought to pay attention to and seek to change. Knowing when to bet on a dark horse is probably as much art as (philosophy of) science but being able to compare track records is a start.

THE TRACK RECORD OF CLIMATE CHANGE

The track record behind the theory of anthropogenic climate change is more than a century old, and begins with theoretical progress and moves on to empirical progress via a succession of models that more accurately predict (and retrodict) climate changes due to greenhouse gas emissions. Theoretical progress occurs when a theory makes a novel prediction—either predicting a completely unknown fact or predicting a fact that was not taken into consideration in the formulation of the theory.

The latter form of "novel fact" was urged on Lakatos by his student and later colleague, Elie Zahar, who suggested that a prediction was novel if the predicted fact "did not belong to the problem-situation which governed the construction of the hypothesis."[45] As such, Zahar's reconstruction captures a dynamic observed in highly valued scientific contributions from the past. The highly valued scientists are used as an initial set in defining criteria for scientific progress. In his classroom lectures, Lakatos calls them the "goodies" rather than the "baddies" peddling questionable or pseudoscientific theories like Marx or Freud. He argues that the best criteria for describing why we value the goodies over the baddies is that they offer theoretical predictions that seem to offer surprising rather than expected accounts of a phenomenon.[46]

As Barrett and Stanford put the point, "the relevant sense of prediction involves not anticipating when and where familiar phenomena will recur but rather discovering the existence of phenomena unlike those that are already familiar."[47] For Lakatos and Zahar, familiarity in this context refers to the background knowledge associated with scientific work in that area. Progressive problem-shifts within a research program occur as successive scientific theories get better at predicting facts about the world without ad hoc maneuvers designed just to fit anomalies to the theory. As a result, "in order to assess the relation between theories and empirical data within a research-programme, one has to take into account the way in which a theory

45. Zahar, "Why Did Einstein's Programme Supersede Lorentz's?"; Lakatos, *Methodology*, ch. 4.
46. Lakatos, "Lectures on Scientific Method," pp. 24, 108–12.
47. Barrett and Stanford, "Prediction," p. 628.

is built and the problems it was designed to solve."[48] On this view, it matters how new ideas are *generated* by heuristics, while they remain fruitful, as well as the relationship between predicted facts and background knowledge. As such, Lakatos sought to rescue Popper's spirit of critical thinking associated with his falsificationist methodology within a more accurate understanding of the actual historical dynamics of science, where empirical corroborations figure much more prominently among history's "goodies" than falsifications.[49]

Lakatos rejected Popper's falsificationist approach because he was convinced that it did not respect "the remarkable tenacity of theories" in history.[50] Successive theories in a research program can always be adjusted by modifications to the *protective belt*, a set of auxiliary hypotheses that are set up to take the force of logical refutation away from the *hard core* of the research program. The hard core is protected from refutation so long as the research program is in use, via a negative heuristic, while the protective belt makes use of positive heuristics that help the scientist interpret the hard core's assumptions and generate predictions on this basis. These positive heuristics often take some time to develop, as attested by the two-century-long history of developing the mathematical techniques that translate Newton's hard-core program of point forces between two isolated bodies into actual predictions of real astronomical bodies.[51]

While Popper, in his more sophisticated moments, did allow that modifications of theories that facilitated novel predictions were legitimate, his conception of novel facts required strict "temporal novelty." That is to say, the scientist could be credited with a novel prediction, rather than an illegitimate ad hoc maneuver designed to evade a falsification, if the predicted fact was not known at the time of the prediction.[52]

For Lakatos, this poorly captured our positive evaluations of the "goodies" in the history of science who did not always predict completely unknown facts. Thus, Copernicus "predicted" the motion of planets already known to the Ptolemaic astronomers, while Einstein predicted observations compatible with Lorentz's ether theory. However, by Zahar's criterion, predictions made by adjusting the parameters of the theory to fit known facts were inferior to

48. Zahar, "Why Did Einstein's Programme Supersede Lorentz's?," p. 103.

49. Lakatos, *Methodology*, chs. 1–2.

50. Ibid., p. 4; Lakatos, "Lectures on Scientific Method," p. 104.

51. Lakatos, "Lectures on Scientific Method," pp. 104–5.

52. Zahar, "Why Did Einstein's Programme Supersede Lorentz's?"; Lakatos, *Methodology*, ch. 4. Lakatos, "Criticism and the Methodology of Scientific Research Programmes" introduced a distinction among three different Popperian positions, the imaginary $Popper_0$, a straw-man caricature of his position, $Popper_1$, the naïve falsificationist who does not allow hypotheses to be modified after failed predictions, and $Popper_2$, the more sophisticated approach discussed above that allows for reformulations that increase conceptual content through temporal novelty.

those that fell out from assumptions that did not already incorporate these known facts into the process of constructing the theory.

While Copernius did still have to engage in parameter adjustment to fit his deferents and epicycles to the observed motion of the planets in the published *De Revolutionibus* (1543), as did Ptolemaic astronomers, he did not need to do so to predict the planets' retrograde motion or their connection to the motion of the sun, evident already in his initial sketch of the system in the manuscript *Commentariolus* (circa 1510). Copernicus's predictions of retrograde motion and the elegant "simplicity" of the model were the unexpected by-product of a new hard core not rigged to produce these results. It is the *conceptual novelty* of the *Commentariolus*, rather than the ad hoc efforts to save all of Ptolemy's facts that served as the starting points of the work of Galileo and Kepler, testifying to the significance of conceptual novelty in generating new knowledge via heuristics that make up a logic of discovery, not just a logic of justification. Similarly, Einstein produced a novel prediction of the known anomalous perihelion of Mercury as "an unintended by-product of Einstein's programme."[53] Conceptual novelty is an important selling point for new research programs and motivates the pursuit of the new program despite the continuing existence of other anomalies. For Lakatos, "[i]t is primarily the positive heuristic of [a] programme, not the anomalies, which dictate the choice of his problems."[54]

The fact that conceptual novelty and confirmations capture more of what we value in the "goodies," while excluding more of the "baddies" in the history of science, suggests that Lakatos's philosophy works better as a reconstruction of the growth of knowledge than Popper's approach. Indeed, it could be argued that attempts to evade falsifications by speculating that there are unaccounted for phenomenon interfering with the result often help found fruitful research programs in the first place. This seemed to be the case, when in 1824, Joseph Fourier calculated what temperature the earth should be based upon a model of the emission of infrared radiation but found that the earth would be much colder than it actually is if something else was not going on. Rather than "falsifying" his model, Fourier then "predicted" that there were gases in the atmosphere that functioned on analogy with glass panes that had the ability to trap heat in greenhouse gases.[55]

This amounted to theoretical progress in predicting the existence of a type of gas about which Fourier had previously been ignorant. The prediction made theoretical progress in Zahar's sense because he made a prediction

53. Lakatos, *Methodology*, pp. 185–88, 76–77.
54. Ibid., p. 111.
55. Weart, *The Discovery of Global Warming*, pp. 2–3. The discussion below draws on Weart's book and a supplemental website tracing the history of global warming at https://history.aip.org/history/climate/index.htm.

about a novel type of gas that was not "transparent" to infrared radiation but would absorb it. In 1859, John Tyndall converted theoretical progress into empirical progress by testing for the existence of greenhouse gases in his laboratory, finding that coal gas (mostly methane) and carbon dioxide (CO_2) did, in fact, absorb heat.[56]

By calculating that too much heat would be dissipated to account for the temperature found on the earth using his simple model, Fourier inferred that something in the atmosphere must be absorbing heat. Finding out that some atmospheric gases did have the property of absorbing infrared radiation corroborated this novel fact. The initial, simple model of the atmosphere was replaced by one that also incorporated consideration of the existence of greenhouse gases, leading to a theoretically and empirically progressive problem-shift.

Fourier's model and Tyndall's corroboration of the existence of greenhouse gases set the stage for understanding the basic causal mechanism contributing to global warming. One thing remained to construct the first global climate model: the concept of feedback that could explain natural variations in climate. An understanding of the possible role of positive and negative feedbacks emerged as scientists tried to make sense of the puzzle of a prehistoric ice age, as evidence surfaced of the geological impact of glaciers in temperate areas in Europe and North America.

In 1896, Svante Arrhenius suggested that changes in greenhouse gases like CO_2 could result from changes in volcanic activity, amplified by the effect it had on the amount of water vapor in the atmosphere. Water vapor is a potent greenhouse gas as well, and a small degree of warming caused by volcanic emissions of CO_2 would allow the warmer air that resulted to hold more water vapor, increasing warming by a process of positive feedback. Likewise, if volcanic activity ceased, the resultant cooling would ensure that the air held less water vapor and cooling would be enhanced, potentially leading to an ice age. According to Arrhenius's rough calculation, cutting the CO_2 in the atmosphere by half would lead to a 5-degree Celsius drop in temperature, which, when further amplified by the reflection of light by snowfall, could in principle lead to an ice age.

His model of how Ice Ages could occur also suggested to him that the emission of CO_2 by recent industrial activities could bring about warming, especially as it seemed from preliminary data put together by Arvid Högbom that human-caused emissions of greenhouse gases were roughly equal to the level of emissions and absorptions of CO_2 that made up the natural CO_2 cycle. Arrhenius calculated that if industrial emissions were to continue for a

56. Ibid., pp. 3–4.

long time and the total levels of CO_2 in the atmosphere doubled, the earth's temperature would rise by about 5 or 6 degrees Celsius.

Arrhenius's model launched a research program by making a novel prediction that would only begin to be corroborated in 1957 when Roger Revelle began measuring CO_2 levels as part of the International Geophysical Year, a Cold War funding program. By 1998–1999, the famous "hockey stick" graph incorporating these and other proxy measurements indicating an uptick in global temperatures after 1900 and significant warming after 1980 offered a corroboration of the novel prediction. The test was subject to significant political and statistical criticism, but ongoing efforts to verify the test supported the warming trend, including a reanalysis of the data carried out by global warming skeptic Richard Muller and funded by Charles Koch.[57]

Arrhenius's model was the first in a series of successive models of the atmosphere extending for more than a century, each of which predicted enhanced warming with the addition of greenhouse gases like carbon dioxide and methane to the atmosphere, predictions which have been confirmed, *ceteris paribus*, via complex measurements of global temperature making use of proxy data like ice cores, pollen, and tree rings.[58] The consistency of the predictions of these models, now incorporating sophisticated computer modeling of feedbacks and interactions, was such that no models predicted cooling and all models predicted warming within a narrow range due to fossil fuel consumption. In fact, the early history of the models from 1896 to 1975 predicted warming resulting from a doubling of CO_2 that varied from Arrhenius's hand-calculated model's predictions of 5–6 degrees Celsius to 0.8 degrees Celsius in 1971, using a variety of assumptions. Increasingly sophisticated computer-based global climate models developed since 1979 have predictions that range between 1.5 and 5.4 degrees.[59] Efforts to track overall warming have required constructing global temperature measures from variable sources and methodologies and are necessarily heavily reliant upon computer models.

While modifications of the models to better retrodict known, rather than unknown, features of the atmosphere would not count as progress in Lakatos's terms, surprising predictions—or even "surprising retrodictions"—made by the model would be (call the surprising retrodictions "unknown knowns"). This means that computer models can be used as a generator of new insights into the climate system even when ad hoc modifications for known effects are

57. Mann, Bradley, and Hughes, "Northern Hemisphere Temperatures during the Past Millennium"; Brumfiel, "Academy Affirms Hockey-Stick Graph"; Ahmed et al. "Continental-Scale Temperature Variability during the Past Two Millennia"; Rohde et al., "A New Estimate of the Average Earth Surface Land Temperature Spanning 1753 to 2011"; Weart, "Hockey Stick and Hiatus."

58. Dörries, "Politics, Geological Past, and the Future of the Earth."

59. Edwards, *A Vast Machine*, pp. 182, 378.

introduced, if they also contribute to explaining aspects of the climate system that were not understood up to that point. The relationship between computer modeling, theoretical "predictions," and "empirical" confirmations remain complex, especially since all empirical measures of climate phenomena are highly mediated instrumentally and theoretically.[60] Neither observation discounts Lakatos's general point about tracking progress. Research programs advance so long as they do make predictions that are corroborated now and again; they do not need to do so at every step.

As for the mediated and artificial character of "observation," in the case of climate models amounting to model-data symbiosis,[61] that too is compatible with Lakatos's system, as Popper had already argued that the facts used to judge predictions were *conventional*, as understood by the assumptions of particular sciences, rather than *given* by direct sense perception. What the facts are taken to be that judge whether predictions are corroborated or not are constituted by social, conceptual, and instrumental assumptions given by the research programs informing the research. There are no such things as pure facts, unmediated by theory and history.

In the case of climate science, even the data are constituted via "data models" that bridge the gap between various instrumental measures of different modalities scattered around the globe and the construction of global temperatures that can then be said to test predictions of warming.[62] If data models are used to test climate models, then there are models all the way down, but this is no ground for rejecting the science. Getting better climate models as well as better data models to test them are both appropriate scientific goals. What one wants to know is if we are making progress in coming up with scientific theories that explain the data painstakingly put together by scientists.

From the beginning, accounting for the positive and negative feedbacks in the models were the most important changes that needed to be made to bring about improved accuracy. Indeed, early accounts of negative feedback support the idea that progressive programs nonetheless are accompanied by an "ocean of anomalies," as early tests suggested that the atmosphere was already saturated with CO_2 and could absorb no more, that the oceans could absorb any excess CO_2, and that water vapor would produce clouds that would reflect light back to space and counter any warming trend.[63] It is precisely accounting for the complex interactions of positive and negative feedbacks like these that have required the construction of incredibly complex computer models of the atmosphere.

60. Galison, *Image and Logic*, ch. 8.
61. Edwards, *A Vast Machine*, ch. 10.
62. Ibid.
63. Weart, *The Discovery of Global Warming*, pp. 7–8.

Climate change skeptics have focused most of their criticism on the "ocean of anomalies," interrogating statistical techniques behind the hockey stick graph and challenging details of the computer models. The climate skeptic and MIT meteorologist Richard Lindzen argued that modelers had engaged in parameter adjustment in a way that would be ad hoc by Lakatos's standards by varying the model to better fit known climate data, leaving future predictions more uncertain than realized. Challenging the treatment of water vapor, Lindzen developed an alternative account that eliminated the positive feedback that accounted for much of predicted temperature increases based on a conviction that, in the long run, the climate acts as a self-regulating system.[64] His specific criticisms were not held to be convincing, just as criticisms of the statistical techniques used in the Hockey Stick graph did not lead to a rejection of the results.

Such anomaly mongering can lead to additional work that tightens the argument, but it does little to advance the climate skeptics' aim of rejecting the claim of anthropocentric global warming. Scientists working within a research program direct their ongoing work to improving their models piecemeal and do not expect to eliminate all anomalies in any one model. Therefore Lakatos urged us to examine the general trend of a series of theories in a research program. Moreover, the simulations of the climate are *models,* which translate the larger theoretical claims of climate change science into provisional predictions that must make simplifying assumptions for reasons of calculative convenience. In fact, computer climate models are now among the largest and most complex computer programs ever created and testify to a willingness to continue to arrive at better models for retrodicting past climate data to improve the accuracy and scope of climate predictions. The scope is increased as models begin to account for more types of interactions and outcomes—for instance, by modeling climate phenomena beyond merely global temperature averages.

By comparison, the positive alternative to climate change science has been left fallow. Most climate skeptics appeal to natural climate variability and/or changes in solar intensity to account for temperature changes. The first thing to notice is that sunspots as cause was long a competing research program. Charles Greely Abbot's work at the Smithsonian Astrophysical Observatory in the 1920s demonstrated significant variations in the brightness of sunlight received by the earth and began connecting sunspot activity and terrestrial temperature.[65] Ultimately, his predictions of temperature change based upon

64. Ibid., pp. 176–77. For a delineation of different levels of dissent to anthropocentric climate change that go beyond political objections to regulation to include cultural differences among scientists, based upon their disciplinary backgrounds in meteorology and weather forecasting that lead them to question climate models, see Lahsen, "Anatomy of Dissent."

65. Weart, *The Discovery of Global Warming*, p. 16.

sunspot activity began to fail. Since then, climate skeptics have also found select correlations between sunspot activity and global temperature, but as new data was collected, the correlations fell apart.[66] Since multiple factors shape temperature, it is still possible to revive the program and try again, but currently the sunspot theory must be judged regressive.

In 1988, the Interplanetary Panel on Climate Change (IPCC) was established and began a more formal process of assessing climate change models in order to advise policy makers about the extent of the risks of anthropogenic climate change. The organization is a hybrid institution, drawing upon "volunteer scientists"[67] who seek to summarize the latest evidence of the likely effects of human fossil fuel use and other greenhouse gases on the climate. Over time, the IPCC's periodic assessment reports incorporated more formal peer review, statements about statistical confidence intervals, and extensive documentation that shifted increasingly from a summary of expert judgments to something that looks more like a brief for a legal argument subject to contestation. This difference reflected the growing politicization of the question of anthropogenic climate change as policy makers increasingly looked to regulate greenhouse gas emissions and encountered significant resistance from affected industries.[68]

Crucial to understanding and evaluating the strength of the evidence that significant anthropogenic climate change may be occurring is to appreciate that science does not work by decisive proof of harm but by an iterative series of models designed to predict climate behavior based upon simplifying models. When U.S. president George H. W. Bush argued that we should wait for more research to be done before acting upon the potential threat of global warming in 2001, he was not only rejecting what has been called the precautionary principle (act now to preempt potential threats even if the evidence is uncertain). He was also drawing upon a discredited view of how science works that Lakatos was at pains to challenge.

Science does not produce "proofs" that particular phenomena behave in a certain way but only highly tentative models, whose strength, if they have any, is that some of them, over time, make predictions that turn out to be corroborated. He did not argue that all successive theories within a research program would unambiguously increase predictive content and empirical confirmation at each step, but only that a research program that showed evidence of theoretical and empirical progress over time should be supported and pursued in preference to one that over time did not do so. There was no "instant rationality," nor should the existence of anomalies—or what Popper and others would

66. Weart, "Changing Sun, Changing Climate?"
67. O'Reilly, Oreskes, and Oppenheimer, "The Rapid Disintegration of Projections," p. 717.
68. Oreskes and Conway, *Merchants of Doubt*.

have called disproof—be taken as grounds for rejecting a research program; it may—or may not—be a motivating factor for arriving at a successor theory within a research program depending upon the particular anomaly and the particular readiness of the development of science up to that point.

These insights can be usefully combined with an understanding how social location affects assessments of uncertainty and their implications within science. Mackenzie has described the operation of a certainty trough, where most supportive public consumers of scientific knowledge presume high levels of certainty. Mackenzie actually maps "uncertainty" on the *y* axis, so these consumers are low on a graph and represent a middle position in terms of social distance from the science involved. Those at a far distance include those who are alienated from the science concerned and posit high uncertainty for the field, but in a negative way, seeing such uncertainty as delegitimating the field and rendering it irrelevant for policy. Passive consumers and motivated outside critics attribute too little and too much uncertainty to science, respectively, based upon a false model of how science handles anomalies.[69]

By contrast, core-set scientists, in general, are well aware of the uncertainties, approximations, and anomalies that exist at any given state in the development of their field—indeed, such an awareness is necessary to understand the work that they do, which may end up reducing select cases of this uncertainty without eliminating the ocean of anomalies as a whole. Thus, thoughtful, reflective scientists are likely to be aware of the provisional and imperfect nature of the state of knowledge in their field. Since a field can make better or worse progress at developing a particular research program, without eliminating uncertainty as a whole or "proving" that some outcome must happen in the future, it is pointless to demand that scientists provide such proof. To be sure, this may leave politicians and the public with less information about potential threats than they would prefer, but that speaks to the limits of our knowledge, not to its non-existence.

To evaluate whether a field is "onto something," the focus ought to be on evaluating its theoretical and empirical progress relative to other, potential or actual, competitors. Here, modern climate science does quite well, even if some of the scientists summarizing climate models may themselves be consumers, rather than producers, of these models and, hence, attribute too much certainty to them.[70] Of course, this can cut both ways, as the drafting of IPCC

---

69. Mackenzie, *Inventing Accuracy*; Shackley and Wynne, "Representing Uncertainty in Global Climate Change Science and Policy." See Lahsen, "Seductive Simulations?" for the argument that Mackenzie's certainty trough is a simplification that ignores that users may sometimes be in a better position to criticize findings oversold by producers of knowledge claims, especially in fields like climate science with multiple sites of production.

70. O'Reilly, Oreskes, and Oppenheimer, "The Rapid Disintegration of Projections"; Lahsen, "Seductive Simulations?"

reports may overstate *or understate* risks relative to what knowledgeable insiders might hold to be the strengths and limitations of particular models.

A good example of this is how a better model produced renewed uncertainty within the IPCC assessment reports about the stability of the West Antarctic Ice Sheet (WAIS) during the twenty-first century and its contribution to sea-level rise. In the third assessment report from 2001, the possibility that the sheet would collapse was considered unlikely before 2100, with estimates of the contribution of WAIS melting due to sea-level rise in the long-term (post-2100) given, despite acknowledging high uncertainty in the estimates.

These projections assumed that physical models that retrodicted successfully glacial-interglacial cycles in the earth's past on a scale of about 100,000 years could predict melting on a scale of a few centuries. In the 1980s, Philippe Huybrechts developed the model used in the IPCC reports based upon a "shallow ice approximation" that allowed tractable calculations to be made given assumptions about the physical forces operating within the sheet and the relationships between ice over land and ice over sea in the WAIS. The approximation was "an educated guess that stands in for a complete solution to the physical equations."[71]

While the fourth assessment report was being prepared, ultimately published in 2007, new satellite observations that tracked streams of water within the ice sheet directly suggested significant melting was occurring and collapse could occur much quicker than anticipated by Huybrechts's modeling strategy. While the situation was in flux before definitive conclusions about the new satellite data were published, the authors of relevant sections of the report abandoned the distinction between short-term stability and long-term decline. Instead, since the IPCC authors were charged with relying upon published, peer-reviewed data, they removed any contribution to sea-level rise from WAIS melting from both short- and long-term projections.[72]

Better data and emerging, dynamical models increased uncertainty in the scientific short term and led to a likely *underestimate* of sea-level rise by 2100, which the authors tried to signal by noting that the estimates excluded the contribution of "future rapid dynamical changes in ice flow." However, the retreat from a more confident prediction of sea-level rise was perceived by policy makers and the media as proof that earlier confidence levels were overstated.[73] A focus purely on confidence levels at one moment in time missed the theoretical and empirical progress that was being made as models

71. O'Reilly, Oreskes, and Oppenheimer, "The Rapid Disintegration of Projections," p. 716.

72. For an argument that the IPCC's exclusive reliance upon peer review should be balanced by a precautionary principle in this instance, employing the new research prior to peer review to avoid underestimating potentially catastrophic harms, see Lewens, *The Meaning of Science*, pp. 131–35.

73. O'Reilly, Oreskes, and Oppenheimer, "The Rapid Disintegration of Projections," pp. 711, 723–24.

were developed that better characterized a fluid situation, pun intended, on the scientific research frontier.

Like science itself, the predictions in the IPCC reports need to be taken as a historical series over time rather than as statements of proof at a moment in time. No one should expect that all scientists should reject consideration of alternative accounts of climate change. The extent of overall consensus in the peer-reviewed literature for anthropocentric climate change is actually something of an anomaly in science, which results from the great effort and resources that have been put into improving these models rather than any grand conspiracy by scientists. The fruitfulness of research on anthropocentric climate change is indicated as well by its use in other sciences as a basis for research programs specific to these fields, such as explaining ecological changes in plant and animal populations. Minority positions such as the appeal to sunspots have not been progressive and should not attract equal time in policy forums or equal funding resources. In other fields, the diversity of positions is greater, and the case for minority views deserves more consideration, as we shall see.

Chapter Two

Generating Challenges to Dominant Ideas

Lakatos and Feyerabend's Dialectical Alternative

In lectures on scientific method to his graduate students, Imre Lakatos sought to develop an alternative to Kuhn and Polanyi's "elitist authoritarianism" that respected the findings of historical sociology of science without granting experts license to define knowledge as the judgment of self-selecting communities. Kadvany has argued that Lakatos's assessment of scientific fields was rooted in a Lukácsian reading of Hegel's dialectic, where reason emerges in history as the long-term outcome of various heuristic logics whose trajectory only becomes evident in retrospect.[1]

Feyerabend shared Lakatos's contempt for elitism, while challenging his proposed criteria for public assessment of knowledge, importing from dialectical materialism consideration of the uneven and combined development of scientific fields. His conception of incommensurability grew out of this sense of the temporary incompatability of distinct theories, judged on their own terms, which operates both as a mechanism for generating qualitative change within history and as a temporary obstacle to hybridization of theories from distinct research programs. In this chapter, I reconsider their ongoing debate and the guidance it offers in constructing an evaluative philosophy of science suitable for an engaged and interventionist Science and Technology Studies (STS).

For Lakatos implicitly and Feyerabend explicitly, the situation is analogous to Leon Trotsky's call for "permanent revolution": knowledge is a productive force but one that the existing social-epistemic relations increasingly impede rather than promote, whereas backward knowledge may leapfrog more advanced knowledge leading to scientific revolution.

Key for both philosophers was the idea that it was necessary to pay more attention to how new and fruitful approaches are *generated* in the practice of

1. Kadvany, *Imre Lakatos and the Guises of Reason*, pp. 287–88, 312–16.

dynamic fields than how they are abstractly justified. Moreover, the disciplinary form that knowledge takes at any moment is as likely to impede future progress as encourage it, a conclusion quite at variance with contemporary historians' reliance upon actor's categories as the framework for their historical analysis.

Following the past five decades of work in the sociology of scientific knowledge and cognate fields, we have a much firmer basis for assessing the social conditions for the growth of knowledge and the possibilities for improving the quality and relevance of our scientific endeavor—but only if we jettison the current tacit philosophy of science studies that exempts scientific communities from outside evaluation on the grounds that all knowledge is local and tacit. On the contrary, one of the key findings of the sociology of scientific knowledge is that the fate of knowledge is in the hands of later (and different) users.

A DIALECTICAL APPROACH TO SCIENCE

Lakatos and Feyerabend developed different versions of a dialectical alternative to Kuhn's view in *Criticism and the Growth of Knowledge*, the proceedings from the 1965 London conference featuring a showdown between the philosophies of Kuhn and Popper. The conference was arranged by Lakatos, where his Methodology of Scientific Research Programs (MSRP) was presented as a synthesis of the two. Feyerabend's own contribution similarly sought to reconcile historical and critical approaches to science, setting the stage for his own dissenting philosophy of science eventually published as *Against Method*.[2]

A dialectical approach incorporates the history of science into the assessment of science in a different way than does Kuhn's appeal to the history of science to undergird a model of alternating stages of normal and revolutionary science. Key to the difference is developing an understanding of the relationship between different, "contradictory" elements within allegedly consensual, normal science that set the stage for scientific revolutions in the first place. Hegel famously argued that the real is rational, which was taken by his early critics as a conservative defense of the status quo. However, Hegel himself clarified in the second edition of his *Encyclopaedia* (1827) that the real is not whatever exists in the present but a part that will end up having played a necessary role in the whole of history; so our own static conception of what we know at a moment in time can be expected to change

2. Lakatos and Musgrave, *Criticism and the Growth of Knowledge*; Feyerabend, *Against Method.*

as the future unfolding of our knowledge reorders the relationships of these incomplete, static parts.[3]

Marx and Engels took Hegel's dialectical philosophy to exemplify how historical change would bring about the transformation of existing historical systems. In *Ludwig Feuerbach and the End of Classical German Philosophy*, Engels pointed out that Hegel's claim that the real is rational meant not that anything that exists is rational but that anything that remains *necessary* within a process of historical change is rational. If the Prussian state relied upon authoritarian measures that spoke to the limitations imposed by the current state of the development of the German people and was a necessary part of political development, that could not be wished away by utopian political thinking.[4]

At the same time, the Prussian state, and the liberal political revolutions unleashed by the French Revolution that made the ancien régime unreal in time, would also become unreal as what once was necessary becomes superfluous, or even an obstacle, to further development. Engels took this to be suitably paraphrased by the slogan that "all that exists deserves to perish," a view that when applied to scientific epistemology implies a thoroughgoing fallibilism, denying that the scientific realist's vision of a complete and final classification of the natural world could be obtained in finite time:

> But precisely therein lay the true significance and the revolutionary character of the Hegelian philosophy . . . that it once and for all dealt the death blow to the finality of all product of human thought and action. Truth, the cognition of which is the business of philosophy, was in the hands of Hegel no longer an aggregate of finished dogmatic statements, which, once discovered, had merely to be learned by heart. Truth lay now in the process of cognition itself, in the long historical development of science, which mounts from lower to ever higher levels of knowledge without ever reaching, by discovering so-called absolute truth, a point at which it can proceed no further, where it would have nothing more to do than to fold its hands and gaze with wonder at the absolute truth to which it had attained.[5]

Lenin echoed this criticism of a dream of a final theory in distinguishing dialectical materialism from metaphysical materialism:

> The "essence" of things, or "substance," is *also* relative; it expresses only the degree of profundity of man's knowledge of objects; and while yesterday the profundity of this knowledge did not go beyond the atom, and today does not go beyond the electron and ether, dialectical materialism insists on the temporary,

---

3. Hegel, *Encyclopedia of the Philosophical Sciences in Basic Outline, Part 1, Science of Logic*, pp. 33–34; Taylor, *Hegel*, pp. 419–27.

4. Engels, *Ludwig Feuerbach and the Outcome of Classical German Philosophy*, p. 10.

5. Ibid., p. 11.

> relative, approximate character of all these *milestones* in the knowledge of nature gained by the progressing science of man. The electron is as *inexhaustible* as the atom, nature is infinite, but it infinitely *exists*. And it is this sole categorical, this sole unconditional recognition of nature's *existence* outside the mind and perception of man that distinguishes dialectical materialism from relativist agnosticism and idealism.[6]

Here Lenin cites Joseph Dietzgen, a contemporary of Marx who helped develop dialectical materialism, who "emphasised that the 'subject-matter of science is endless,' that not only the infinite, but the 'smallest atom' is immeasurable, unknowable to the end, *inexhaustible*, 'for nature in all her parts has no beginning and no end.'"[7]

THE INEXHAUSTIBLE ATOM

Lakatos echoed this view as early as 1947, when he published a critique of an appeal to (what would later be called) intelligent design by a biologist, Vilmos Csiszár, who had argued that cell division could not be explained in purely materialistic terms. Drawing on Lukács's view of Marxism as a method not wedded to any particular content, Lakatos defended materialism, but rejected "vulgar materialism," which equated matter with our current scientific conception of it. Since "[e]ach new discovery modifies our concept of matter," Lenin's concept of real, but "inexhaustible" atoms should be preferred over Csiszár's belief in a "well-arranged and comforting *final* explanation, which makes our existence meaningful and purposeful."[8] In the surviving article, "Modern Physics, Modern Society," incorporated into his missing 1947 Hungarian dissertation, Lakatos describes "the only thesis of materialism" capable of surviving the historical transformation of scientific concepts to be the "infinitely broad Leninian axiom postulating an existence *independent of our mind*, of which our mind reflects constantly more and more, without ever exhausting it."[9]

Key to Lakatos's 1947 dissertation is the idea that dialectics does not just describe a relationship between concepts abstractly considered; it shows the inherently historical dialectic within science that gives rise to those concepts in the first place and which eventually exhaust themselves as science develops further. This Lukácsian extension of the sociology of knowledge to

6. Lenin, *Materialism and Empirio-Criticism*, p. 269. On debates within physics at the time about the structure of the electron, attacks on materialism by Poincaré, and Dirac's "dialectical" solution, see Bunge, "The Inexhaustible Electron."

7. Ibid., p. 267.

8. Cited in Kutrovátz, "Lakatos' Philosophical Work in Hungary," p. 121 (emphasis added by Lakatos).

9. Kutrovátz, "Imre Lakatos' Hungarian Dissertation," p. 359.

science, referred to by Lakatos as "planting historicism into natural science," itself is directly opposed to classic internalism, even if "the objective aspects [of scientific knowledge] play an ever larger role in the development of science" over time.[10] Rejecting either a cumulative view of scientific advance or a logical reconstruction of the relationship between scientific representations and objective reality, Lakatos reflects on how to reconcile Lenin's picture theory of scientific representation with his notion of the inexhaustible character of natural reality, and hence, the sociologically delimited character of all scientific knowledge:

> But what should we mean by "more and more" exact reflection? Conceptual development is not a quantitative development that renders ever larger concentric circles of exactness. Nor is it "logical," for conceptual development changes logic as well. So what guarantees the "more and more" exact system of concepts? Does the sociological sphere not penetrate into scientific concept-building itself?[11]

Even the elements of scientific culture that contribute to a more exact representation of nature reflect the sociological basis of scientific work, both facilitating scientific advance and inhibiting further progress until social conditions allowed.

In Lakatos's version of a Marxist sociology of scientific knowledge, science itself was a productive force that would create economic, conceptual, and organizational contradictions for the bourgeois society that created it. Thus, Lakatos saw the discovery of nuclear energy as transforming the mode of production in a way that challenged bourgeois property law since the promise of limitless energy would introduce the threat of a society "not regulated by the law of value." The result was that the stage was set for a socialist mode of production but resisted by capitalist society, turning nuclear power toward destructive ends.[12] At the same time, the transformation of our conception of matter by quantum mechanics, an exemplification of Lenin's inexhaustible atom, was resisted by idealist and subjective philosophies put forward by scientists working in a capitalist society. Organizationally, the drive toward international scientific cooperation, the concentration of instrumentation into ever more centralized laboratories, and the introduction of greater planning within scientific work was resisted as the specific "hermetically isolated" national research schools reflected the same contradictions driving conflict between nations under capitalism.[13]

---

10. Ibid., pp. 359, 368.
11. Ibid., p. 368.
12. Ibid., p. 357.
13. Ibid., pp. 367–68.

Dating to the height of Lakatos's ideological polemics against alleged enemies of Marxism in Stalinist Hungary, Lakatos's notion of a dialectically based fallibilism in science and mathematics remained central to Lakatos's thinking throughout his career. Lakatos's mature philosophy abandoned Marxism but continued this early belief that the dialectical method "unmasks the seemingly unchanging and eternal things as historical categories revealing their birth, flourishing, and demise."[14]

Aside from the extension of Trotsky's concept of uneven and combined development to the history of science, in his contribution to *Criticism and the Growth of Knowledge*, Feyerabend had already begun to develop some of the key themes of his mature philosophy of science by developing Lenin's dialectical materialism. Feyerabend believed that there were few contemporary historians of science "who are as well acquainted with contemporary science as was Lenin with the science of his time, and no one can match the philosophical intuition of that astounding author."[15] In assessing the impact of historical materialism on quantum mechanics, Feyerabend sought to develop key insights from dialectical materialism as an explicit methodology for scientists to follow. According to Feyerabend, the key constructive themes introduced to scientific theorizing by dialectical materialism were the following:

> (1) the emphasis on the fact that in nature there are no isolated elements but that everything is related to everything else; (2) the emphasis on the existence of discontinuities, indicating essential limitations of our knowledge; (3) connected with this, the emphasis on the approximate character of knowledge; (4) the demand to unite practice with theory, so that there is neither unreflected practice nor empty theory; (5) the change (motion) of concepts in the course of the development of our knowledge.[16]

Feyerabend uses these points to argue that Bohr followed a dialectical approach in developing the Copenhagen interpretation of quantum mechanics.

He defends Bohr against the charge that he was an idealist, promoting the role of consciousness in constituting reality, arguing instead that the first point "was now extended to include the human observer and his peculiarities." This extension was thoroughly materialist in Lenin's "inexhaustible" sense, since Bohr "provides objective criteria for the implications of the fact that in theo-

14. From a 1946 book review cited in Kutrovátz, "Lakatos' Philosophical Work in Hungary," p. 117. On the influence of Lukács on Lakatos, see Dusek, "Lakatos between Marxism and the Hungarian Heuristic Tradition"; Kadvany, *Imre Lakatos*, pp. 287–316. For Lakatos's retention of an interest in dialectics after abandoning Marxism, see Kutrovátz, "Lakatos' Philosophical Work in Hungary," pp. 129, 131.

15. Feyerabend, "Dialectical Materialism and the Quantum Theory," p. 414; Feyerabend, "Consolations for the Specialist," pp. 205, n. 1, 211.

16. Ibid., p. 415.

rizing we are restricted by our own (material) organization."[17] The false view that Bohr was an idealist resulted from "secondary philosophies, that is, from philosophical points of view which, though not effective in research, are used afterwards for explaining the results in general terms." Notably, Feyerabend mentions physicist James Jeans, who had come in for similar criticism by Lakatos in the surviving articles incorporated into his Hungarian dissertation.[18]

Both Lakatos and Feyerabend explicitly built their philosophies of science upon a dialectical approach, and it is dialectical philosophy's combination of a historical and critical perspective that distinguishes them from Kuhn. A fundamentally historical approach to knowledge does not imply that scientific experts possessed and controlled the knowledge that they produced, as Kuhn and, later, constructivist sociologists of science believed.

Kuhn, Lakatos, and Feyerabend all endorse the existence of social or conventional elements in science straight through to its core. Where they differ is in evaluating what this difference means to the practice of science. Lakatos and Feyerabend, despite their differences with each other, agreed on the following three criticisms of Kuhn's philosophy of science, which animated their contribution to *Criticism and the Growth of Science* and their subsequent philosophical work.

First, they held that the typical state of science was one involving intense disagreement about foundations, not the operation of homogeneous paradigms accepted without dissent. Second, they argued that the state of consensus within science was overstated by proponents of dominant paradigms by selective presentation of the actual current and past views of scientists in a way that is *dysfunctional* for science (as opposed to Kuhn's view that a paradigm's rewriting of the past served scientific progress within normal science). Third, politics, religion, or other cultural belief systems, whether judged good or bad, reactionary or progressive, could enable positive contributions to science, even if those contributions were not recognized as such at the time. Their conclusion is that the history of science can and should continue to be a source for ideas in the present even if those ideas are at odds with dominant perspectives or have been rejected as refuted or mistaken.

THE SURVIVAL OF MINORITY VIEWS: THE CASE OF LAMARCKISM

In this sense, minority views never disappear within science and can be re-established as science changes, making formerly implausible ideas plausible

17. Ibid., p. 416.
18. Kutrovátz, "Imre Lakatos' Hungarian Dissertation," pp. 361–62, 354–55.

again, albeit in new form. A good example of this is the growing reconsideration of non-germ line inheritance in biology. "Lamarckian" inheritance of acquired characteristics has long been a textbook example of a minority view that was seen as not credible and displaced by the advancement of science, first as August Weismann distinguished germ line from somatic influences, then again, as the modern synthesis in biology put forth its "central dogma" that genes drove the construction of proteins and not vice versa, blocking a path for phenotypes to influence genotypes. The role of the environment is strictly to "select" adaptive genes over those that will not be passed on (or to cause random mutations), but environmental modifications of phenotypes could not be passed on genetically and genetic inheritance was the only kind of inheritance that was considered possible.[19]

With the growth of epigenetics, the recognition that at least some examples of "epigenetic inheritance" exist, whereby environmental modifications of the "epigenome"—physical attachments to the genome that can turn on or off specific genes—can be passed on to successive generations in some cases.[20] Some scientists now speak of epigenetic inheritance as its own level of evolutionary change.[21] Although most scientists generally avoid speaking of the inheritance of acquired characteristics, it can be argued that epigenetic inheritance is exactly that and that study of molecular modifications of the genome—through processes such as methylation and histone acetylation or deacetylation—can provide causal specificity for epigenetic change and epigenetic inheritance in just the way that molecular biology did for genetic evolution via the study of DNA.[22]

The fact that developmental biology was left out of the modern evolutionary synthesis facilitated this neglect of epigenetics, as focus on the developing embryo by scientists such as Hans Dreisch and Conrad Waddington had already highlighted the fact that different genes must be turned on or off, once it was realized that the entire genome was copied by each cell even as cell differentiation occurred.[23] Neglect of this proto-epigenetic finding within developmental biology was made possible by developmental biology's disciplinary exclusion from the modern synthesis in biology, despite the clear demonstration that something like epigenetic modifications of the genome were needed to explain cellular differentiation in development, once it was realized that the genome was not divided and distributed to distinct body parts in development. In short, a viable research program developed apart from

19. Graham, *Lysenko's Ghost*, pp. 15, 19–22.
20. Moore, *Developing Genome*.
21. Jablonka and Lamb, *Evolution in Four Dimensions*; Jablonka and Lamb, *Epigenetic Inheritance and Evolution*.
22. Graham, *Lysenko's Ghost*, pp. 12, 15–16.
23. Moore, *Developing Genome*, ch. 3.

dominant approaches to biology as a whole and has only recently begun to be incorporated into a potential new synthesis.[24]

In developing the new science of epigenetics, biologists have found themselves struggling against orthodox representations of scientific consensus that were always mistaken. Darwin himself, and many other contributors to "orthodox" evolutionary thinking—like Weismann—were actually supporters of the inheritance of acquired characteristics as one mechanism accompanying strict "Darwinian" selection. Other scientists supportive of the inheritance of acquired characteristics are remembered more for their other contributions to science, with their "Lamackianism" passed over in silence, as with Ernst Haeckel and Ivan Pavlov.[25]

In the case of Paul Kammerer, once well known as a defender of the inheritance of acquired characteristics, accusations of fraud and his resulting suicide may have led to the continued neglect of experimental results for which replication has not been attempted and which could potentially have an epigenetic explanation.[26] Cultural differences between different national scientific communities may also have been obscured by orthodox claims of scientific consensus, as Russian scientists both before and after the Russian revolution remained sympathetic to the inheritance of acquired characteristics. While most scientists condemned the suppression of genetics by Lysenko and Stalin, contemporary Russian scientists have raised the question of whether Lysenko was right (scientifically) after all. In truth, his own particular contributions were neither original nor based upon a strong experimental foundation, but the metonymic collapse of all historical developments of the concept of the inheritance of acquired characteristics to a discredited "Lysenkoism" obscured the extent of dissensus that actually existed historically and obscured new approaches building upon these suppressed views.

JUSTIFICATIONISTS VERSUS CONJECTURALISTS

In his lectures to his graduate students, Lakatos makes clear that Kuhn and Polanyi's philosophies were problematic not because they incorporated history or sociology, but because they evidence an epistemic politics he called "elitist authoritarianism," the view that self-selecting experts know best what to believe and who is to be believed.[27] Research programs are defined by

---

24. Pigliucci and Müller, *Evolution: The Extended Synthesis*.
25. Graham, *Lysenko's Ghost*, ch. 2.
26. Ibid., ch. 3.
27. Lakatos, "Lectures on Scientific Method," p. 94. Despite Lakatos's criticisms of Polanyi's elitist view of science, Lakatos was influenced by his views on the role of heuristics, and invited him to the London conference debating the philosophies of Kuhn and Popper, over the objections of Pop-

Lakatos in a way that makes possible an understanding of science's conventional component and its change over time. It does so because the conventional components provide rules for producing new scientific work that may or may not prove fruitful.

Metaphysical assumptions about the phenomena under study are incorporated into a conventional hard core of the research program, which is generally exempt from critical scrutiny. A positive heuristic based upon the hard core provides guidance for producing predictions that may or may not be corroborated. We can compare the success of research programs in generating novel confirmed predictions to assess which programs have been successful and—crucially—to offer an unexhausted positive heuristic that gives scientists something to do with their training.

While most scientists will tend to pursue programs that offer the chance for fruitful predictions, there is no instant rationality or decisive refutations in Lakatos's judgment. Fruitful knowledge produces *new* knowledge, and it is this *productive* element of scientific culture that distinguishes it from other cultural forms that may nonetheless include representations of nature. In other words, science is not just defined as any representation of nature but is confined to a new kind of dynamic cultural practice promoting rapid discovery and specialization.[28]

As a result of the dynamic quality of modern science, ideas are not judged primarily by how justified they are by empirical evidence but by the extent to which they continue to help generate additional cutting-edge research that leads somewhere. In dialectical terms, knowledge is not a static representation but a moment in an ongoing transformation, which is why Lakatos focused on a *series* of theories within an overarching research program.

The most important idea shared by both Lakatos and Feyerabend is that unjustified ideas are welcome in science and deserve the opportunity to make their case. In a letter to Lakatos, Feyerabend distinguishes "justificationists" from their own position as "conjecturalists" or "those who admit that unjustified hypotheses are fully fledged members of the domain of knowledge" (explaining, for instance, Lakatos's defense of Arthur Jensen's racial IQ research and Feyerabend's defense of acupuncture or witchcraft).[29]

---

per. See Dusek, "Lakatos between Marxism and the Hungarian Heuristic Tradition," p. 71. It should also be noted that Polanyi defended a form of realism, arguing that scientists' judgments tracked an underlying metaphysical reality. He saw this as fundamentally opposed to Kuhn's relativism. See Moleski, "Polanyi vs. Kuhn," pp. 18–22.

28. Lakatos's implicit sociology of science has been developed as a model independently by the sociologist Randall Collins in *The Sociology of Philosophies*. See also Lynch, "Cultural Evolution and Social Epistemology," pp. 67–68.

29. Feyerabend to Lakatos, September 24, 1970, in Lakatos and Feyerabend, *For and Against Method*, p. 216.

While they share the conjecturalist position with Popper (see table 2.1), they differ from his commitment to "abstract" epistemology, favoring the historical view that "pay[s] attention to the (historical, cosmological, physiological etc.) conditions which must be satisfied if a certain epistemological ideal . . . is to be realized in the form of a science."[30] Their conjecturalist quality distinguishes them from Hegel, whose historical approach they share, leading to Feyerabend's joking reference to Lakatos as a fusion of Popper and Hegel: Pegel, Hopper, or Pop-Hegelian.

Table 2.1. Feyerabend's Classification of Philosophies of Science, from a Letter to Lakatos

	justificationist	*conjecturalist*
abstract	Descartes	Mill (*On Liberty*)
	Plato (*Republic*)	Popper
historical	Hegel	Vico
	Plato (*Laws*)	Lenin
		You
		I

Source: Lakatos and Feyerabend, *For and Against Method*, p. 216.

As for Feyerabend, his dialectical approach took a distinctly Marxist form, though coupled with a skepticism that fits better his epistemological anarchism and political distaste for authoritarian radicals.

> All *I* need is Xenophanes; frg. 189: "no man knows, or will ever know, the truth about the gods and about everything I speak of; for even if one chanced to say the truth, yet one knows it not; but seeming wrought over all things." This [asserts] the hypothetical character of all our knowledge. Frg. 190: "Let these things be opined as resembling the truth . . ." That is, although everything is hypothetical, let us not give up altogether (irrationalism) but deal with the hypotheses. Frg. 191: ". . . by seeking, men find out better in time" (against skepticism: the hypotheses can be improved). That is all that is needed, and if we want to take care of *science*, we add *dialectical materialism* to it. Had Popper been content simply to defend this tradition (as he does in *Conjectures and Refutations—without mentioning dialectical materialism*) . . . we would not now be faced with the battle of the technicians.[31]

30. Ibid.

31. Feyerabend to Lakatos, July 27, 1968, in Lakatos and Feyerabend, *For and Against Method*, p. 148. On the influence of "Red Vienna" on Popper's philosophy, see Hacohen, *Karl Popper, the Formative Years, 1902–1945*, chs. 2–3.

Interestingly enough, they share the square with Lenin, whose anti-Machianism helped define the limits of conventionalism for Lakatos. Feyerabend, for his part, defines his position as building upon Greek skepticism, supplemented by dialectical materialism, a tradition Popper abandoned for sterile analytical distinctions that Lakatos and Feyerabend believed led to a sterile, empty approach to the philosophy of science.

Upon hearing of Feyerabend's "conversion to Marxism," Lakatos refers him to his manuscript of *Proofs and Refutations*, for his own "Marxist" take on mathematics. In the relevant section, Lakatos summarizes his philosophy of mathematics as the view, evaded by most mathematicians, that "reason is too weak to justify itself." In support, he cites the Russian mathematician Aleksandr Aleksandrov, who argued that "the rigour of maths is not absolute; it is in process of continual development: the principles of maths have not been congealed once and for all, but have a life of their own and may even be the subject of scientific quarrels."[32]

Feyerabend echoes back that he can fully subscribe to Trotsky: "A political struggle is in its essence a struggle of interests and forces, *not of arguments*": if *astronomy* can't get on on the basis of argument only, *how should politics be able to?* Therefore—back to Marx, I say to myself, and re-examine what can be preserved of 'formal democracy,' which plays the same role in politics that argument plays in astronomy."[33] Cultural forms do not stand in need of formal justification but can be judged by their role in a developmental process over time.

A DIALECTICAL PHILOSOPHY OF SCIENCE

Kadvany argues that the relevant context to understand Lakatos's Methodology of Scientific Research Programs is his hidden engagement with Hegelian philosophy rather than his explicitly analytical style of philosophy of science.[34] Feyerabend's own continuing engagement with Lakatos's philosophy similarly reflects a more sophisticated understanding of Lakatos's project than is evident in Anglo-American philosophy of science. As a result, Lakatos's philosophy of science may speak to contemporary Science and Technology Studies more than would be apparent from focusing too literally on the "internalist" character of his writings. Lakatos was less concerned to subject scientific practice to the externally imposed rules of the philosophy of sci-

32. Lakatos to Feyerabend, August 1968, in Lakatos and Feyerabend, *For and Against Method*, pp. 150–51; Lakatos, *Proofs and Refutations*, pp. 54–55. On Lakatos's application of Popperian fallibilism to mathematics, see Glas, "The 'Popperian Programme' and Mathematics," parts I and II.
33. Ibid., pp. 151–52.
34. Kadvany, *Imre Lakatos and the Guises of Reason*, ch. 7.

ence than to recover grounds for judging the success of research programs in advancing on their own terms, while simultaneously comparing their progress to competing perspectives.

Lakatos's approach has been understood as part of the philosophy of science's opposition to sociological approaches to science. In particular, his emphasis on reconstructing an internal dynamic within scientific research programs looked to most sociologists like a rejection of a sociological approach for classic internalism. Internalism, on this view, is the thesis that science develops from its own autonomous dynamic of knowledge that should remain insulated from outside, sociological influence. Lakatos, however, was putting forward his own form of a sociological approach to knowledge that shared Kuhn's view that science was driven by the conventional approaches of distinct expert communities but denied his view that single paradigms dominated a field in mature science.

Key to reconsidering this verdict on Lakatos is to recognize that his "internal" history is a retrospective reconstruction of knowledge that *includes within it* social dimensions of epistemic interest that drive science forward, but excludes *accidental* psychological or sociological factors. Historical process is key here: an attempt is made to explain the *dynamic* character of scientific change.[35] Notice that this means that he is not just walling off science from external influence (for macrosociological or cultural elements could in principle be key in explaining the intellectual origins of the hard core and positive heuristic that drives the research program). The perspective is also internal for *us*, for analysts of science, in the sense that it puts forth a candidate explanation for reconstructing the growth of science that can succeed or fail in generating novel insights in the history of science.

Here we might briefly consider what the "ontology" of knowledge is for Lakatos—that is, what kind of entity knowledge is taken to be. Analytical philosophers have been obsessed with relating knowledge to beliefs of a certain form—those that are justified, true, and otherwise characterized by their relationship to the outside world. On this view, there is a correspondence between a psychological faculty and a state of the world. Sociologists like Harry Collins take this conception and make it collective, with knowledge understood as *shared* beliefs. Both views accept that the kind of entity knowledge is involves individual or collective beliefs.

The picture of what the entity called "knowledge" is actually like in Lakatos's philosophy is actually *more* social than much SSK (sociology of scientific knowledge) because it doesn't just reify personal or culturally shared *beliefs* but treats knowledge as an *alienated product of scientific*

35. Nickles, "Justification and Experiment."

labor.[36] When Lakatos advocated putting the "real" history in the footnotes—always a sore spot among historians—it reflected his argument that the psychological attitude of a scientist to his or her theory may or may not reflect the theory's place in a dialectical succession of theories driven by a positive heuristic (this has affinities with poststructuralist approaches that deconstruct the author).

Lakatos's intimate linkage of the context of justification with the context of discovery, evident especially in his work on the history of mathematics, allows for a very strong sociological account of scientific knowledge, so long as the sociological analysis tends to illuminate the dynamic character of science—its ability as an institution to promote the *growth* of knowledge via the generation of new theories emergent from failed theories. This dynamic theory of scientific change amounts to an application of a sociological account of the *internal* dynamics of science as driven by available research programs, which themselves are socially situated. Thus, Lakatos's mature philosophy of science represents not an anti-sociological internalism, but a sociology of knowledge that is dialectical in being driven by the contradictions between existing scientific concepts and the current state of knowledge.

Lakatos's lost 1947 Hungarian doctoral dissertation, "On the Sociology of Concept Building in the Natural Sciences," demonstrates his long immersion in debates within the sociology of knowledge in the Hungarian context, as documented by surviving reports of its content and articles published reflecting its content. Including obligatory discussion of the sociology of knowledge of Max Scheler and Karl Mannheim, Lakatos extended the ambit of the field to the natural sciences, making his dissertation a contribution to a fully general sociology of scientific knowledge comparable to Ludwig Fleck's *The Genesis and Development of a Scientific Fact* (1935). It is no wonder that Lakatos saw himself in a position to synthesize the approaches of Kuhn and Popper, given that his intellectual background in the sociology of knowledge was deep where Kuhn's was shallow and serendipitous. One of the readers of his dissertation, Sándor Karácsony, summarized its central claim as the argument

> that scientific concept-building is not objective, not the alphabet of nature, but it is a relation and a function of the prevailing social structure. Therefore, natural science, during the course of its progress, occasionally outgrows its proper system of concepts, as long as socialist society has not yet reached its static rest

36. See the discussion of Lakatos's Lukácsian sociology of concept formation in science from his (missing) 1947 Hungarian dissertation, "On the Sociology of Concept Building in the Natural Sciences"; Kutrovátz, "Lakatos' Philosophical Work in Hungary," pp. 125–29. For similar arguments, see Feyerabend, "Consolations for the Specialist," p. 228, as well as Bartley, "Alienation Alienated," and Bartley, *Unfathomed Knowledge, Unmeasured Wealth*, for a Popperian account of alienated knowledge.

> state, at which it will force scientific concept-building into a form more independent of time and facts.[37]

It is interesting to note here that it is socialist society that functions to ultimately exempt science from historical determination, a view Lakatos no doubt jettisoned when he abandoned Marxism. Marxism itself functions as a historically determined limit on the dialectical philosophy to which Lakatos remained committed.[38]

What remains evident in Lakatos's London career, given this background, is his emphasis on the necessity of confronting "contradictions" rather than evading them. The other reader, Ottó Varga, emphasized this aspect of Lakatos's dissertation, describing Lakatos's commitment to the view "that the dissolution of contradictions is made possible only when we view any scientific statement as a creation of the society which created that science—that is, truth is always a function of history."[39] In his fallibilist and empirical account of mathematics, Lakatos objected to *monster-barring* in nineteenth-century mathematics—revising a flawed proof by changing the definition of the mathematical entity to eliminate the counterexample and preserve the proof from challenge. Lakatos noticed that the most progress was made historically in generating new, more refined proofs by *lemma incorporation*—proofs that show why the counterexample doesn't fit, even if that involves *concept-stretching* where the meaning of the original theorem was unavoidably changed. Only a historical approach to mathematics and not a focus on formalism could illuminate how mathematical discoveries are generated.

He shows this through a close examination of the counterexamples proposed to Euler's formula, $V-E+F=2$, relating the vertices, edges, and faces of polyhedra, showing that the existence of apparent counterexamples to the theorem led to the incorporation of revised lemmas used in the proof of the theorem. Because this back and forth between deductive reasoning and quasi-empirical observation of counterexamples undercut the idea of strict deductive necessity, Lakatos used the history to argue that the project of mathematical formalism did not make mathematics certain.[40] As one commentator put it, for Lakatos, "*nothing* in mathematics is self-evident. Self-evidence in mathematics is an

37. Quoted in Kutrovátz, "Lakatos' Philosophical Work in Hungary," p. 125. For a translation of the reader reports and the surviving article, "Modern Physics, Modern Society," which was incorporated into the dissertation, see Kutrovátz, "Imre Lakatos' Hungarian Dissertation."

38. Compare Lukács on Marxism as method in *History and Class Consciousness*, a work Lakatos cites in the "Modern Physics, Modern Society." The dissertation reader, Sándor Karácsony, tacitly invoked Lukácsian influence in noting that the dissertation was "based on dialectic Marxism, but in its modern and not its orthodox form." See Kutrovátz, "Imre Lakatos' Hungarian Dissertation," pp. 369, n. 13, 372, 374, n. 21.

39. Quoted in Kutrovátz, "Lakatos' Philosophical Work in Hungary," p. 126; Kutrovátz, "Imre Lakatos' Hungarian Dissertation," p. 373.

40. Lakatos, *Proofs and Refutations*.

illusion."[41] Feyerabend observed that Lakatos's argument "remove[d] the last Aristotelian element, the element of necessitation, from modern science."[42]

Lakatos's philosophy of science—with its attention to the role of positive heuristics in promoting theoretical dynamism—extended this approach from mathematics to science. The idea that methodological ideals in science, like formalism in mathematics, were themselves fallible and changeable guides to inquiry linked his 1947 dissertation straight through to a planned work at the time of his death, titled *The Changing Logic of Scientific Discovery*.[43]

In light of this scholarship on Lakatos's Hungarian intellectual background, it challenges an identification of Lakatos's philosophy of science with anti-sociological internalism.[44] In his philosophy of science, what is external is any element of culture or biography that may shape some scientist's activity but that fails to be connected to our best reconstruction of the dynamic growth of knowledge over time. Elements internal to a research program are not culture-free, but highly conventional approaches to explaining specific phenomena. Different research programs are also competitors who nonetheless cannot easily be compared except retrospectively in terms of their fruitfulness in generating discoveries over time.

Just as Lakatos has been pigeonholed as a standard internalist, Feyerabend's epistemological anarchism has been treated as a kind of free-standing, wild interpretation of science, when it would be better to take notice of the way his position is dialectically tied to Lakatos's effort to explain the growth of knowledge within and across research programs. In this sense, Feyerabend's philosophy of science is a successor theory in the Lakatosian historiographical research program. Thus, Lakatos can explain novel predictions (with modifications introduced by Zahar) in the Copernican case but only if he ignores the question of *how* the earth moves. Feyerabend follows the rules of Lakatos's program by picking out Galileo for examination since he fits our current scientific intuition of a great scientist. He then argues that Galileo's work will not fit the pattern of progressive improvement; instead, he only achieves the importance he has for us by violating these rules and engaging in propaganda. In one sense, this is a falsification of Lakatos's program. However, they both reject naïve falsification as a criterion for evaluating research programs, so Feyerabend's intervention can be seen as closer to a lemma incorporation of counterexamples to Lakatos's philosophy of science.

41. Koetsier, *Lakatos' Philosophy of Mathematics*, p. 19.

42. Cited by Bağçe and Başkent, "An Examination of Counterexamples in *Proofs and Refutations*," p. 19.

43. Kadvany, *Imre Lakatos and the Guises of Reason*, pp. 393–95; Motterlini, "Reconstructing Lakatos"; Kutrovátz, "Lakatos' Philosophical Work in Hungary"; Dusek, "Lakatos between Marxism and the Hungarian Heuristic Tradition."

44. Shapin, "Discipline and Bounding."

Indeed, in his contribution to *Criticism and the Growth of Knowledge*, Feyerabend cites Lakatos along with Lenin and Trotsky to underwrite what he later called counterinduction: holding on to a theory in the face of refuting facts, since "a 'phase difference' between theories and the corresponding auxiliary sciences" may be the cause of the apparent refutation. This principle of *tenacity* must be supplemented by one of *proliferation*, borrowed from Mill and Mach. Tempered by "dialectical materialists who objected to extreme flights of fancy," proliferation and tenacity find their synthesis in the writings of Engels, Lenin, and Trotsky, who help answer what Kuhn cannot: how a transition from normal science to revolution can actually take place.[45] Just as the seeds of the new society grow in the old for the Marxists, proliferation does and should exist within "normal science" to make scientific revolutions possible. It should be clear that Feyerabend's concept of tenacity contrasts with Merton's norm of disinterestedness, corresponding closely to Mitroff's counternorms of interestedness and emotional commitment and fitting with Kuhn's description of paradigm-guided reasoning in science. Unlike Kuhn, such tenacity must be coupled with the proliferation of approaches within science to avoid dogmatism and stagnation.

We can conclude that Feyerabend supplements Lakatos's approach in two ways. First, he abandons the distinction between research programs and theories, since theories ought to compete without separation into distinct, relatively autonomous programs and since research programs often shield themselves from criticism from nearby fields, as his Galileo argument showed.[46]

Second, he argues for the use of incommensurability as a tool of *criticism*, which makes possible "jumps from quantity to quality in the language of dialectical materialism."[47] This makes sense of Feyerabend's use of incommensurability not just as an obstacle to scientific communication, as it was for Kuhn, but as a constructive tool for subjecting a stultifying observation language to criticism. Galileo's account of motion is incommensurable with Aristotle's, for Feyerabend, because this allows him the freedom to explore how a rejection of "established" facts might free up the theoretical imagination. This in turn accounts for the curious lessons that Feyerabend gives us from his discussion of Galileo: the Church was justified in rejecting Galileo's view as incompatible with known facts, and Galileo was right to ignore these facts.

45. Feyerabend, "Consolations for the Specialist," p. 211.

46. Ibid., p. 203, n. 2; p. 211, n. 1.

47. Ibid., p. 211, n. 1. See, also, Feyerabend's posthumously published *Philosophy of Nature*, p. 185, where he follows Hegel in arguing for "motion" in our concepts: "Stability in our concepts and ideas does not show that we have found truth. Such stability is evidence that we have not succeeded in getting beyond a random state in our knowledge process to reach a higher state of consciousness and understanding. The question even arises whether we actually possess any knowledge in such a state. We adopt the existing categories, we learn the available alternatives until we can repeat them without any difficulty, and our thinking loses its spontaneity; it turns into 'bestial staring at the world.'"

Not only is the introduction of a new, incommensurable, theory a positive thing for the new theory, it also allows for a better engagement with the world by the *old* theory. Indeed, in a letter to Kuhn providing criticism of the draft of *The Structure of Scientific Revolutions*, Feyerabend argued that the introduction of a new perspective in a scientific revolution improved the empirical content of the old theory by disassociating it from the tendency to seek the fitting of the natural world into normal science boxes by allowing criticism from an external point of view.

> The practice of repressing novelties, the attempt to let the theory penetrate every part of nature (and of thinking) usually leads to a situation where people are psychologically unable to imagine alternatives, where procedures such as transcendental deduction will be very successful and where the impression arises that "we are always chasing our own tails in a formal system" (as Eddington has expressed it). This practice therefore finally leads to the establishment of a language which can be used for the description of every and any situation and a situation which does not fit the theory cannot even be imagined. This means, however, that this method of excluding novelties, this attempt to press nature into the boxes of the theory will gradually decrease the empirical content of the theory until it is finally almost zero. A scientific revolution which shows the limitation of the theory and which points out very plainly where it is wrong therefore gives back empirical content to the theory, i.e. makes it again a theory of nature. Conclusion: better live in permanent revolution than in the state of normalcy.
>
> Revolution in Permanence should be the battle cry of every empiricist. The "close fitting" between theory and nature which normal science tries to achieve decreases the empirical content of the theory and thereby also decreases its ontological significance. "Normal Science" therefore tries to achieve something which is very undesirable and its undesirable aftereffects are eliminated only by a revolution. The closer we try to get to nature (by the method of trying to fit nature into boxes designed by us) the further we get away from it.[48]

This idea could be called *empirical overshoot*, a situation where the mutual adjustments between observation and theory within a paradigm leave little room to reconsider our language for describing nature.

For Feyerabend, incommensurability represents a tool of criticism that is useful for bringing back an older theory into a better connection with nature than Kuhnian-style normal science allows. Feyerabend was clear in this letter and throughout his career that Kuhnian-style normal science was undesirable, whatever its prevalence in the history of science. Feyerabend objected particularly to the conflation of "is" and "ought" in Kuhn's work. While he drew his lessons from historical examples, his emphasis was on maximizing the productive consequences of criticism for advancing new scientific

48. Hoyningen-Huene, "Two Letters of Paul Feyerabend to Thomas S. Kühn," p. 358.

perspectives and better forging a connection to the natural world through a reconsideration of its conventional content on a regular basis.

Keeping in mind Lakatos's terminology above, Feyerabend shares Lakatos's focus on producing new conjectures rather than ensuring logical justification. For Lakatos, "reason is too weak to justify itself."[49] What interested both Lakatos and Feyerabend was not whether a particular theory was justified at a particular moment in time but whether the practice of science was tending to generate new knowledge from old, transforming in dialectical ways the older theoretical language, and establishing new possibilities repressed by the tendencies to scientific inertia in dogmatic training and practice valorized by Kuhn.

In a sense, Feyerabend improves upon Lakatos's account by introducing new concepts that amplify and extend the analysis of research programs. In the long run, we understand that Galileo was right about the earth's motion, so the question is whether we can show he is right in the short run. Lakatos and Zahar's attempts to show this fail. On Feyerabend's analysis, there is a sense in which *both* older and newer theories can be justified, as facilitated by their respective observation languages (but, again, justification is not the right focus). Both Lakatos and Feyerabend were more concerned with the generation of novel insights than the preservation of the old. As such, the real import of incommensurability is not that it allows each perspective to *ignore* the other, but that it forces a choice where dogmatism had reigned before.

Revolutionary perspectives often require "breathing space" to work out a new observation language when the old is so infected by the older theory and flawed common sense. In Feyerabend's analysis, incommensurability is functional for the development of Galileo's program, not only by allowing breathing space for a seemingly falsified view (that the earth moves) but by allowing evaluation of the established view from the perspective of a *new, external* standpoint. This differs radically from Kuhn's view of incommensurable paradigms as occupying separate worlds that do not interact. Instead, since the natural interpretations are so seamlessly integrated with Aristotle's Tower argument against the motion of the earth, they would *never break down of their own accord*, and require the construction of an alternative natural interpretation in order to see their limitations.

Thus, Feyerabend rejects the view that normal science will uncover serious anomalies on its own. He encourages the construction of alternative standpoints from which to evaluate the adequacy of a normal, disciplinary scientific practice rather than treating normal science only in terms of the normal scientist's own intentions. This approach contrasts with the reigning orthodoxy in science studies—in this sense, actor's accounts are just as

49. Lakatos, *Proofs and Refutations*, pp. 54–55.

external for Feyerabend as they are for Lakatos, what matters is how a rival perspective displaces another through insinuating a new common sense.

In Feyerabend's judgment, Galileo's astronomical discoveries may be progressive, but overall his approach appears to be degenerating by Lakatos's own criteria (this can be seen as an accusation that Lakatos falls back from the conjecturalist position to a justificationist perspective). The reason that Galileo can be seen to be degenerating—and this is not a bad thing in the long run—is that science exhibits "uneven development," following Trotsky's political concept, and so propaganda defending Copernicanism from a premature negative judgment was justified.

This is where the much misunderstood slogan, "Anything Goes!" comes into play: it is not that scientists should be able to do whatever they want without consequence, alternately focusing on or ignoring facts, but that the significance of a scientific contribution has to do with its role in transforming science's past into its future. It is the dialectical role Galileo played in getting science out of a (completely justifiable) tight spot that accounts for his status as one of Lakatos's "goodies"—those scientists we credit with significant contributions—not his establishment of a new, unquestioned theoretical perspective. Incommensurability is a tool of criticism not of quietude.[50]

One might gloss the differences between Kuhn and Feyerabend on scientific revolutions as a difference between "Townsendian" and "Trotskyite" views of revolution. In the lyrics to the Who song "Won't Get Fooled Again," Pete Townsend laments that following his support for a revolution, the revolutionary leaders' "beards have all grown longer overnight." A revolution is the mobilizing of the people to overthrow an old orthodoxy and establish a new one, and at best one can be wistful about the betrayal involved, praying that one won't get fooled again (but knowing that one will).

By contrast, Trotsky promoted a view of Marxist revolution that was predicated on following a revolutionary seizure by moving against the new order before it could consolidate a new orthodoxy. Trotsky had seen the promise of the Petrograd Soviet of 1905 to be that Russia's overall backwardness was an advantage since it experienced only the latest, most concentrated form of capitalist production without any of the earlier political forms that inhibited socialist revolution in the rest of Europe. Weak feudal relations, an autocratic czar, a feeble domestic bourgeoisie, and a mostly foreign ownership of Russian capital meant that none of the religious or abstract democratic character of the English or French revolutions would apply, while Russian peasant traditions

50. Oberheim, "On the Historical Origins of the Contemporary Notion of Incommensurability"; Oberheim, *Feyerabend's Philosophy*, ch. 5; Feyerabend, "Explanation, Reduction, and Empiricism." Contrast Kuhn's approach discussed in Fuller, *Thomas Kuhn*, pp. 6–18.

of collective ownership of land and the workers' soviets allow them to skip the bourgeois revolution and set the stage for worldwide proletarian revolution.[51]

Uneven development in science means that the lack of methodological justification for Galileo's ideas given the dominant science of the time can be a means for launching a revolutionary leap unavailable to those currently possessing the most empirically justified theories. Once again, dialectical potential matters much more than temporary empirical justification, which is why counterinduction can advance science in the right hands, and more specifically, at the right time.

Feyerabend's "uneven development" builds upon Lakatos's identification of the historical dynamic of research programs, since if different fields did not have their own internally driven patterns of growth, there would be no sense in talking about the different paces that development takes in different fields. Lakatos already allows that programs may degenerate for a time only to recover and vanquish their opposition. Feyerabend provides more content to this position by specifying the reason for this, albeit a reason that itself is only justifiable in the long view of the history of science. In the terms of Lakatos's history of mathematics, Feyerabend does more than monster-barring—pointing to a counterexample—he gives a lemma-incorporation, which specifies the reason why this counterexample exists in a manner that allows the theory in question to be modified.

> The synthesis [of Popper and Kuhn] consists in Lakatos's assertion . . . that proliferation and tenacity do not belong to *successive* periods of the history of science, but are always *copresent*. Lakatos's analysis . . . can be further improved by . . . allowing for incommensurability (jumps from quantity to quality in the language of dialectical materialism). Improved in this way it would be a truly dialectical account of the development of our knowledge.[52]

If we connect Feyerabend's argument to the Hegelian context for Lakatos's work that Kadvany provides, we can see Feyerabend's accomplishment as a rejection of MSRP (Methodology of Scientific Research Programs) as a static summary of the historiography of science, resituating MSRP as the first stage in the same kind of dynamic pattern for the historiography of science that Lakatos has shown characterizes science itself. Lakatos would agree with this need to see MSRP as a first step to be improved upon by successive programs that better explain the growth of science, or more precisely, that lead to fruitful case studies that corroborate some of our evolving historiographical insights. MSRP, for Lakatos, represents the best reconstruction (if temporarily so) of the internal dynamic of science available at this stage in

51. Trotsky, *The Permanent Revolution.*
52. Feyerabend, "Consolations for the Specialist," p. 211, with text from footnote 1 appended.

the history of science, an understanding that can be expected to grow as a dynamic interaction of historiography and empirical research in the history of science proceeds and as science itself evolves, thereby offering new data and new endpoints for our reconstructions.

In this sense, MSRP is presentist, though in a fashion that I would argue is unavoidable and not Whiggish, since Lakatos's fallibilism insists that our contemporary judgments about great science (e.g., Galileo) may change, though there always remains an unavoidable endpoint in a process of directed change. We don't know what the future will bring, but at the same time we must not reject any tentative standpoint at all. Also, we may *begin* from contemporary judgments of great science, but our examination may question some of these judgments as is evident in his analysis of Michelson's ether experiments—where Lakatos came to the conclusion that his work was rather pedestrian and confused and does not form the basis for the development of relativity.

If Lakatos is right, Science and Technology Studies (STS) should consider the evolving relationship between our historiographical or theoretical understanding of the growth of knowledge and empirical research in the history of science. If we consider how theory and empirical research relate, we could only conclude that STS is degenerating, since we have more or less tabled the discussion of the longer-term development of knowledge or, worse, overlaid our preferred narratives of science on a case study in an ad hoc fashion. Park Doing demonstrates this in his discussion of laboratory studies by showing that the theoretical account of constructivism does not follow from the empirical research that was carried out in laboratories. Rather, constructivist language is used as a narrative overlay for describing what was going on in the laboratory without showing that such a perspective led to new insights in explaining laboratory practices.[53] As history of science has increasingly disengaged itself from a relationship with philosophy of science, and has grown closer to political and cultural history and cultural studies, the field has degenerated.

Just as in the case of Galileo, we may need to rely upon the long run of history before we can work out how our improved understanding of the history of science should inform a mature philosophy of science. But we should look about for neglected research programs with promise and spell out if we can a picture of the growth of knowledge that is not only compatible with recent history of science (allowing for the inevitable ocean of anomalies) but also can guide further investigation that helps us understand science as a knowledge-producing enterprise. It is not enough, however, for the critic to emphasize the existence of anomalies in the dominant approach to the field. Such anomaly mongering ought to be replaced by the development of approaches that generate new knowledge. This, too, is a corollary of the

53. Doing, "Give Me a Laboratory."

emphasis on favoring the production of useful and fruitful conjectures over abstract justification of a static view by means of existing facts.

Reflexively, empiricism is not only a poor philosophy of science but it is a poor basis for understanding how the history of science should inform STS research. One of the biggest obstacles to a sustained, dialectical relationship between theoretical historiography and history of science is the persisting disciplinary tendency among historians of science to adopt a naïve empiricism as the tacit epistemology of their own research. Lakatos was clear about this deficiency in the practice of historians, and his use of reconstructions coupled with corrections in the footnotes was an effort to call attention to the theory dependence of historical observation, so it does little good to respond with pious platitudes about getting the facts right or understanding history "in its own terms."[54]

For a Hegelian, we might add that there is an elementary difference between what history is "for us" and what it is "for them" (the participants, on their own terms), which may be synthesized at a higher level of research but cannot be wished away. To illustrate this limitation of historians' epistemology, he included Pearce Williams as a contributor to *Criticism and the Growth of Knowledge* to document the problem. Despite his traditionalism vis-à-vis contemporary history of science, Williams's words still find echoes in our field:

> Both Kuhn and Popper *really* base their views of the structure of science on the history of science and the main point of my remarks here is that the history of science cannot bear such a load at this time. We simply do not know enough to permit a philosophical structure to be erected on a historical foundation. . . . For example, there could be no better illustration of "normal" science than the experimental researches in electricity of Michael Faraday. . . . Here was puzzle-solving with a vengeance. . . . Yet the minute one moves behind the published papers to the *Diary* and the manuscript notes and letters, a strange Faraday emerges. From 1821 on he was *testing* fundamental hypotheses on the nature of matter and force. How many "normal" scientists (as defined by their published papers) are really revolutionaries at heart?[55]

Lakatos's point is that just as science is not built on induction from facts, neither is the history of science. Williams continues with a classic example of monster-barring by suggesting that Faraday may have been acting as a revolutionary rather than normal scientist if one looks closely at his unpublished documents. We can get a sense of why Lakatos would dismiss as external Faraday's own attitudes toward his work if we question whether Faraday's

---

54. See Kadvany, *Imre Lakatos and the Guises of Reason*, p. 343, n. 25, on Lakatos's frustration that Kuhn and other critics "never grasped his point about the codependence of historical narrative and methodology."

55. Williams, "Normal Science, Scientific Revolutions and the History of Science," p. 50.

private, revolutionary thoughts actually shaped the ongoing, public development of science. The point is not that archival research is unilluminating but that an argument must be made, not assumed, that it matters what Faraday thought he was doing if no one else took it that way. What Faraday was for himself is not what he is for us. Williams's concluding remarks about the need for more research is a still common refrain that Lakatos would refer to often when describing his dissatisfaction with atheoretical history of science. Williams explicitly weds the call for more research to an inductivist historiographical methodology:

> As a historian, then, I must view both Popper and Kuhn with a somewhat jaundiced eye. Both have raised issues of fundamental importance; both have provided deep insights into the nature of science; but neither has amassed sufficient hard evidence to lead me to believe that the essence of the scientific quest has been captured. I shall continue to use both as guides to my researches, always keeping in mind Lord Bolingbroke's remark that "history is philosophy teaching by example." We need a lot more examples.[56]

Contemporary emphasis on case studies continue this call for more examples, as if the mere proliferation of case studies written from the standpoint of science studies meant that science studies was more well founded. This rhetorical strategy was an important part of the field's early legitimation, ignoring the point that more studies on the basis of SSK does not mean that the interpretive framework of sociology of scientific knowledge (SSK) has been vindicated.[57]

In the sociology of scientific knowledge, we also see a tendency to see sociology as empirical and sensible and philosophy as normative and illegitimate. Lakatos's response would be the same as his response to Williams:

> [M]any historians will abhor the idea of any rational reconstruction. They will quote Lord Bolingbroke: "History is philosophy teaching by example." They will say before philosophizing "we need a lot more examples." But such an inductivist theory of historiography is utopian. *History without some theoretical bias is impossible*.[58]

A related point is made by Lakatos about the tendency to appeal to a participant's view of science, rather than an analyst's theoretical account of the history of science. Such a view invites the easy extension of the category of science to any human representation of nature, an approach that muddies the

---

56. Ibid.
57. This is the explicit strategy of legitimation in Shapin, "History of Science and Its Sociological Reconstructions."
58. Lakatos, "History of Science and Its Rational Reconstructions," p. 107.

water of understanding the historical contingency of the specific elements of modern science that drive discovery in a novel fashion:

> Long texts have been devoted to the problem of whether, and if so, why, the emergence of science was a purely European affair; but such an investigation is bound to remain a piece of confused rambling until one clearly defines "science" according to some normative philosophy of science. One of the most interesting problems of external history is to specify the psychological, and indeed, social conditions which are necessary (but, of course, never sufficient) to make scientific progress possible; but in the very formulation of this "external" problem some methodological theory, some definition of science is bound to enter. History of *science* is a history of events which are selected and interpreted in a normative way.[59]

If one simply emphasizes how actors define science, we cannot even identify science or discuss the historical contingencies that have led to the cultural evolution of science.[60]

Part of the problem with the way that Lakatos was received lay in the fact that we took this emphasis on historiography as if it implied that the direction of influence between theory and historical fact was one way, top down. Yet if we notice that his exposition of his perspective enacts an emerging criticism of philosophical standards by reference to history, and we register his later interest in authoring a book titled *The Changing Logic of Scientific Discovery*, we can see that criticisms of philosophy of science by historians are welcome, so long as they enact the equivalent of lemma incorporation and avoid monster-barring. One should not just find exceptions to philosophical generalizations about science, but show how these exceptions suggest new explanations for the growth of science.

The continued attitude toward philosophy of science in our field is a compulsive and outdated criticism of a simplistic rationalism or empiricism. Instead, we should incorporate an understanding of just where Lakatos's system breaks down in order to be able to show that a renewed understanding transcends rather than just ignores its past. Feyerabend can be read as initiating such a criticism, though his adoption of cultural relativism tended to inhibit a recognition of this and looks more like monster-barring by leveling all claims to epistemic authority. Nevertheless, Feyerabend's belief that all perspectives need to remain open to criticism ultimately pushed him away from strict epistemic relativism. This led to his own lemma incorporation of relativism in his *Conquest of Abundance*.

---

59. Ibid., pp. 107–8.

60. Lynch, "Darwinian Social Epistemology"; Lynch, "Complexity, Natural Selection, and Cultural Evolution." On the necessary use of hindsight in the history of science, see Tosh, "Anachronism and Retrospective Explanation"; Lynch, "The Utility of the Present in Reconstructing Science's Past."

Relativism tended to treat different cultures as their own universes, as Wittgenstein's take-it-or-leave-it approach to rule-following demonstrates. For Barnes, Bloor, and Henry, a deviant rule follower *might* be able to find solace in a *different* philosophical community, as the critic of Millikan's oil drop experiment Ehrenhaft was to find in being judged a poor experimentalist by the experimental physics community but finding favor among Machian, positivist philosophers. For Michael Lynch, the whole notion of alternative rule-following is a mistake: if a new regime of rule-following emerges, it does so on its own in a new context, not as the result of deviant understandings brewing beneath the surface.[61]

For Lakatos, by contrast, concept stretching can change meaning *retrospectively*: interpretive flexibility exists, but it is not monopolized by a community as it exists at a particular moment of history. A scientific community's own attitude to its creations may not capture the dynamic of knowledge it has unleashed. Feyerabend extends this notion to question whether cultures ever have clear boundaries and whether it makes sense to treat forms of life in a holistic fashion.

Drawing on the work of anthropologist Renato Rosaldo, Feyerabend suggests that cultural forms do not have a predefined identity and are not limited in principle by the semantic horizon of their discourse.

> I am not denying differences between languages, art forms, customs. But I would ascribe them to accidents of location and/or history, not to clear, unambiguous, and immobile cultural essences: *potentially every culture is all cultures*.[62]

In Lakatosian language, what is internal and external may change and research programs can also be hybridized.[63]

This is something that the latter Feyerabend believed was an insight blocked by "the trap of relativism" since relativism sees only the distinct cultural assumptions of diverse groups and not how they can be modified by human action and by the influence of nature. Speaking of the nature-culture connection as "an exchange between two rather unequal partners," cultures can be selected for or against by the natural world even more than culture can impose itself on nature:

> Not all exchanges produce beneficial results. Like unfit mutations, the actors of some exchanges (the members of some cultures) linger for a while and then disappear (different cultures have much in common with different mutations living

---

61. Barnes, Bloor, and Henry, *Scientific Knowledge*; Lynch, *Scientific Practice and Ordinary Action*. See chapter 4 for more discussion of the limitations of Wittgenstein-inspired SSK.
62. Feyerabend, *Conquest of Abundance*, p. 33.
63. Lynch, "Between Kin Selection and Cultural Relativism," pp. 280, 306–7.

> in different ecological niches). The point is that there is not only one successful culture, there are many and that their success is a matter of empirical record, not of philosophical definitions. . . . Relativism, on the other hand, believes that it can deal with cultures on the basis of philosophical fiat: define a suitable context (form of life) with criteria etc. of its own and anything that happens in this context can be made to confirm it. As opposed to this, real cultures change when attempting to solve major problems and not all of them survive attempts at stabilization. The "principles" of real cultures are therefore ambiguous, and there is a good sense in saying that every culture can in principle be any culture.[64]

Within science, interaction of incommensurable perspectives is encouraged to maximize critical perspectives on taken-for-granted views. Comparing the pragmatic effects of adopting incommensurable perspectives, rather than imposing a common logical framework, challenges the tendency within modern science of taking a narrow, reductive approach as exhausting reality.[65]

Lakatos's own philosophy encompassing one element of how science has gained knowledge over the past 300 years or so is one such narrow approach, but its essence is no more fixed by his intentions than any other knowledge product. Indeed, Feyerabend is able to clarify more clearly whether "the scientific view of the world [has] a special status compared with other views," because he doesn't just define science as what scientists do: he has a preliminary, tentative, sense of the accomplishments and methods of science to use as a touchstone.

By contrast to Feyerabend's view that "potentially every culture is all cultures," contemporary science studies tends to reify research programs in terms of the intentions and tacit understanding of a self-selecting group of experts. We may compare what different expert-communities have to say, but we rarely examine how they change either other's minds, or how they could change each other's minds if certain "accidental" features of the situation did not block them. Most important, in this context, is the university's reinforcement of disciplinary turf as the basis for judgments about the quality of work, evident even in interdisciplinary fields like STS. In the next chapter, we will examine an interdisciplinary synthesis—the modern synthesis in biology—and interdisciplinary challenges and responses to weak points in the program.

---

64. Feyerabend, "Art as a Product of Nature as a Work of Art," p. 98. For discussion of Feyerabend's post-1990 turn away from relativism, developing from his "semantic nihilism" that denies that well-defined contexts of meaning can be delineated, see Preston, "Science as Supermarket," pp. 87–88, 91–94.

65. Oberheim, *Feyerabend's Philosophy*, pp. 70–76, 123–65, shows that Feyerabend intended his concept of incommensurability to be a tool of criticism for dislodging well-confirmed theories that blocked progress. "Incommensurability does not imply incomparability. To the contrary, the best way of testing the merits of some theory is by comparing it to incommensurable competitors because they offer the most fruitful critical comparisons" (p. vii). See also Lynch, "Imre Lakatos and the Inexhaustible Atom," p. 29.

Chapter Three

Making Anomalies Work

Dialectical Engagement versus Anomaly Mongering in Evolutionary Biology

The distinctiveness of Lakatos and Feyerabend's philosophical approaches to science can be seen in the way they sought to understand science as a process developing over time, where the "parts" do not make sense except in the context of an evolving whole. This "dialectical" approach allowed each to engage the history of science in ways that did not confine attention to the past as if it were a foreign country, self-contained and self-warranting. Seen most explicitly in Lakatos's approach to mathematics, the idea is also evident in his Methodology of Scientific Research Programs (MSRP) in its focus on a *series* of theories constituting an evolving research program rather than focus on a static moment in time in the logic of justification.

If science can be described as a dialectical process, it is because the history of science can only be understood as a totality, meaning as the sum of its theoretical moments in series. Since science proceeds in part by subsuming or transcending past developments in new forms, it bears similarities with what Hegel and Marx said about historical change. It is not sufficient to focus on a momentary snapshot of the state of knowledge in the history of science, analyzed for its correspondence to the world or, for that matter, its correspondence with a static culture (see chapter 4). Connections to the world—or to forms of life—are developed over time, and changes are made not by refining the fit of theory and fact in isolation but by changing successively the character of the theory and the facts as well, as Feyerabend emphasized with his concept of counterinduction.[1]

Lakatos likewise understood that facts are conventions generated by developing traditions of inquiry whereby empirical corroboration is driven by ever-

1. See the discussion of the "One-Pass Fallacy" in Nickles, "Reconstructing Science: Discovery and Experiment"; and Nickles, "Justification and Experiment."

changing conceptions of how to uncover the facts in the first place. This is why it makes little sense to Lakatos to compare the fit of competing theories directly to the empirical facts. Rather, progress in generating and corroborating predictions over time is a sign that a research program is developing in ways that are guiding further understanding, that one is on the right track, and that the theoretical approach is continuing to generate fruitful new work. It also means that new theories are also generated dynamically in part from creative engagement with previous theories. Research programs can be fruitful to the extent to which they continue to generate new theories that are connected to the older ones but that surpass them by generating novel predictions.

Feyerabend took the approach further, understanding the concept of incommensurability not as a permanent block to communication or comparability—which is to remain stuck at the level of static comparisons—but as a way to transform quantitative change into qualitative change. Innovative work like Galileo's proceeds not by directly challenging the older approach on its own terms but by altering the language for observing the world in a way that generates a transformation in the way that scientific work is done. Significant breaks with past scientific understandings are possible, but they must be capable of generating their own ongoing solutions, not simply sniping at the older theory's anomalies.

Likewise, anomalies make a poor basis for comparison of competing theories since each side will have anomalies that the other side feels are fatal but that are safe to ignore so long as they do not impede ongoing work. Lakatos's criteria of progress are designed to generalize a common trait of past progressive science in the face of Kuhn's "ocean of anomalies." Feyerabend's "breathing space" and the understanding that competing approaches might not be mature enough for direct comparison as the result of "uneven development" make a similar point. What this means above all is that progressive science for Lakatos, or creative, transformative scientific work for Feyerabend, must be capable of generating new work, not simply cataloguing the faults of competitors. This also means that new developments are dialectically tied to past developments, even when significant challenges to past ideas are incorporated.

What this means for the tension between scientific liberty and the social character of knowledge is that even "scientific revolutions" depend upon what went before to provide the basis against which the new approach defines itself. We should not believe in what sociologists call "voluntarism," whereby social structures are created at once from the free actions of individuals. At the same time, we should not become so fixated on the role of training and social constraint in science (the "oversocialized" conception) that we miss the ways in which traditional elements are extended and transformed in practice (this theme is taken up in chapter 4).[2] Indeed, if Lakatos and Feyerabend are right, the best science will proceed by reconceptualizing past science in creative ways that advance and deepen our understanding without starting from scratch.

The contrast between the introduction of transformative alternatives and approaches created de novo can be highlighted by considering different ways orthodox Darwinian ideas have been challenged and modified over the past half century—including inclusive fitness, group selection, and the extended evolutionary synthesis—with the ideas of intelligent design. The former cohort of very different theories can be seen as part of the ongoing process of debate in evolutionary theory over the past century and a half, and the latter represent an attempt to opt out of this scientific history. The former includes attempts to engage core anomalies in thinking about natural selection and evolution, anomalies that stimulate the generation of alternative perspectives, while the latter are more likely to ask whether the overall picture generated by evolutionary science is justified or compatible with other beliefs, such as religious beliefs. We can show this by contrasting intelligent design as a contemporary hypothesis salvaging theism against a perceived materialist onslaught with a dialectical history of changing understandings of what evolution by natural selection entails. Both offer minority views that often struggle to get a hearing, but both cannot be said to engage and transform the past in generating the future in quite the same way.

SCIENCE AND RELIGION

To be sure, Feyerabend himself felt that scientists' dismissal of the creationism of his time proceeded more from scientific dogmatism and prejudice than any justified demonstration that creationism (or witchcraft or Aristotelian physics) failed to explain the world. Recall that Feyerabend's considered judgment on Lakatos's project was that at best Lakatos had shown how the modern institution of science had generated its own standards of rationality, which were consequently inappropriate for evaluating alternative knowledge systems outside its borders. For him, these distinct approaches to knowledge should have no reason to feel bound by science's standards. Feyerabend's own later criticism of relativism as a conception of distinct and separate communities that fail to interact may seem to temper this judgment, but it is clear that Feyerabend would err on the side of allowing competing approaches outside the materialist ambit of modern science to proceed unhindered.

However, if we do want to understand what is distinctive about science as a way of generating knowledge about the world, and we do not simply want

2. Wong, "The Oversocialized Conception of Man in Modern Sociology."

to identify science with any human representation of the natural world, then understanding something distinctive about the difference between being a maverick *within* science and being one *outside* science may be important. Considered purely in terms of justifying ideas based upon premises each community has chosen to adopt, it may be that materialist science and religious apologetics can hardly engage each other. Science may generate knowledge claims about the world that are modifiable in peculiar ways compared to other cultural means of producing and modifying belief. Whether that means that its best representations are truer of reality than that of the theologian, mystic, or artist is probably a pointless question.

What it is possible to show, however, is that certain kinds of criticisms of scientific theories just do not have what it takes to be real contenders for generating viable scientific research programs. My argument is that intelligent design is one example of that and the emergence over time of (ever-changing) "materialist" conceptions of the world are unlikely to be reversed for dualist or spiritualist conceptions within science, no matter how much that might alleviate anxiety about science among some students or citizens.[3]

Static emphasis on justification based upon premises supplied by other traditions of belief are not sufficient for doing science, even if they might turn out to be true for all we know.[4] Take Pope John Paul II's conciliatory engagement with evolutionary theory, where he learned the lesson from the Galileo affair by conceding to scientists that evolution by natural selection was well established. He argued that Catholics should defer to scientists on

3. Fuller, "The Dissent over Dissent over Descent," pp. 485–86; Lynch, "Darwinian Social Epistemology."

4. Fuller, "The Dissent over Dissent over Descent," p. 480, has argued that incorporating intelligent design into school curriculums can be justified by a prospective reading of its promise as a research program, which philosophers of science ignore because they focus on a forced choice at a moment of time: "Philosophers (still) tend to operate with a much more backward-looking and static view of the task. In other words, they presume that competing scientific theories are in a race that ends at the moment a philosopher is taking the decision, in which case the theories' track records are used to justify their respective fates. In contrast, a school board—or a court considering an appeal to a school board decision—is focused on what the next generation should learn to deal effectively with the life challenges they are likely to face. In that case, the theories' track records are read for clues to future developments." I agree that philosophers too often employ a static view of theory choice, which is why I have drawn on the neglected approaches of Lakatos and Feyerabend, who do try to use track records to think about the potential contributions of promising new approaches, what Fuller, following Laudan, calls the context of pursuit rather than the context of justification. (On this point, see Feyerabend, "More Clothes from the Emperor's Bargain Basement," pp. 241–45.) However, I argue that intelligent design does not offer compelling reasons to motivate us to consider developing the theory further except if one has a perspectivist agenda external to science. Moreover, if it is important for alternative research programs to develop their own internal, dynamic programs rather than create one's alternative program from suggesting alternative hypotheses for shelved anomalies within dominant research programs, what I have called anomaly mongering. While one should support scientific liberty and allow scientists to pursue research programs of their choice, judgments that it has little promise to emerge as a viable research program are necessary for making decisions about funding and inclusion in the curriculum.

all properly scientific questions, reserving only the interpretation of theological concepts like the soul to Church doctrine. Rather than choosing conflict with evolutionary theory as young-earth creationists did, he saw science and religion as complementary. As long as Catholics realized that at some point the evolving primate was invested with a soul by God, the explanation of the evolution of new forms in biology was left to Darwinian explanation.

Some scientists likewise choose coexistence over conflict. The pope's approach might well be said to fit Stephen Jay Gould's understanding of the proper relationship between science and religion, particularly as he developed his thoughts on the question after observing that aggressive attacks on John Paul's position by scientists would serve science poorly and lead scientists to speak outside their competence on moral issues. Gould makes the reasons for this division of labor clear:

> Religion is too important to too many people for any dismissal or denigration of the comfort still sought by many folks from theology. I may, for example, privately suspect that papal insistence on divine infusion of the soul represents a sop to our fears, a device for maintaining a belief in human superiority within an evolutionary world offering no privileged position to any creature. But I also know that souls represent a subject outside the magisterium of science.[5]

Gould put forth an approach for making sense of this division of labor called NOMA, for non-overlapping magisteria, where science and religion each have distinct aims and cannot be said to be in conflict, properly understood. Science attends to representations of the natural world, while religion is concerned with the care of the soul.

As critics have pointed out, however, this division of labor is itself historically dependent, with science claiming more and more of the territory of religion as it generates more knowledge that impinges upon what the Church believes. We can see this process already in Asa Gray's 1860 review of Darwin's *Origin of Species*, where what had formerly been conceded to be the realm of religion, how animal species came to be, is shown to be explained by a new scientific hypothesis. Failing the discovery of intermediaries between animals and humans, however, the origin of the human species is left to divine creation. For Gray, "put the genealogy of the brutes upon what footing you will, the four-handed races will not serve for our forerunners;—at least, not until some monkey, live or fossil, is producible with great-toes, instead of thumbs, upon his nether extremities. . . . Until some testimony of the sort is produced, we must needs believe in the separate and special creation of man, however it may have been with the lower animals and with plants."[6]

5. Gould, "Nonoverlapping Magisteria," p. 21.
6. Gray, "Darwin on the *Origin of Species*."

Until such a discovery, Gray concludes that the transmutation of animal forms alone can be explained by his theory. Indeed, since Gray tells us that Darwin accepted the special creation of life itself as a prelude to biological evolution and was opposed to the concept of spontaneous generation of life, two cases of divine intervention are supported by the evidence: one the original emergence of life itself, and one the origin of humans. Of course, Neanderthals had already been identified by 1859 and the discovery of Australopithecines would come by 1924 that more directly delivered on the prediction of an emerging bipedal primate.

As for the origin of life itself, Darwin himself had speculated privately in 1871 that abiogenesis, the emergence of life from non-living chemicals, could explain the first event:

> It is often said that all the conditions for the first production of a living organism are now present, which could ever have been present. But if (and oh! what a big if!) we could conceive in some warm little pond, with all sorts of ammonia and phosphoric salts, light, heat, electricity, &c., present, that a proteine compound was chemically formed ready to undergo still more complex changes, at the present day such matter would be instantly devoured or absorbed, which would not have been the case before living creatures were formed.[7]

This proposal represents theoretical progress in Lakatosian terms since it develops a novel prediction of the origin of life itself from non-living matter in situations where the interference of already existing life is removed.

While he recognized that science was in no position to confirm this account at that time, his speculation underscores the generative capacity of new research programs, the tendency for the scope and explanatory power of proposed explanations to be extended over time, challenging both of Gray's special creation events. By now, abiogenesis is a developed research program, building on the primordial soup hypothesis of Alexander Oparin and J. B. S. Haldane by the 1920s. A key prediction of the theory was confirmed in 1952, when Stanley Miller and Harold Urey produced organic molecules from non-organic ones in conditions similar to those proposed by Oparin and Haldane.[8]

Intelligent design has increasingly targeted Gray's transition between non-living and living matter, as evident in Stephen Meyer's work, a former oil company geophysicist who returned to school for a Cambridge PhD in history and philosophy of science and now heads the Center for Science and Culture at the Discovery Institute, the predominant think tank promoting intelligent design. With the growth of fossil and genetic evidence supporting the evolutionary transition from non-human to human species, the confirmation of

7. Darwin, *Life and Letters*, III, 18,
8. Rauchfuss, *Chemical Evolution*.

transitional fossils in the record as well as new methodologies tracing genetic divergences between species, intelligent design has focused on the origins of the information-bearing system of DNA itself as something that cannot be explained through Darwinian mechanisms. Here the issue is not just that biological systems are complex but that they have a specified complexity that serves a biological function. That functional specificity is "profoundly mysterious" insofar as the DNA/RNA/protein molecular machinery for building living systems depends upon the prior existence of the machinery itself to build new living forms and to replicate DNA itself. Without the prior existence of "some physical medium of information transfer," the whole process could not get off the ground. Here, Meyer draws upon Richard Lewontin's criticisms of genetic determinist views like that of Richard Dawkins, a key forerunner of the extended evolutionary synthesis discussed below.[9]

However, what Meyer sees as justifying an inference to intelligent design, Lewontin sees as the grounds for interrogating misleading metaphors relating genes and adaptations without considering the influence of the precise mechanisms of the molecular machinery that link genetic information, the physical environment of the cell, and a developmental process incorporating a variety of feedbacks from the environment. If Lewontin asks "what makes the proteins that are necessary to make the protein," he is articulating a core anomaly for a field that presents a challenge that different scientists will try to explain. Familiar concepts may have to be jettisoned as the problem is addressed.[10]

We might see this process as the dialectical motion of concepts that Feyerabend speaks about, typical for how early attempts at explaining a new problem within science act as scaffolding for constructing better answers. Thus, early attempts to explain how DNA was implicated in constructing living cells presumed that DNA would act as a direct template for producing proteins based upon the physical geometry and chemical affinity of the molecule, a well-established technique that had informed the discovery of the structure of DNA itself. It is only with the identification of a complex, RNA-based, intermediary mechanism between the information stored in DNA and its expression through protein synthesis that the relevance of a model of the transfer of *information*, encoded in the equivalent of the digital codes then animating emerging information science and computer technologies that a more complete solution was discovered. It is the existence of a functionally specified complexity within the apparatus of cellular replication that motivates Meyer's appeal to intelligent design, as its characteristics resemble nothing in nature so much as the products of human design in modern culture and technology.

9. Meyer, *Signature in the Cell*, pp. 110, 133–34.
10. Quoted in ibid., p. 134.

On this view, Darwin may have overturned the need for special creation of species by an intelligent designer, but his approach, extended to the attempt to explain the emergence of biological forms from a prior chemical evolution will fail because the machinery of life indicates a systematic interrelationship of parts that make it hard to see how the process could ever get off the ground in the first place. If proteins already need to exist to build proteins, then how could proteins emerge in the first place?

To accept this core puzzle as necessarily blocking explanations emerging from materialistic explanations depends upon stopping the "motion" of concepts that is key to science's development historically. Scientists have not only extended their understanding of the evolutionary process from biological to chemical evolution, they have generalized the mechanism of natural selection to apply to any circumstance where we encounter reproduction, heredity, variation, and selective advantage. Such an approach was first corroborated in experiments with RNA molecules undergoing mutation and selection in laboratory conditions in the 1960s and 1970s.[11] In the 1980s, ribozymes were discovered, RNA molecules that could function as enzymes like proteins, suggesting that they were "a relic of an ancient RNA world in which RNA molecules could have functioned as both chicken and egg."[12] In the past decade, progress toward laboratory production of synthetic cells and a growing understanding of how RNA molecules can replicate prior to the existence of the protein enzyme machinery of modern cells demonstrate that molecular biology and an understanding of the emergence of Darwinian replicators are closely linked.[13]

The specific answers that science has proposed for "evolutionary transitions" like this are undoubtedly incomplete at the least, or profoundly mistaken like the direct template model. But the hypothesis that systems can emerge that store and transmit information from lower-level processes that do not has been the basis of ongoing research.[14] The development of such

---

11. Note that Darwin appealed to special creation of the first life forms in the *Origin of Species* to placate Church of England clergyman Charles Kingsley, the first to reconcile Darwinian evolution and Christianity. Ernst Haeckel's objection implicitly recognized this prospective motion of concepts: "The chief defect of the Darwinian theory is that it throws no light on the origin of the primitive organism. . . . When Darwin assumes a special creative act for this first species, he is not consistent, and, I think, not quite sincere." See Fry, "The Emergence of Life on Earth and the Darwinian Revolution," pp. 323–24, 327.

12. Ibid., p. 327.

13. Blain and Szostak, "Progress toward Synthetic Cells." Szostak won the Nobel Prize in Physiology or Medicine for 2009 for research on how telomeres protect chromosomes from degradation, a key finding for understanding how chromosomes accurately replicate. Szostak's work is only the latest refutation of the claim in Fuller, *Dissent over Descent*, p. 204, that no Nobel Prizes have been given "to a Darwinist for work that requires being a Darwinist." See Lynch, "Darwinian Social Epistemology," pp. 56–57.

14. Maynard Smith and Szathmáry, *The Major Transitions in Evolution*; Szathmáry, "Toward Major Evolutionary Transitions Theory 2.0"; Hazen, "Emergence and the Experimental Pursuit of the Origin of Life."

research programs over time can elicit new answers that may eventually explain more naturally what was initially puzzling when analyzed according to existing scientific research programs. In this case, scientists may even do so by reconsidering the information model that was crucial to the discovery of the DNA/RNA/protein system itself, as the discussion on the extended evolutionary synthesis below will make clear. Conceiving of genes as equivalent to a computer program may be a progressive move at one moment in the history of science and a regressive move at a later moment of time. Therefore a focus on a dialectical motion of concepts should lead one to expect that seemingly disabling anomalies may not be so in the future.

The Catholic Church has come to modify its claims in turn, not only conceding over the course of time that Galileo and Darwin were right, but incorporating more and more limitations and qualifications on its spiritual beliefs. The Church has not abandoned all belief that humans can be possessed. However, exorcists within the Church must now collaborate with mental health experts to be sure that true possessions are distinguished from mental illness. The psychology and evolutionary science behind moral thinking or religious practices also threaten to challenge established beliefs in religion.[15] Over time, the boundary between science and religion shifts as scientific attention expands and previously "non-scientific" questions are treated by new programs in science. Religion is left to claim the residual elements of this process as its special province. Such a claim only looks plausible when considering the history of science in a specific moment of time, not when understood as a developing temporal process.

The key point here is to notice not so much that the pope is provably wrong about souls but to see that an emphasis on whether a given phenomenon can be explained by science *now* does not establish that current research programs in science will not be able to explain them further down the road. Within archaeology, explaining the transition between less sophisticated, albeit tool-using, primates in the historical record and our own species, capable of speech, artistic representation, and burying the dead is one of the key unsolved anomalies within modern science. What accounted for the seemingly sudden, explosive development of the human mind and culture, the emergence of "behavioral modernity"?[16] While some archaeologists question whether the transition was quite so sudden, we still do not have a clear understanding of exactly how and why this transition took place.

An obvious explanation that could be proposed is precisely that of Pope John Paul II, considered now as a scientific hypothesis since it explains

15. See the responses to John Paul II, "Message to the Pontifical Academy" in the *Quarterly Review of Biology*, especially Dawkins, "Obscurantism to the Rescue."

16. Klein, *The Human Career*.

observable outcomes better than they are currently understood by "materialist" evolutionary science. Evolving primates could have been given souls by a supernatural process that brought about behavioral modernity. Considered purely in terms of who has a viable explanation, the pope can even be seen to come out ahead, since evolutionary biologists and archaeologists can only promise that further study may elucidate this transformation.

And yet it would be foolish to bet against evolutionary explanations of behavioral modernity based upon the track record of modern science. Focusing on the anomaly of behavioral modernity does not provide a good basis for thinking that proposed alternatives outside of science will prevail in the long run. The weakness of the pope's view as a scientific hypothesis is clear: nothing in what the pope says can easily be made into a viable, dialectically evolving, research program within modern science.

Catholics are free to believe that their Church's doctrine is right, but turning it into a progressive research program requires more than an alternative explanation targeted at an ongoing anomaly within science. At some point, we can expect that an anomaly that is ignored or treated speculatively will become amenable to direct treatment by science when the conditions are right within the edifice of science across a number of disciplines. It will likely become possible to explain how new traits evolved and were selected for favoring behavioral modernity. (One likely option is that biological evolution gave us biological traits necessary for behavioral modernity, but only by facilitating cultural evolution, which took time to develop historically as practices favored by group selection slowly accreted.)[17] It is this process of transforming an anomaly from a source of annoyance to a basis for a satisfying explanation that occurs often in science and that should be looked to for signs that minority positions within science may emerge and prevail.

Part of the unavoidable conflict between elements of science and religion, even for moderates like John Paul II and Gould, stems from the fact that key metaphysical components of each institution differ in ways that are hard to reconcile. Modern science, by and large, builds explanations on materialist terms, while all but the most metaphysically diffuse religious traditions adhere to some kind of alternative. Sometimes the overwhelmingly materialist character of modern science is challenged by equating materialism with Newtonian conceptions of space and causality and observing that the theories of relativity and quantum mechanics challenge these ideas.[18] However, most scientists reject wooly interpretations of consciousness creating realty and

17. Sterelny, *The Evolved Apprentice*; Wilson, *Darwin's Cathedral*.

18. Fuller, "Dissent over Dissent over Descent," p. 482. See also Hands, *Cosmosapiens*. The most sympathetic aspect of Sokal and Bricmont's *Fashionable Nonsense*, their critique of postmodernism and science studies following on the heels of Sokal's hoax publication (discussed in chapter 5) is their understandable frustration with questionable interpretations of relativity and quantum physics put

the like familiar from some popular treatments of quantum physics, as challenges to localism and the experimental demonstration of spooky causality hardly imply a rejection of material causality within physics. The key point to observe is that a commitment to materialism has acted as an overarching metaphysical commitment associated with a variety of highly successful research programs spanning all areas of science. Evaluated in terms of its track record, materialism can expect little challenge from alternative worldviews as a basis for constructing progressive research programs within science. As it relates to evolution, science takes it a step further by positing explanations that dispense with the need for a designer.[19]

Steve Fuller has challenged methodological naturalism by pointing out how belief in God has motivated the pursuit of science in the Western tradition. Historically, this has been the case with the idea of the book of nature as authored by a divine creator informing conceptions of natural philosophy during the scientific revolution, as seen in the work of significant scientists like Kepler, Boyle, and Newton. While this has been the case, it is one of the metaphysical elements within modern science that has been jettisoned from field after field, particularly since Darwin's work initiated a direct alternative to conceptions of design in nature. Natural theology has been a degenerating program and materialist approaches have been more fruitful since Darwin.

"Accidental" metaphysical aspects within science help drive research programs before being exhausted and needing to be replaced by more accurate ones. Uneven development plays a role here, since prior to Darwin's work, there was no way for Newton to know, for example, that the evidence of design he and other scientists identified had another explanation. Scientists may use all kinds of metaphysical hypotheses as stepping-stones, and they have not been confined to materialism. Moreover, there has been a special motivation to consider natural laws created by God. However, the point is that these metaphysical stepping-stones, when viewed in the context of the arc of scientific history, look like dispensable aids for generating new theories as each of the assumptions is challenged in turn.[20] In dialectical terms, they become unreal and irrational as their relationship to the accumulated body of scientific knowledge changes.

Newton relied upon God to intervene and maintain the rotation of the solar system to avoid running down with time, and his flawed understanding of the necessity of such an intervention was only rectified later by those developing

---

forth by their objects of critique, misunderstandings that figure prominently in Sokal's spoof (Sokal, "Transgressing the Boundaries").

19. Fry, "Philosophical Aspects of the Origin-of-Life Problem," pp. 70–77.

20. Both Oparin and Haldane's theories of the origin of life were influenced by Marxist dialectical materialism, which may mean that dialectical materialism has been fruitful for science but not that it is true. See Fry, "The Emergence of Life on Earth and the Darwinian Revolution," pp. 324–26.

the new approach to mathematical natural philosophy that he had initiated. Before Darwin, a concept of God was relied upon both to create biological forms in the first place, as well as to create humans as creatures possessed of a soul that distinguished them from other animals. All of these were once essential to complete scientific explanations at a moment in time, but no longer remain so. They have become unreal.

Meyer, in his defense of intelligent design, emphasizes that the assumption of intelligent design, the belief that only a rational intelligence was capable of producing an intelligible order in the world, motivated the explosion of knowledge arising from the scientific revolution.[21] Without this key motivating belief, such discoveries would never have been sought and so the hypothesis of intelligent design can be understood as a particularly fruitful one. In 1999 testimony before a U.S. government panel, Meyer used this point to challenge the claim of Eugenie Scott, head of the National Center for Science Education, that appeal to design violated the scientific norm that explanations must be naturalistic. Meyer pointed out that Newton had specifically invoked design to explain the arrangement of the planets, something he held that his law of universal gravitation could not do. Naturalism has not always been a norm of science.

What Meyer misses, however, is that we now know that Newton was mistaken in his beliefs that physical laws explaining the motion of planets needed to be supplemented by appeal to a designer. Scientists have developed naturalistic explanations of the origins of the solar system to replace Newton's invocation of God's intervention. The experience of the last few centuries has been that apparent limitations to naturalistic explanation have been surpassed again and again.[22]

That tendency has taken over science after Darwin showed that natural selection could produce the same effects that were previously explainable only through appeal to design. Darwin had produced a competing explanation for design and the extension of that explanatory scheme to new fields has accounted for the development of numerous, fruitful research programs since that time. We have learned that the initial, motivating assumption for the discovery of lawlike behavior in the universe is mistaken and dispensable.[23] It is a scaffold that we have dispensed with, and this accounts for the prevalence of naturalistic explanation in science today. Meyer is banking on the fact that naturalistic scientific investigations of the initial origin of life will fail and we will need to rethink the entire edifice of modern science. This is an extraor-

21. Meyer, *Signature in the Cell*, ch. 6.
22. Fry, "The Emergence of Life on Earth and the Darwinian Revolution," p. 213.
23. Contrast Fuller, "What Has Atheism Ever Done for Science?" and Lynch, "Cultural Evolution and Social Epistemology."

dinarily unlikely possibility, which is why betting on it based on science's overall track record is a bad idea.

Indeed, it seems that Meyer's core, motivating assumption is an objection to scientific materialism for its perceived erosion of ethical values and human rights, a view shared by both intelligent design theorists and young-earth creationists. Pennock argues that Meyer's objections to naturalism follow from a foundational assumption that naturalism would erode Christian theism and ethical value. On this view, only belief in theistic creation ensures that humans have dignity and rights, while the Soviet Union's hostility to human rights arose from its commitment to materialism.[24] Meyer concludes that "without a purpose-driven universe, there can be no 'purpose-driven life.'"[25]

Moreover, there are grounds for suspecting that Meyer seeks not only to overturn scientific materialism in science but also to restore philosophical idealism, though not in its Hegelian form! Unlike Hegel's God, who was in some sense immanent within the evolving universe itself,[26] Meyer's preferred understanding of the intelligent designer of the information system encoded in DNA is a God that stands apart from creation. He responds to Richard Dawkins's objection that appeal to DNA's intelligent designer is question-begging because it leaves unanswered the question of who designed this designer. For Meyer, such a view presupposes that only natural causes could explain the emergence of intelligence. Dawkins sees our own intelligence as a product of Darwinian evolution, so Meyer's appeal to the design encoded in human artifacts *is itself the outcome of natural causes that have produced our species through Darwinian evolution*. Therefore, if the DNA/information system is attributed to an intelligent designer, we should still seek to explain the natural processes that led to the emergence of that intelligent designer.[27]

Meyer will have none of that and ultimately rejects an infinite regress of intelligent designers by appeal to an unmoved mover who is also *non-material* —not part of this universe—and *indivisible*, so not explainable via component parts and mechanisms like we are as embodied beings.[28] The possibility is broached that human intelligence, too, is immaterial, and that this accounts for its ability to impose a specified complexity of the material world:

> ID advocates do not necessarily conceive of minds as complex (and functionally specified) collection of parts (or even symbols). Some would argue that complexity . . . is a feature that characterizes material effects, not self-conscious minds. Indeed, a long tradition in Western philosophy conceives of mind as

24. Pennock, "DNA by Design?," pp. 133–37.
25. Meyer, *Signature in the Cell*, p. 449.
26. Wallace, "Hegel's God."
27. Meyer, *Signature in the Cell*, pp. 388–95; Dawkins, *The God Delusion*, pp. 147, 185–88.
28. Meyer, *Signature in the Cell*, p. 545, n. 33.

> an immaterial entity consisting of no material parts, in which case it cannot be characterized as complex at all.[29]

So much for scientific psychology, evolutionary or otherwise, not to mention broad areas of cognitive science, biology, and the social sciences. Like Michael Cremo's Vedic archaeology discussed in chapter 1, Meyer's presumption of privileged access to the correct metaphysical view of the world allows him to dismantle modern science's pretensions as built upon sand, requiring starting again virtually from scratch. Intelligent design theory "poses a serious challenge to the materialistic worldview that has long dominated Western science and much of Western culture."[30]

It would be a mistake to say that science *proves* that the world is purely material and that God does not exist. What Meyer calls scientific materialism is best understood as methodological naturalism: only naturalistic explanations help advance scientific research, while supernatural ones block its advancement.[31] Arguments by scientific atheists like Dawkins are best seen as putting forth philosophical arguments that build upon scientific explanations but are not themselves scientific explanations. Dawkins uses the success of methodological naturalism to argue for the existence of only naturalistic, material entities. One can find reasons to deny that such strictly materialist explanations of the world must be true and make use of arguments like the pope's to suggest that the materialist metaphysical assumptions underlying scientific research programs are simply instrumentally useful. Much like the tension in Collins's methodological relativism discussed in the first chapter, it may be that treating the natural world *as if* it evolved without the intervention of a prime mover is particularly well suited to generating fruitful investigations into how the world works without this metaphysical assumption being true.

However, while it would be glib to suggest that science refutes religion in toto, what has been said so far does bring to light that scientific research programs possess the characteristics of being capable of dynamically generating new problem-shifts and novel predictions about the world, which may not be possible based on just any metaphysical starting point whatever.

29. Ibid., p. 545, n. 32; p. 555, n. 43.

30. Ibid., p. 449.

31. Pennock, "DNA by Design?," pp. 133–34. Meyer, *Signature in the Cell*, p. 434, objects to Pennock's account of methodological naturalism for rejecting non-material causes. Pennock's argument was endorsed by Judge Jones in *Kitzmiller v. Dover Area School District* ruling against teaching intelligent design in science classes. Meyer's objection amounts to the claim that only purely formal methodological rules should be invoked to exclude intelligent design ("a theoretically neutral norm or definition of science"), rather than the overall track record of modern science where non-material causes (what Meyer calls "intelligent" causes) are rejected as non-starters. Contrary to Meyer's insinuation, when scientists search for extraterrestrial life in the universe, they are not positing intelligent design in Meyer's sense because they are not presuming that intelligent life, extraterrestrial or terrestrial, is non-material.

Whether considered realistically or instrumentally, the materialist and blind-evolutionary dimensions of modern science may have the characteristics of Lakatosian progress that they do because this kind of metaphysical machinery helps make it possible. By contrast, explanations based upon an unexplicated intelligent designer and irreducible or specified complexity may offer little basis for producing progressive problem-shifts.

Similar problems may plague other alternative explanations. A good example might be good-faith investigations of psychic phenomena lacking any underlying physical mechanism to describe and investigate. Collins and Pinch examined the science of parapsychology and concluded that most investigators were methodologically circumspect, so that they could claim that "nothing unscientific is happening."[32] Parapsychology's lack of integration into modern science may flow less from failure to follow proper experimental protocols than from an inability to specify and investigate physically understandable bases for the alleged phenomena. In short, not just any picture of the world can be made into the basis for a viable scientific research program. Should psychic phenomena actually exist, science would probably require preliminary identification of a causal basis for such phenomena before a progressive research program in parapsychology could develop.[33] Should God exist, science would probably require understanding of God's present activity in the world—Charles Lyell's "causes now in operation"—to support an abductive inference that the historical origin of life can be accounted for by design.[34]

Meyer argues that the existence of a putative "intelligent cause" that can account for the emergence of life from chemical conditions on the early earth makes it a viable scientific explanation because it is either the only conceivable cause that can do the job or, at the least, "a known (presently acting) and adequate cause for the origin of explanation." Design accounts for "the difference between a chaotic prebiotic environment in which simple chemical constituents interacted with each other in accord with natural entropic (randomness-producing) processes and an environment in which systems with highly specific arrangements of chemicals arose that stored information for building proteins and protein machines. Intelligent design could have made a causal difference between continuing chemical chaos and the origin of

32. Collins and Pinch, "The Construction of the Paranormal," p. 262.

33. It is instructive in this instance that Vienna Circle positivists Moritz Schlick and Hans Hahn were actively involved in researching paranormal activity, open to the possibility that the phenomena could be real. My point is that lacking viable, naturalistic accounts of how paranormal activity could be possible, open-ended "testing" will necessarily fail to establish its existence. See Josephson-Storm, *The Myth of Disenchantment*, pp. 259–60.

34. It is precisely this appeal to a distinctive pattern of historical explanation based upon known causes in the present that Meyer endorses, drawing on Lyell's uniformitarian geology. See Meyer, *Signature in the Cell*, p. 160; Meyer, "Of Clues and Causes."

information-rich molecules."[35] On this view, our experience of "intelligent agents" as the cause of information satisfies Lyell's criterion of present "causes now in operation" by our understanding of *human-created* design, not present knowledge of divine or intelligent creation of a superior being.

That is a questionable interpretation of Lyell's criterion,[36] and Meyer ignores the alternative explanatory scheme originated by Erwin Schrödinger in 1944, before the discovery of the structure of DNA and anticipating an understanding of it as a carrier of information. Meyer invokes the role of entropy in the default prebiotic chemical environment, a situation deemed unlikely to produce an information generating and transmitting system. He then invokes an "intelligent cause" to make up the deficit. What Schrödinger pointed out was it did not make sense to think about the thermodynamics of life by reference to thermodynamic equilibrium and the closed conditions in which entropy can only increase. Life exists as the result of open conditions in which the infusion of outside energy allows the organism to maintain itself homeostatically, decreasing entropy locally. Information in the biological organism is equivalent to "negative entropy," made possible by the overall increase in entropy that occurs when we factor in the infusion of energy from the sun as part of a closed system.[37] Research in "far from equilibrium" thermodynamic processes is ongoing and depends upon a less equivocal conception of causes now in operation than appeal to intelligent design.[38]

Even if we were to concede the case that Meyer's appeal to intelligent design is a conceivable cause that modern science lacks, Meyer still has not shown how a research program addressed to non-human intelligent design as a cause could provide the basis for a scientific research program. Minority positions introduced since the Darwinian synthesis within modern evolutionary biology engage the history of their science in a different way than intelligent design or the invocation of souls to explain behavioral modernity. Above all else, such approaches introduce a theoretical reconfiguration that launches a new basis for puzzle solving. In the process, anomalies are recast in ways that preserve elements of the prior approach, showing that evolution by natural selection can explain apparent anomalies by linking the broad terms of Darwin-

---

35. Meyer, *Signature in the Cell*, p. 171. Note that intelligent design defenders routinely treat complex structures as irreducible because they assume that its component parts had to emerge randomly rather than from prior, repurposed, functional components. See Miller, "The Flagellum Unspun," pp. 88–93.

36. Fry, "Philosophical Aspects of the Origin-of-Life Problem," p. 76. Note also that Meyer begs the question by assuming that humans, as the cause of intelligent design of artifacts known to us, are not themselves products of a naturalistic process. See Pennock, "DNA by Design?," p. 142.

37. Schrödinger, *What Is Life?*, ch. 6. See also Prigogine, *Introduction to Thermodynamics of Irreversible Processes*. Meyer, *Signature in the Cell*, ch. 12, restates his claim that information cannot emerge from deterministic physical laws in objecting to Prigogine's approach.

38. Chen and de Vries, "From Underwear to Non-Equilibrium Thermodynamics"; Machrafi, *Extended Non-equilibrium Thermodynamics*.

ian explanation to a new positive heuristic that generates new explanations for answering anomalies that had previously been set aside. The shelving of anomalies gives way to their recasting on new terms.

THE CORE ANOMALY OF DARWINISM: ALTRUISM

The significance of certain anomalies that are recast as theoretical explanations are redeveloped, after a period where the anomalies tended to be ignored or papered over, makes them *core anomalies* for the scientific transformation in view. The fact that such anomalies reveal unexplained problems at the heart of progressive research programs also points to the futility of fixating on the limitations of a research program's explanatory power at any one moment in history. As long as one is continuing to generate progressive problem-shifts and corroborates them on a regular basis, one is not only free to ignore the challenge of core anomalies, but well advised to do so. Only when conditions ripen in ways that allow a recasting of the theoretical tradition to show that the core anomaly was not the problem it seemed does it make sense to turn our focus to the anomaly.

This process can be seen most clearly in accounting for cooperation and altruism among living organisms within a theory where natural selection is seen as focused on the competition for resources. The prima facie demonstration that animals, especially highly social animals like ourselves, have cooperated in ways that impose costs on individual fitness presents challenges to standard treatments of natural selection. While some forms of cooperative behavior may be mutually beneficial, altruistic acts are understood by biologists as acts that increase the fitness of others at the cost of decreasing the fitness of the individual who carries out the act. In the extreme case, individual animals engage in self-sacrifice, for instance, by warning calls that alert group members to the presence of a predator while, in the process, calling the attention of the predator to the signaling animal. Under a strict reading of natural selection as understood in the modern synthesis, such altruistic traits should not spread as they are selected against in the population. Attempts to appeal to Darwin's solution of group selection were rejected by biologists as inconsistent with natural selection, as selfish individuals would leave more offspring than altruistic ones no matter what the benefit to the group.[39] Consequently, explaining observed altruistic behavior in living organisms became a *core anomaly* for Darwinism.

---

39. See Williams, *Adaptation and Natural Selection*, p. 93. Williams criticizes the group selection claims in Wynne-Edwards, *Animal Dispersion in Relation to Social Behaviour*.

Large overarching organizing ideas for constructing research programs like Darwinism can be seen as hierarchically organizing more specific research programs that translate its principles in ways that can guide more focused investigations. Such overarching frameworks can be described in part by core anomalies that are almost defiantly opposed to common sense and direct observation, just as heliocentrism was at the time of Galileo. Altruism is a good example of a concept that Darwinism addresses with counterinduction. Cooperative behavior exists within the animal kingdom in ways that often go against the genetic interests of individuals. Altruistic acts up to and including self-sacrifice have been observed in many species, including humans. This creates a puzzle for how such traits could ever emerge via a process focused on individual fitness. As Bourke observes, for Darwinism, "the expectation of selfishness has frequently been taken to epitomize the theory."[40]

How are core anomalies handled in science? Core anomalies can be ignored or evaded until new theories within the research program are developed that offer an explanation for the anomaly. Initially, the manifest exceptions to selfishness were glossed over, treated as in principle reducible to individual selection, connected to supplementary, underdeveloped explanatory principles (as in Darwin's appeal to group selection), or otherwise dealt with in ways that do not necessarily give satisfying explanations for phenomena that previously had them. For instance, most explanations prior to Darwin of cooperative behavior in humans would argue cooperative behavior exists because humans were created in imitation of God, which makes possible true altruism for those who resist sinful temptations. The introduction of a new theory (evolution by natural selection) to replace traditional views of creation and the great chain of being were accompanied by what has been called "Kuhn loss": the abandonment of previously satisfying explanations and the loss of content with the transition to a new paradigm in a scientific revolution. In Feyerabend's terms, the new approach engages in counterinduction by introducing an approach that seems to be incompatible with known facts—in this case, that humans do often sacrifice themselves for others, especially during wars or natural disasters.

This way of looking at things, developing an observation language that recasts the very way that we look and describe natural phenomena, offers promissory notes that such a core anomaly will be dealt with via the explanatory structure of the overarching research program at some point. When such an explanation is finally given, the history of the research program is retold with the claim that the new explanation was tacit within the larger program all along.

40. Bourke, *Principles of Social Evolution*, p. 21.

In the case of explaining cooperation by natural selection, the key breakthrough came with Hamilton's inclusive fitness theory.[41] Inclusive fitness theory has been described as "the only significant elaboration of Darwinian fitness since the nineteenth century."[42] Darwin's own attempts to explain cooperation focused on group selection, but such an explanation did not fit well with the modern evolutionary synthesis and an emphasis on genetic underpinnings of natural selection. Group selection is a view that has been marginal to evolutionary explanations since the development of an understanding of the genetic underpinnings of evolution following the discovery of the structure of DNA by Watson and Crick in 1953, although there has been a revival of this theory recently, in part transforming Hamilton's theory in turn.

Hamilton's theory can be seen as taking its starting point from a quip by one of the founders of the modern synthesis, J. B. S. Haldane. Haldane suggested that he would sacrifice his life for two siblings or eight cousins, based on the average percentage of genes shared by each. For Hamilton, rather than focusing only on an individual's direct reproduction (*personal fitness*), passing on one's traits to offspring, we need to consider *indirect* reproduction (*inclusive fitness*) as well, meaning the presence of identical traits in other individuals. If an individual sacrifices his or her own life to save the larger group from predation, evolution may select for this if the actions tend to ensure the survival of genes for altruism shared by others in the group, typically kin. Maynard Smith employed the term "kin selection" for this process, though it is possible for inclusive fitness to extend to those who share genes but are not close kin.[43]

In Dawkins's elaboration on this idea, natural selection does not directly select for individual fitness but for that of genes—his "selfish gene" theory.[44] In either case, what looks like actions that would tend to select against the genes of the individual engaging in them can select for them: counterinduction accomplished through a progressive problem-shift in the underlying explanatory structure of the theory. Altruism is in a sense "explained away" as an artifact of selection at the level of genes. More important, the new theory offers new types of explanations that can be used to explain how cooperative

---

41. Hamilton, "The Evolution of Altruistic Behavior"; Hamilton, "The Genetical Evolution of Social Behaviour," I and II.

42. Costa, "Hamiltonian Inclusive Fitness."

43. Hamilton's view is an expanded conception of inclusive fitness over earlier ideas of kin selection tacit in the modern synthesis, such as Haldane's quip that he would sacrifice his life for two brothers or eight cousins, since copies of genes may be prevalent in a population that does not only include close relatives. See Hamilton, "Innate Social Aptitudes of Man," pp. 140–41, for criticism of Maynard Smith's discussion of the theory in terms of kin selection in Maynard Smith, "Group Selection and Kin Selection."

44. Dawkins, *The Selfish Gene*. See, also, Williams, *Adaptation and Natural Selection*; Wilson, *Sociobiology*, p. 3.

behavior takes place, whether in ants or humans. The theory, in turn, helped spur the development of the controversial theory of sociobiology.[45]

Lakatos's research programs constitute a succession of theories that potentially make theoretical progress by making novel predictions compared to predecessor theories, and empirical progress if these predictions turn out to be corroborated. The predictions can be of previously unknown facts or of facts that were not previously explained by means of a theory's positive heuristic—in this case, altruistic traits unexplained by individual selection.

Consider the following series, which begins from an incomplete sketch of an idea (labeled with subscript 0) through Hamilton's formulation to application of the formula to tests of kin selection in anthropology:

> KS_0: Altruistic acts on behalf of kin can be selected for in proportion to the extent of shared genes (Haldane).
>
> KS_1: Altruistic acts can be selected for if the number of offspring of the beneficiary, B, multiplied by the degree of relatedness, r, is greater than the decrease in offspring of the donor, C:
>
> $$r \times B > C$$
>
> Altruistic acts can be directed to kin (or others sharing genes) by evolving traits for *kin recognition* or because kin remain close together and are more likely to benefit by default (*low dispersal*) (Hamilton).

It is worth noting that research programs are often built, tested, and criticized via *typifications* that are not strictly accurate to the theory under consideration. For example, frequent assumptions found in the literature include that only kin are selected for, rather than others who may share genes, or that altruists must be able to recognize each other, when conditions of low dispersal of kin may serve the same function.[46] Arguably, Maynard Smith's substitution of the term "kin selection" for "inclusive fitness" is an example of a formulation that facilitated such typifications.[47]

As a result, versions of "kin selection" theory may represent a slight theoretical regression while facilitating a potential test that could demonstrate empirical progress. Kin selection had an important influence in a section of the anthropology discipline. For instance, Napoleon Chagnon applied kin selection theory as follows to ethnographic data collected on the Yanomamö:

> KS_2: Individual men in human groups who compete for resources of land, food, or women will be more likely to be related to individuals with whom

45. Wilson, *Sociobiology*.
46. Park, "Persistent Misunderstandings of Inclusive Fitness and Kin Selection."
47. Maynard Smith, "Group Selection and Kin Selection"; Hamilton, "Innate Social Aptitudes of Man," pp. 140–41.

they cooperate than they are to individuals with whom they engage in conflict (Chagnon).[48]

Chagnon's attempt to wed anthropological interest in kinship to Darwinism was fiercely resisted by cultural anthropologists, who described human cultural constructions of kinship as "fictive" rather than genetic, driving cooperation beyond the limits of genetic relationships. Marshall Sahlins made this point in his criticism of Wilson's sociobiology, and the objection has been made based upon a research program of *cultural relativism* launched by Franz Boas:

> CR_0: Cultures organize cooperation in radically different ways that cannot be sorted into hierarchies ranked by success or rationality. Cultures are autonomous creations of the societies that create them (Boas).[49]
>
> CR_1: Cultural conceptions of kinship organize cooperation among genetic non-kin (Sahlins).[50]

Tests of Chagnon's kin selection theory (KS_2) have been mixed. Chagnon has confirmed the results among the Yanomamö, while the effect seems to break down in larger groups.[51] Likewise, tests of the extent to which real or fictive kinship (CR_1) organizes cultures has received mixed results.[52] In both cases, methodological criticisms of alleged corroborations of the theory have focused on whether the simplifications necessary for arriving at a "testable" result were appropriate.

By contrast, theoretical progress can be made by the development of new mathematical formalisms that generalize the theory. George Price developed a formalism that was able to derive Hamilton's formula as a special case while specifying separate terms for individual and group selection (and multilevel selection more broadly). Price's equation blurred the differences between kin selection and group selection. Each side now claimed that the formulation of Price's equation reduced their opponent to a special case of their own approach.[53]

---

48. Chagnon, "Kin-Selection Theory, Kinship, Marriage and Fitness among the Yanomamö Indians."
49. Boas, "Museums of Ethnology and Their Classification"; Li, "Marshall Sahlins and the Apotheosis of Culture"; Lynch, "Beyond Kin Selection and Cultural Relativism."
50. Sahlins, *The Use and Abuse of Biology*; Sahlins, *What Kinship Is and Is Not.*
51. Chagnon and Bugos, "Kin Selection and Conflict"; Alvard, "Kinship, Lineage, and an Evolutionary Perspective"; Alvard, "Kinship and Cooperation."
52. Etter, "Sahlins and Sociobiology"; Shapiro, "Contesting Marshall Sahlins on Kinship."
53. Nowak and Allen, "Inclusive Fitness Theorizing Invokes Phenomena That Are Not Relevant for the Evolution of Eusociality"; Nowak, Tarnita, and Wilson, "The Evolution of Eusociality"; Sober and Wilson, *Unto Others*; West, El Mouden, and Gardner, "Sixteen Common Misconceptions about the Evolution of Cooperation in Humans."

The Price equation describes the change in the average value of a trait between two generations and shows that if the difference between groups is large enough it can overcome selection within the groups. The Price equation has consequently been appropriated by group selection theorists, while inclusive fitness theorists claim it for their own as well, showing that group selection is rare compared to kin selection. Group selection theorists point to warfare, as well as processes that suppress phenotypic variability within groups such as conformist bias in cultural transmission or enforced egalitarianism, as cases common among humans where group selection effects are large.[54] Incommensurability between kin selection and group selection gives way to formal, mathematical commensurability satisfying instrumentalists, while realist proponents in either camp continue to debate what is the "real" underlying cause.

Thus, Andy Gardner's argument for the equivalence of the two perspectives nonetheless concludes that kin selection provides a more natural way to derive and understand consequences of the theory:

> [B]ecause kin selection and group selection theory are both based upon the same Price equation, it is easy to show that the two approaches are mathematically exactly equivalent, and are simply alternative ways of carving up the total selection operating upon the social character. Irrespective of the approach taken, individual organisms are expected to maximise their inclusive fitness—though this result follows more easily from a kin selection analysis, as it makes the key element of relatedness more explicit.[55]

By contrast, David Sloan Wilson concedes that the theories are essentially equivalent, with the exception that kin selection cannot easily explain equilibrium selection, community-level selection, and human cultural evolution.[56] A survey of evolutionary anthropologists has shown a generational shift to supporting group selection, in marked contrast to an earlier generation where one professor is reported to have advised his students to avoid three forbidden ideas: Lamarckism, phlogiston, and group selection. (I have discussed reconsiderations of two of these but will not advocate reviving phlogiston theory!)

Group selection theory was also correlated with liberal, rather than conservative, political opinions in keeping with the political significance of cultural

---

54. Bowles, "Did Warfare among Ancestral Hunter-Gatherers Affect the Evolution of Human Social Behaviors?"; Richerson, Boyd, and Henrich, "Cultural Evolution of Human Cooperation"; Boehm, "Emergency Decisions, Cultural-Selection Mechanics, and Group Selection," p. 775.

55. Gardner, "The Price Equation," R200. Van Veelen et al., "Group Selection and Inclusive Fitness Are Not Equivalent" and van Veelen, "On the Use of the Price Equation" argue that the two approaches are not mathematically equivalent on a rigorous derivation of Price's equation with group selection and inclusive fitness giving different results for some cases.

56. Wilson, "The Tide of Opinion"; Wilson, *Does Altruism Exist?*

evolution as a more progressive alternative to sociobiology. At the same time, liberal respondents were less likely to see warfare as the driving force behind group selection in humans, even though that is the predominant mechanism proposed in the literature.[57] In response, evolutionary psychologists reflecting a different political perspective challenge group selection theory as flawed and question funding for Wilson's work by the Templeton Foundation, which sees its charge as reconciling science and religion.[58] In an inversion of the sociobiology debate—or better, a new counteroffensive—Steven Pinker, Richard Dawkins, and Jerry Coyne have accused group selection advocates of imposing politics and moral ideals emphasizing cooperation, harmony, and religious belief onto human society, distorting science through ideological bias.[59]

With this in mind, we represent the Price theory here as a hybrid theory supporting both kin selection and group selection in the reconstruction of the group selection research program:

> GS_0: Traits can evolve by group selection when they favor the survival and reproduction of groups (Darwin, Wynne-Edwards).
>
> GS_1: To the extent that distinct groups differ in their possession of altruistic traits, and the altruistic groups survive and reproduce better (for instance, by avoiding predation), the greater survival of altruistic groups can overcome individual selection favoring non-altruist "cheaters" when altruists disperse to other groups (David Sloan Wilson, Maynard Smith).[60]
>
> $KS_3 = GS_2$: Change in the frequency of traits between generations are related to the combined result of between-group and within-group selection, as described by the sum of two terms that together equal the covariance between a trait and fitness, corresponding to the natural selection term in Price's equation.[61]

As we have seen, inclusive fitness is not the only research program that carried out a reconfiguration of the core anomaly of altruism in Darwinism. The alternatives that have been introduced testify to the presence of multiple, competing research programs within evolutionary biology, while their interaction and hybridization show that research programs do not remain self-contained and distinct from competitors. The most significant alternative research programs for explaining altruism were brought about by (a) the revival

57. Yaworsky, Horowitz, and Kickham, "Gender and Politics among Anthropologists in the Units of Selection Debate"; Sober and Wilson, *Unto Others*, p. 40.

58. Coyne, "Big Dust-Up about Kin Selection"; Barash, "Genes Are Selfish; People Are Not."

59. Pinker, "The False Allure of Group Selection"; Dawkins, "The Descent of Edward Wilson."

60. Wilson, "A Theory of Group Selection"; Maynard Smith, "Group Selection and Kin Selection."

61. Price, "Extension of Covariance Selection Mathematics"; Hamilton, "Innate Social Aptitudes of Man." The natural selection term is added to a term that delineates the change in traits resulting from transmission between parent and child to arrive at the change in frequency of traits between generations. For the actual equations and discussion, see Gardner, "The Price Equation."

of group selection theory by David Sloan Wilson and Elliott Sober, exploiting the formalism of George Price that converted inclusive fitness theory into an account of distinct levels of selection (also called multilevel selection),[62] and (b) the development of cultural evolution theory by Boyd, Richerson, Cavalli-Sforza, Feldman, Durham, and others, where human cooperation, in particular, is explained via a separate inheritance system, available to animals who possess and accurately transmit culture.[63] In both cases, minority positions treated as degenerating or obsolete were revived, subject to theoretical reconfiguration and empirical progress.

Related developments within evolutionary biology have also reinforced the idea that there are levels of selection, focusing on how evolution at lower levels can lead to the emergence of new types of entities subject to selection. When higher levels begin to act in a coordinated way, a new "individual" subject to selection emerges. On this view, there have been multiple transitions in the history of life where higher-level individual units emerge. Besides cultural evolution, transitions included those from prokaryotic cells to eukaryotic cells, from single-celled organisms to multicelled organisms, from asexual to sexual organisms, and from solitary organisms to eusocial species organized as colonies.[64]

Multilevel selection and the identification of major transitions in evolution can be used to explain not only how cooperation emerges but how it breaks down, acting as novel predictions in the Lakatos-Zahar sense. A good example of this is the reversion of some cells within a multicellular organism to pursuing selfish interests rather than the good of the whole organism: cancer. The equivalent process for group selection is the emergence of cheaters who benefit from opting out of the altruistic traits shared by the group. Shelved anomalies are taken off the shelf when conditions allow a creative transformation of existing theoretical elements within science to accommodate them.

Minority positions within evolutionary biology continue to develop by finding new ways to creatively reinterpret what is going on in cases of altruistic behavior. Building on social evolution's discussion of the emergence of structures of greater complexity over the course of evolution, biologists have proposed alternatives to the selfish gene view of evolution, views that have not displaced the selfish gene model at this point. The revival of group selection as an explanation fits with the larger emphasis on emerging complex structures that become subject to natural selection at a different level than genes. E. O. Wilson's research on eusocial insects, insects with an "altruistic" division of sexual labor where male worker drones are sterile, represent

62. Sober and Wilson, *Unto Others*.

63. Boyd and Richerson, *Culture and the Evolutionary Process*; Cavalli-Sforza and Feldman, *Cultural Transmission and Evolution*; Durham, *Coevolution*.

64. Maynard Smith and Szathmáry, *The Major Transitions in Evolution*.

an extreme form of cooperation going against personal fitness. Moreover, recently Wilson has become convinced that inclusive fitness fails to explain the emergence of eusociality and has begun to speak of colonies of insects being selected for directly.[65]

Wilson, collaborating with mathematicians Martin Nowak and Corina Tarnita, has developed an alternative theory that denies equivalence with kin selection theory, arguing that a gene-centered view bypasses the need for inclusive fitness. They argue that Hamilton's theory requires that personal and inclusive fitness are additive components applied to individuals, an assumption that does not hold generally. They also argue that alleged empirical confirmations are not based on predictions of actual inclusive fitness. Their alternative hypothesis argues that eusocial species can emerge by standard natural selection when coherent, long-lasting groups emerge in freely mixing populations. The theory has been fiercely criticized by defenders of inclusive fitness theory, who have called it "obviously incorrect" and suggested that it "involves, and I suspect requires, close kinship," while others support the theory as an alternative to address weaknesses of an excessive emphasis on kinship and inclusive fitness.[66]

The controversy has been so intense that 137 scientists signed a letter in *Nature* objecting to their paper.[67] Their claim that genetic relatedness was not the cause of the evolution of eusociality was challenged by arguing that Nowak, Tarnita, and Wilson's own theory depended upon relatedness; their response in turn is that current defenders of Hamilton's rule tacitly incorporate population structure in a model described as focused on genetic similarity. Common to these disputes is the concern to delineate the "real," causal basis for selection in models that necessarily incorporate both population structure and relatedness, as well as the relationship between approximations of Hamilton's rule developed for tractability and more general, correlational treatments of Hamilton's rule that abstract away from a strict genealogical focus.[68]

Group selection approaches are implicated in most theories of the development of cultural evolution. Explanations of large-scale human cooperation among non-relatives has been explained as the effects of the emergence of

65. Nowak, Tarnita, and Wilson, "The Evolution of Eusociality"; Wilson and Nowak, "Natural Selection Drives the Evolution of Ant Life Cycle."

66. Nowak, Tarnita, and Wilson, "The Evolution of Eusociality"; Keim, "E. O. Wilson Proposes New Theory of Social Evolution." For critical discussion, see Queller, "Kin Selection and Its Discontents"; Birch and Samir, "Kin Selection and Its Critics."

67. Abbot et al., "Inclusive Fitness Theory and Eusociality."

68. Liao, Rong, and Queller, "Relatedness, Conflict, and the Evolution of Eusociality"; Hughes, Oldroyd, Beekman, and Ratnieks, "Ancestral Monogamy Shows Kin Selection Is Key to the Evolution of Eusociality."

cultural evolution in our species.[69] Once the development of human "hyper-cultural" learning systems are in place due to biological evolution, cultural practices that help groups survive and thrive are selected for at another level than genetic evolution. Groups possessing favorable cultural adaptations can pass on these practices to the next generation. It is also possible that other groups will imitate these practices, allowing "horizontal" transmission to supplement "vertical" transmission typical for genetic evolution.

These separate levels can in turn interact, understood under the label of culture-gene coevolution, as the spread of cultural practices that favor group survival in turn influence the selection environment for biological evolution. For an example of this process, consider the introduction of the cooking of food as a cultural practice that facilitates greater extraction of nutrition. The behavior spreads through the human population via the evolved social learning process at the basis of human cultural abilities (a biological adaptation of our species), which facilitates the survival of those cultural practices by cultural selection (those groups who have learned to cook their food extract more nutrition from the environment than those who do not). This, in turn leads to biological adaptations to these cultural practices, with a consequent shift from large guts and smaller brains to smaller guts and larger brains, allowing them (and their cultural products) to survive. Cultural practices that predigest food make possible biological adaptations that speed the development of larger brains to further support cultural learning.[70]

It is worth mentioning as well, in this context, that this more complex view of evolution has important implications for attempts to model science itself in evolutionary terms. The more complex understanding of evolution that has emerged in biology in recent decades as the result of methodological critiques of overly "adaptationist" approaches shows that mere survival of a trait does not mean that it is an adaptation, since it may well be the result of genetic drift or a by-product of selection for other traits. So, too, the survival of theories in science does not automatically imply fitness, and fitness does not imply truth. The survival of theories in science does not imply that the theories are optimally adapted by the process of science to representing the world.

Even if surviving theories in science are understood as adaptations generated by the process of cultural evolution, and not by-products or other non-adaptive changes, it does not mean that these adaptations are necessarily true, approximately or otherwise, as developments in the evolutionary understanding of religion particularly make clear. Religious practices and beliefs may have evolved by cultural evolution by promoting cooperation and group

69. Boyd and Richerson, *Culture and the Evolutionary Process*; Richerson and Boyd, "Complex Societies"; Richerson, Boyd, and Henrich, "Cultural Evolution of Human Cooperation"; Henrich and Henrich, *Why Humans Cooperate*.

70. Wrangham, *Catching Fire*.

survival, without the beliefs constituting true representations of supernatural phenomena.[71] Fitness, in cultural evolution, whether understood as memes merely using us as vehicles for reproducing themselves at our expense or as complex adaptations that facilitate group survival does not equate to truth. Truth, like religion, may be an occasional by-product of cultural adaptations that serve other functions in enhancing individual and group survival.[72] Alternatively, different "ecological" contexts for scientific work can be expected to lead to knowledge that functions in specific niches, rather than bringing about correspondence to nature as a whole.[73] The challenging of "just so" stories in evolutionary epistemology can suggest new questions neglected by standard models of science.

What should we conclude about the "function" of theoretical diversity in science? Clearly, from a viewpoint seeing science as a form of cultural evolution where scientific groups compete to reproduce themselves, the existence of theoretical diversity in science has, at times, provided needed variation that can reemerge to be selected for as the environment changes.[74] There remains the question whether or not the transformation of prior frameworks was a necessary step in the history of science, whether one could skip to later "integrated" or "extended" syntheses, or whether there exist multiple paths that science could have followed. What is clear is that the advance of science "reframes" the meaning of past science. There may be essential dogmas that help facilitate advance but ultimately block later insights. When these blocks are finally overcome (again when "ripe"), it calls for a reconsideration of everything that was once "known" within the facilitated perspective.

This dialectical process is quite different from imposing an outside dogma unconnected to the current state of science. Science tends to block the importation of random, outside metaphysical building blocks as a basis for science. Instead, a dialectical reversal of a crucial dogmatic, metaphysical building block within science leads to a transformation of knowledge that is revolutionary in rewriting the established picture of the world. Science may not adopt an utterly different picture of the world de novo but rather incorporate qualitative transformation of prior science.

The incorporation of developmental biology, and affiliated perspectives excluded or neglected by the modern evolutionary synthesis, back into an "extended evolutionary synthesis" and the development of the science of epigenetics in ways that transforms genetic reductionism are examples of this kind of process.[75] To paraphrase Feyerabend, *potentially every scientific*

71. Wilson, "Species of Thought"; Wilson, *Darwin's Cathedral.*
72. Lynch, "Cultural Evolution and Social Epistemology."
73. Wilson et al., "The Nature of Religious Diversity."
74. Compare the literature on the "Baldwin effect": Weber and Depew, *Evolution and Learning.*
75. Pigliucci and Müller, *Evolution: The Extended Synthesis*; Moore, *The Developing Genome*.

research program is all scientific research programs; that is to say, scientific development belies any essential identity to research programs in the long run, even if incommensurability can be functional in defining distinct niches where different research programs do well. Disunified science is only temporary, however, and integration occurs when temporary obstacles to unification are overcome in the course of time. A dialectical view of science is not compatible with an emphasis on permanent disunity and the autonomy of self-warranting expert communities.

Just as Lakatos showed that monster-barring and other ad hoc ways of avoiding contradictions would have inhibited the development of approaches that advanced mathematics, a similar dialectical point can be made about science. If one only focuses on whether a thesis or its opposite can be justified or not at a particular moment in time, the emphasis will be on rhetorical justification and current assessments of existing evidence. Where evidence is not definitive for a dominant approach, one can be tempted to demand that "by right," alternative views should be considered. Metaphysical commitments of science are then to be subject to a kind of emphasis on democratic representativeness. Everyone should have their favorite big metaphysical beliefs about the world incorporated by scientists within their research programs.

If science works through dogmatic perspectives, exhausting them of their fruitful potential, before new perspectives emerge that reframe its insights, then one's metaphysical commitments should not be seen as a "free choice" subject to the requirements of political representation. Another way to put this—even scientists who radically challenge some accepted view in their field usually do so by drawing on other elements of the research programs within science.

POLITICS AND SCIENCE ON A DIALECTICAL VIEW

There is one addendum we need to add to our discussion about the viability of research programs and that is the role of politics within science. On the standard view, science incorporates a method for overcoming politics and one is advised to leave politics at the door. In the case of sociobiology and its criticism within evolutionary biology, the role of political beliefs has mutually reinforced distinct philosophies of science shaping scientific work. Politics, if nothing else, can serve like the metaphysical stepping-stones we have discussed.

In the case of the "Marxist" school of evolutionary biologists, political critique was also wedded to a view of scientific truth that was in some ways more demanding than that put forth by Lakatos, or for that matter, the average scientist. As we will see, there are distinct advantages and disadvantages to "political" critique of science. The critique of the "adaptationist" program in evolutionary biology carried out by Lewontin and Gould, based in part on Marxist conceptions of dialectical philosophy, the critique of ideology, and a commitment to human diversity, led to an extended debate in biology.

The critique of "adaptationism" highlighted an excessive faith within biology that adaptations arising from natural selection could be identified and distinguished from other causes and processes. The critique was carried out by biologists influenced by Marxist politics, explicitly political groups like the Boston Sociobiology Study Group (SSG) of Science for the People, and, not incidentally, a number of historians and philosophers of science. The critical program was productive *for science* in some important ways, but it also has some significant limitations both as epistemology and as moral and political philosophy. At the same time, the critics of adaptationism were working within the research programs of evolutionary biology in ways that eventually led beyond criticism to constructive bases for new research programs. A number of such programs resulted, among them punctuated equilibrium theory, niche construction theory, developmental systems theory, evolutionary developmental (Evo-Devo) biology, and the extended evolutionary synthesis.[76]

The Marxist aspect of the criticism of science was developed in ways that wedded it to an epistemological critique of reductionist science and to the political history of the abuse of biology.[77] Aside from objecting to the social consequences of sociobiology, the criticism of Lewontin, in particular, favored scientific realism over empiricism. If Lakatos argued that science required multiple research programs, Lewontin treated truly scientific research programs as more like gold bullion than promissory notes. Scientific theories should engage underlying causes; research programs needed to deliver the goods rather than develop hypothetical models.[78] To the extent that this body of writing was purely negative, it could be argued that the critique of adaptationism came close to anomaly mongering, this time to defend society against the potentially pernicious consequences of human sociobiology.

The political critique advanced by Gould and Lewontin involved arguing that sociobiology's emphasis on human biological differences would be used to justify inequality as social policy. Rather than confront this issue purely at the level of social policy and democratic politics, Gould and Lewontin argued that this was also an epistemological and moral issue for scientists themselves. In a sense, they took the lesson from physicists like Leo Szilard,

---

76. Gould, *Punctuated Equilibrium*. Lewontin's impact is traced in Singh at al., *Thinking about Evolution*. The influence of Gould's *Ontogeny and Phylogeny* on Evo-Devo is discussed in Love and Raff, "Knowing Your Ancestors."

77. Levins and Lewontin, *The Dialectical Biologist*; Gould, *Mismeasure of Man*.

78. Segerstråle, *Defenders of the Truth*, ch. 14.

who worked for disarmament after the development of atomic weapons during World War II, arguing that scientists had a moral responsibility for the social consequences of their work. As Kitcher argued, we may legitimately demand a higher threshold of evidence for claims that have potentially negative political effects, just as we use more caution with pharmaceutical drugs that may have serious negative effects (see chapter 6 for discussion of the precautionary principle).[79]

The political critique was also connected to an epistemological critique insofar as they argued that E. O. Wilson and other sociobiologists were not *justified* in their claims that human social behavior could be connected to genetic causes. While they did not deny that humans were shaped by their genetic heritage, the identification of even hypothetical genetic causes for human behavior was too explosive a claim to not demand *proof*, while alternative hypotheses existed that explained human behavior in terms of autonomous cultural causes.

In Lewontin's view, the critique of provisional scientific modeling demanded that scientific work be oriented to the identification of *true causes* underlying biological phenomena and not *plausible models* that could justify research programs that successfully solved scientific puzzles.[80] Lewontin had honed his approach to critique in the controversy over Arthur Jensen's claim that group differences in IQ scores testified to differences in intelligence among races. Lewontin argued that IQ scores did not in fact measure actual intelligence, that Jensen conflated the causes of within-group variation with between-group variation, and that the scientific case for racial differences did not exist.[81] As such, Jensen was abusing science for purposes that must be based on an ideological agenda. In his critique, Lewontin was able to rely upon his own research that showed that only 15 percent of the variance in human blood groups could be accounted for by group differences, with 85 percent of the variance accounted for by within-group variation.[82] This supported the post–World War II consensus in biology that races were not biologically meaningful categories and that biological differences among humans could not be mapped onto racial categories.[83]

Lewontin argued that Jensen, and later Wilson, were corrupting science by promoting unjustified scientific claims in ways that would harm society. The

79. Kitcher, *Vaulting Ambition*, p. 9. For discussion of the extent to which acceptance of a hypothesis in sociobiology ought to entail corresponding beliefs and actions in the political realm, see Holcomb, *Sociobiology, Sex, and Science*, pp. 64–82.

80. Segerstråle, *Defenders of the Truth*, ch. 14.

81. Lewontin, "Race and Intelligence."

82. Lewontin, "The Apportionment of Human Diversity."

83. Belief in this alleged consensus against the use of race in biology ignores continuing disagreements. See Maio and Santos, "Antiracism and the Uses of Science in the Post–World War II" and Obasogie et al., "Race in the Life Sciences."

psychological study of racial differences in intelligence was not one legitimate research program among many but an ideological program built upon a false biological premise. Lewontin later argued explicitly that the mere existence of "progressive" research programs did not prove anything, as it was possible to establish make-work programs for scientists who could solve to their own satisfaction puzzles built upon a false foundation.[84]

Lewontin's argument, founded as it is upon high standards for causal reasoning in science, testifies to the significance of a *realist* philosophy of science for the practice of science, an approach not accidentally connected to Marxism (see chapter 5). At the same time, the success of the critique in influencing scientists potentially shows that political critique can have a legitimate place within science. It is not hard to see how this contrasted with Lakatos's approach, with its emphasis on scientific liberty and the legitimacy of a pluralism of research programs in any scientific field. As we have seen, Lakatos believed that Jensen's claims were legitimate scientific claims whose suppression would damage science and encouraged his graduate student, Peter Urbach, to demonstrate that the research program was progressive in generating new, corroborated facts. Lakatos also used the realist and Marxist-influenced philosopher of science Hilary Putnam's proposal that the American Philosophical Society condemn Jensen's work as a critical foil for the view that knowledge should be evaluated by its social consequences, a view he traced back to the Catholic Church's opposition to Galileo.[85]

A defense of racial differences in IQ was probably not justified under the criteria of Lakatos's philosophy of science. Urbach had ignored the possibility that individual variation in mental faculties existed but not group differences, a view that constituted the hard core of geneticists critical of Jensen such as Lewontin.[86] On this view, inferring intergroup heritability from intragroup heritability is a logically fallacious prediction in the first place. By contrast, even granting the objections to claims about human behavior associated with sociobiology, it is clear that the field built on a hard core based on kin selection and inclusive fitness that proved a productive guide to research in biology and anthropology.[87]

---

84. Segerstråle, *Defenders of the Truth*, pp. 276–80, 346.

85. Lakatos and Feyerabend, *For and Against Method*, pp. 348–49, n. 200; Urbach, "Progress and Degeneration in the 'IQ Debate'"; Lakatos, "Lectures on Scientific Method," pp. 21–22.

86. Deakin, "On Urbach's Analysis of the 'IQ Debate.'"

87. See, however, Matthijs van Veelen's argument that the Price equation has been misused in predicting empirical tests, some of which were nonetheless corroborated. Van Veelen, "On the Use of the Price Equation"; van Veelen et al., "Group Selection and Inclusive Fitness Are Not Equivalent." Compare Lakatos's argument that progressive research programs can be built on inconsistent premises, as evidenced by Bohr's model of the atom (Lakatos, "History of Science and Its Rational Reconstructions," p. 142). See, also, Meheus, *Inconsistency in Science*.

The revival of group selection theory and the development of cultural evolution theory did show that multiple research programs based upon different postulated causal processes can coexist in science, with the question of realism a muddled one, as each camp argued that its competitor's success was attributable to their tacitly borrowing from their superior theory. Thus, inclusive fitness proponents argue that group selection theory only works as a result of shared genes among proximate organisms, while group selection proponents claim inclusive fitness theorists smuggle in group selection by the way they calculate fitness.[88]

While it is probably easy to see that Lakatos's view of the scientific status of IQ research reflected his move to the right politically, Lewontin's view has its own problems as a lasting philosophy of science. An emphasis on scientific realism as a criterion for scientific theory evaluation sets impossibly high demands on scientific reasoning, begs the question as to which research programs track underlying reality in the flux of scientific debate, and sets a bad precedent for policing scientific claims ideologically.

One long-term consequence of Lewontin and Gould's criticisms of adaptationism within evolutionary biology is that their critique has been taken on board in the manner of pro forma qualifications of claims to identify adaptive function in biology.[89] In this sense, biologists have become trained to acknowledge all the exceptions to a research program tracking adaptation by natural selection, citing the paraphernalia of critical concepts including exaptations, spandrels, hierarchical selection, and the like as needed to dispel potential criticism while continuing to propose adaptive functions. To the extent that the effect of Gould and Lewontin's arguments remains at this level, acknowledging the limitations and lack of proof of current approaches without substantially changing their methods, they could be seen as offering a useful methodological critique without going much beyond anomaly mongering. However, their criticisms do go beyond anomaly mongering to the extent that they have become the basis for distinct research programs in contemporary biology, as we will see.

We will consider arguments for scientific realism within science studies in chapter 5. Suffice it to say that anyone arguing for the underlying truth of a scientific claim will need to make their case in part by demonstrating that the research program producing the claim has generated novel predictions (or retrodictions) of fact that have been corroborated over time. So, for example,

---

88. Park, "Persistent Misunderstandings of Inclusive Fitness and Kin Selection"; Nowak and Allen, "Inclusive Fitness Theorizing Invokes Phenomena That Are Not Relevant for the Evolution of Eusociality"; Nowak, Tarnita, and Wilson, "The Evolution of Eusociality"; Wilson, "Hunting, Sharing, and Multilevel Selection." See also Sober and Wilson, *Unto Others*, pp. 31–50, on the "averaging fallacy."

89. Laland and Brown, *Sense and Nonsense*.

the best evidence that the theory of anthropocentric global warming is true and the sunspots theory of its critics is false is that the former continues to generate fruitful work, thereby suggesting that it is on to something, while the latter does not. Alas, research programs do not bear fruit in perpetuity, so the issue of what the best scientific representations will turn out to be await the sorting out of the relationship between current contenders and their unknown future replacements at the end of inquiry.

As Hacking argued, we can also have confidence in our causal claims to the extent that we manipulate the underlying causes to generate new effects.[90] It should be noted that if the theory of anthropocentric global warming is correct, we *have* manipulated the postulated causal mechanisms without knowing it, but, of course, that is exactly what is questioned by critics. Perhaps large-scale geoengineering solutions to global warming, if adopted, will involve active manipulation of underlying atmospheric causes. Even in that case, sorting out the effects of our intervention and other natural and artificial processes will depend upon research programs that can only be evaluated pragmatically by their consequences.

Understood as a purely corrosive force, the employment of a realist approach can aid scientists in identifying causal claims that may fail to be justified. As a positive program, realist scientists need to develop positive programs that can generate fruitful research before they can contribute substantially to the ongoing development of science. In this sense, evaluating research programs still leaves us to judge progress by something like Lakatos's criteria. That being said, the political-cum-epistemological critique of Lewontin has been adopted as a basis for research programs that have shown initial signs of progress.

Burghardt and Laland credit Lewontin with inspiring niche construction theory to make good on Lewontin's claim that the modern evolutionary synthesis erred in treating the environment as a static entity shaping natural selection of the organism. The organism is seen as the passive recipient of environmental influence. Evolutionary adaptations function as the "key" to environmental "locks" challenging the organism.[91] By contrast, Lewontin argued that organisms shape their environments actively in ways that in turn influence evolutionary change. Laland et al. apply this insight in developing a program of "niche construction." In particular, they postulate that repeated modifications of the environment by organisms can lead to directed change along lines shaped by the organism's own actions.

---

90. Hacking, *Representing and Intervening*.

91. Burghardt and Laland, "Deconstructing Niche Construction"; Lewontin, "Organism and Environment"; Lewontin, "Gene, Organism and Environment"; Lewontin, "The Organism as the Subject and Object of Evolution"; Lewontin, *Triple Helix*.

They develop tests of the program that put them into competition with other research programs. For instance, they predict that shared constructed environments will tend to reduce temporal and spatial variability in evolutionary pressures and that clusters of traits will coevolve with animal artifacts such as birds' nests or spider webs. They also argue that niche construction in the past may have served to buffer "adaptive lag" postulated within evolutionary psychology, where a mismatch between current social practices and evolved characteristics of the human mind shaped by Paleolithic foraging is expected. Instead, selective forces brought about by human niche construction may shape both genetic evolution and cultural evolution, such as adult lactose tolerance or technological adaptations to urban lifestyles.[92] Niche construction has recently contributed to a number of works integrating humanistic and scientific methods in explaining the biological-cum-cultural emergence of human culture, including capacities for technological innovation, musical composition, ecological transformation, symbolic thought, and scientific discovery.[93]

A related program of developmental systems theory (DST) argues that genes should not be privileged as the unit of selection, since all kinds of material structures are passed on to future generations, including modifications of the environment emphasized by niche construction theory and the results of developmental processes that affect how genes are expressed. The theory "applies the concept of inheritance to any resource that is reliably present in successive generations, and is part of the explanation of why each generation resembles the last."[94] This can include passing on advantageous symbiotic bacteria, chromatin markers involved in epigenetic inheritance, prions, maternal milk, culture, and ecologies.[95]

All of these resources can contribute to an expanded conception of differential fitness as "[c]hange over time in the developmental system of a lineage is driven by the differing capacity of variant developmental systems to reconstruct themselves."[96] Thus, variation in the lead present in water or changes in availability of discarded shells inhabitable by hermit crabs can be incorporated into accounts of the reproduction of individual life cycles with

92. Laland and Brown, "Niche Construction, Human Behavior, and the Adaptive-Lag Hypothesis"; Odling-Smee, Laland, and Feldman, *Niche Construction*. See, also, the related arguments for inheritance of behavioral traditions shaping the environment in ways that reinforce those traditions in Jablonka and Lamb, *Evolution in Four Dimensions*, pp. 176–80; Avital and Jablonka, *Animal Traditions*.

93. Rendell, Fogarty, and Laland, "Runaway Cultural Niche Construction"; Sterelny, *The Evolved Apprentice*; Tomlinson, *A Million Years of Music*; Tomlinson, *Culture and the Course of Human Evolution*; Boivin et al., "Ecological Consequences of Human Niche Construction"; Barker, Desjardins, and Pearce, *Entangled Life*; Rouse, *Articulating the World*.

94. Griffiths and Gray, "The Developmental Systems Perspective," p. 411.

95. Ibid., p. 413.

96. Ibid., p. 422.

the expectation that that will better predict differential reproduction than a change in gene alleles alone. A novel prediction in Zahar's sense of DST is that organisms can become "addicted" to the environment when available resources in the environment replace or substitute for genetic capability to supply the resource. Hermit crabs are reliant upon transmission of shells from crabs that produce their own shells, while primates living in fruit-rich environments lost the ability to produce vitamin C endogenously.[97]

These examples suggest that the critical program launched by Gould, Lewontin, and other critics passed beyond the threshold of anomaly mongering in setting up research programs that put themselves into competition with research programs associated with the modern synthesis. The negative and harsh rhetoric directed at scientists for incorporating racial ideology within their scientific work—and the negative replies accusing the critics of Marxist bias and corruption of science—has set another questionable precedent for scientific debate.[98] While it is important that scientists incorporate ethical and political responsibility for the impact of their work on society,[99] scientific claims and their moral and political import are all in flux. What may seem politically objectionable in a moment of time may be less so as science and society adjust (and vice versa). Moreover, the conflict of scientific and politically incommensurable viewpoints may give way over time to new, hybrid perspectives.[100]

Still, in some ways, the persistence of questionable claims about human nature have only proliferated since the 1970s in ways that have tended to make the speculative biological reductions of Wilson and Hamilton seem quaint. The carrying out of the human genome project and the development of new methods for correlating bits of the human genome with human behaviors, implicitly treating these traits as genetically fixed, pathological, and subject to potential cure has exploded, despite severe criticism by many of the scientists involved.[101] Just-so stories have given way to just-so correlations, badly interpreted from weak evidence and ignorant of challenges from the fields of the extended evolutionary synthesis.

There is no question that the practice, interpretation, and use of modern biological science are fraught with ideological consequences. The question is how should scientists sort out which programs are worth pursuing, how they should be described and assessed within science, and how their public meaning

---

97. Ibid., p. 426. This line of argument is also developed in Jablonka and Lamb, *Evolution in Four Dimensions*.

98. Segerstråle, *Defenders of the Truth*, ch. 11.

99. Marks, *What It Means to Be 98% Chimpanzee*, ch. 12, makes the case for scientific responsibility for the impact of new biological knowledge. For discussion of science and engineering ethics, see the discussion below in chapter 6.

100. Lynch, "Between Kin Selection and Cultural Relativism."

101. Marks, *What It Means to Be 98% Chimpanzee*, ch. 5; Sarkar, *Molecular Models of Life*, ch. 4.

and use should be monitored. What is clear is that the adage "the enemy of my enemy is not my friend," applies to political critiques of allegedly bad science. The bulwark against ethnocentric and racist science, cultural relativism, held out against sociobiology's emphasis on biological explanations of human behavior, has itself come under criticism for its tendency to treat traditional societies as locked into static cultures incapable of change. The cure for colonialism and cultural prejudice has become a suspect cause of it.[102]

In the meantime, the collateral damage from assessments of political correctness in science is high. A new genre of political critique of science has gone for the jugular. Chagnon's research arguing that kin selection explained patterns of cooperation and conflict among the Yanomamö and that violent killers had more offspring attracted criticism as the primary exhibit for biological accounts of human nature. Rather than just subjecting his ideas to criticism, Patrick Tierney sought to censure him, accusing him of inciting violence among the Yanomamö and contributing to their deaths from measles by a eugenics-style measles vaccine experiment. Tierney's attack was ultimately judged fraudulent but not before his criticism of Chagnon was endorsed by the American Anthropological Association in a controversy recently analyzed by historian of science Alice Dreger.[103] Dreger also examines similar controversies that sought to censor ideas and punish the individuals who propagate them, including an alternative theory of transsexualism developed by Ray Blanchard and popularized by Michael Bailey. The strength and weaknesses of views judged oppressive to marginalized identities are not debated so much as ruled out of court by a self-selected portion of the affected community.

Despite their political and philosophical differences, Lakatos and Feyerabend shared an emphasis on scientific and political fallibilism. Feyerabend's complaints about the moral certainty and manipulative approach of many political activists at Berkeley is to the point here.[104] In a letter to Lakatos, Feyerabend objected to the effort at his Berkeley campus to have Arthur Jensen fired after the publication of his article arguing for innate racial differences in IQ. Feyerabend made the point to his class that removing him did nothing to dispose of his ideas, which he would likely just take up elsewhere. More to the point, the proper response was criticism:

---

102. See the discussion of the Sahlins-Freedman debate in Lynch, "Between Kin Selection and Cultural Relativism."

103. Dreger, *Galileo's Middle Finger*. On the scientific and ethical debate over Chagnon's ethnography, see Borofsky, *Yanomami*.

104. Feyerabend, *Science in a Free Society*. See also his complaints about Berkeley student activists in his letters to Lakatos in Lakatos and Feyerabend, *For and Against Method*. Feyerabend was broadly supportive of the New Left, though critical of some of its tactics, while Lakatos saw the movement as a threat to universities and free inquiry similar to the one he spearheaded as a Stalinist activist in Hungary. See Martin, "The Battle Is On," pp. 5–6, 12–17, 21–26.

> Assume the guy was misled in his research by his ideology, assume that he experimented with racism in his mind. Wouldn't it be interesting to see, to study from close-up how allegedly "objective" science is influenced by very subjective ideological considerations? Is it not absolutely necessary to keep such people at hand close by in order to learn from a living example, that the "objectivity" of science is just moonshine? . . . Getting rid of somebody does not mean getting rid of his ideas. The proper way is to keep him here and, if one thinks he is using a faulty method, to criticize him *in class*. But for this the splendid *New Left* is too cowardly—they might get an F![105]

Above all else, Feyerabend rejected a patronizing view that the public needed to be protected from dangerous ideas, agreeing with Lakatos that an ideological litmus test for science was a bad idea.[106]

Lakatos had made a similar point in 1956, during his period as an anti-Stalinist activist in Hungary, in distinguishing a truly dialectical approach fostering the criticism of views with the "extermination of fact" associated with "a 'Marxist' fight against empiricism":

> The history of science indicates that we ought to teach the future scholar to be modest, to be humble in his scientific claims, to be averse to all kinds of fanaticism. He ought to learn that what he does not understand, or what he disapproves of, still has a right to exist, and that no scientific theory, no theorem can conclude anything finally, in the history of science.[107]

Strong criticism is a vital ingredient driving the dialectical progress of science, but it is vital to couple this with a fallibilistic attitude toward claims of absolute truth, whether scientific or moral.

---

105. Feyerabend to Lakatos, July 25, 1969, reprinted in Lakatos and Feyerabend, *For and Against Method*, 169–70, p. 169. Feyerabend also considers the possibility that Jensen was right but draws different normative implications. He asks whether "[i]t is so bad to live in a society where different people have different talents and learn in different ways, or is it not rather an advantage? Conversely, does the reaction of the lefts not show that they want to be like everyone else, that is, they want an absolutely boring society?" Aside from reflecting his move toward cultural relativism at this time, the thought anticipates in some ways Howard Gardner's theory of multiple intelligences (Gardner, *Frames of Mind*), although Gardner does not accept Jensen's emphasis on racial differences.

106. See Lakatos, "Lectures on Scientific Method," pp. 21–22, 64–67, for criticisms of Hilary Putnam's proposal that the American Philosophical Association should condemn Jensen's theory.

107. Imre Lakatos, "On Rearing Scholars," p. 379; Bandy, *Chocolate and Chess (Unlocking Lakatos)*, p. 221.

Chapter Four

Members Only

Wittgensteinian Sociology and the Debate about Carbohydrates and Fat

> "But the other benefit [of admitting all professions to the Royal Society] is, that by this equal Balance of all Professions, there will no one particular of them overweigh the other, or make the *Oracle* onely speak their *private* sence: which else it were impossible to avoid. It is natural to all Ranks of men, to have some one Darling, upon which their care is chiefly fix'd. If *Mechanicks* alone were to make a Philosophy, they would bring it all into their Shops; and force it wholly to consist of Springs and Wheels, and Weights: if *Physicians*, they would not depart farr from their Art; scarce any thing would be consider'd, besides the *Body* of *Man*, the *Causes, Signs*, and *Cures* of Diseases. So much is to be found in Men of all conditions, of that which is call'd *Pedantry* in Scholars: which is nothing else but an obstinate addiction, to the forms of some private life, and not regarding general things enough."
>
> —Thomas Sprat, *History of the Royal Society*, pp. 66–67

What relevance does the history of science have for the ongoing conduct of science itself? Should philosophers or Science and Technology Studies scholars use their understanding of science to criticize scientific theories and practices? How much leeway is there for scientists themselves to criticize their own fields or the fields of others? When scientists say that they applied particular methods to their research, should we believe them? Can we understand what difference commitment to competing methodologies or research programs makes to the ongoing trajectories of scientific disciplines? Can we assess the extent to which scientists have correctly followed the methods they claim to have followed or interpreted the evidence they have produced appropriately? Or should outsiders butt out and let the experts get on with their work, trusting that the scientists' tacit knowledge will point them in the right direction?

If one were to closely follow the overheated public debate labeled the "science wars," where "postmodernist" science studies scholars squared off against scientific defenders of the truth, one might conclude that Science and Technology Studies (STS) as a field offered a challenge to accepted beliefs in science. One might expect that defenders of the thesis that science is a social construction would leverage this insight to criticize the practice of science, suggesting alternative constructions are possible (and possibly desirable), while those developing an account of the proper methodological and epistemological structure of scientific reasoning would defend the scientific status quo. While this might fit the reigning stereotype of the debate between "enlightenment" scientists or philosophers and "postmodern" or "academic left" cultural critics,[1] it makes little sense of the root development of a social account of the internal practice of science associated with the sociology of scientific knowledge (SSK) or its interdisciplinary offshoot, Science and Technology Studies.[2]

As we have seen, this "internal" sociology of knowledge looked less to broad, social causes of the origin of science and its normative structure, as the "external" sociology of science of Hessen or the early Merton did, than to a close, social-interactional retelling of the internal reasoning of science.[3] In order to carry out this retelling, SSK appropriated Wittgenstein's later philosophy to talk about traditional epistemological topics like induction and experiment in collective, sociological terms. This made sense because Wittgenstein had abandoned his representational theory of meaning in the early *Tractatus* (1921) for a contextual theory in the *Philosophical Investigations* (posthumously published in 1953). For the later Wittgenstein, meaning is determined by the various actual uses of words in concrete situations rather than its reference to an external reality.

The sociological interpretation of Wittgenstein allowed sociologists to explain the content of science through a sociological focus on different forms of life within science. Different forms of life will apply distinct meanings to experience based on the accepted meanings reinforced by participation in

---

1. Gross and Levitt, *Higher Superstition*; Koertge, *A House Built on Sand*; Ross, *Science Wars*.

2. All the major schools of SSK were represented in the early anthology, Knorr-Cetina and Mulkay, *Science Observed* (1983), among which only the dissenting approach of Chubin and Restivo, "The 'Mooting' of Science Studies," evidenced an interest in normative questions about science. Handbooks surveying the field include Jasanoff et al., *Handbook of Science and Technology Studies* (1995), still SSK-centric, and Felt et al., *The Handbook of Science and Technology Studies* (2017), reflecting a wider diversity of interdisciplinary approaches, still predominantly symmetrical or anormative in approach, but with a growing minority engaging normative questions in feminism, postcolonial studies, environmentalism, science policy, or ethics.

3. Hessen, "The Social and Economic Roots of Newton's *Principia*"; Merton, *Science, Technology and Society in Seventeenth-Century England*.

social life. Science is seen as tied to cohesive forms of life that are taken to have their own integrity and to be immune from outside criticism.

In this chapter, I will examine and criticize this Wittgensteinian "sociologizing" of epistemic topics and suggest a more promising alternative that allows for the possibility of criticism from those who do not share taken-for-granted ways of going on. The expertise model of Collins and Evans is only the most explicit outcome of this "members only" rule for science, with its ultimate roots in sociologists' interpretation of Wittgenstein's later philosophy. Expertise depends upon internalizing the right kind of tacit knowledge about how to do science, which "can be acquired only by immersion in the society of those who already possess it."[4]

The idea that you must share a form of life to have a "right" to criticize that form of life underpins the concept of "elitist authoritarianism" Lakatos rejected. Feyerabend admired Wittgenstein's appeal to concrete cases rather than abstract principles, and had initially planned to study under Wittgenstein before his death forced a switch to Popper. He also authored an influential, early treatment of Wittgenstein's *Philosophical Investigations*. Nonetheless, and despite some similarities between Wittgenstein and Feyerabend's mature philosophies, Feyerabend rejected Wittgenstein's philosophy for its antipathy to criticism.[5]

In some ways, Wittgenstein's approach still acts as a firewall against normative approaches within Science and Technology Studies today. After the critical approaches of Popper, Lakatos, and Feyerabend gave way to empirically and contextually oriented philosophy and sociology of science in the 1980s, a variety of normative approaches to science reemerged in the late 1980s and early 1990s, growing out of earlier feminist, Marxist, radical, or philosophical approaches to science; however, these approaches were marginalized in the coalescing new field of Science and Technology Studies. Some have gained belated recognition in recent years, not least as graduate students attracted to critical perspectives have gone on to populate faculty positions and expand the field.[6]

---

4. Collins, "Three Dimensions of Expertise," p. 254.

5. Feyerabend, "Wittgenstein's Philosophical Investigations." Contrasting views on the influence of Wittgenstein on Feyerabend can be found in Kidd, "Reawakening to Wonder" and Oberheim, *Feyerabend's Philosophy*, ch. 2.

6. Rose and Rose, *The Political Economy of Science* (1976); Harding, *The Science Question in Feminism* (1985); Harding and O'Barr, *Sex and Scientific Inquiry* (1987); Rouse, *Knowledge and Power* (1987); Fuller, *Social Epistemology* (1988); Haraway, *Primate Visions* (1989); Haraway, *Simians, Cyborgs, and Women* (1990); Longino, *Science as Social Knowledge* (1990); Harding, *Whose Science? Whose Knowledge?* (1991); Harding, *The "Racial" Economy of Science* (1993); Sclove, *Democracy and Technology* (1995). For the belated recognition of critical approaches, see the list of the Society for Social Studies of Science's Bernal Prize at https://www.4sonline.org/prizes/bernal (a prize named for the Marxist scientist and historian of science John Desmond Bernal).

In the words of Harry Collins, SSK "leaves pure science very nearly as it is; in the main it simply redescribes it," an approach that reflects Wittgenstein's rejection of philosophy as an external standpoint for criticism.[7] Collins rejects the traditional image of science that science depends upon the application of a general method or methods applied to scientific research. With a focus on scientific method, it might be possible for outside observers to judge whether scientists have followed the methodological rules properly. Rules, however, do not determine their application for Wittgenstein. When a pupil learns to continue the number series, "2, 4, 6," with "8," it is not an abstract rule that is determining that conclusion but the internalization of a form of life in which "8" is the customary way to carry on the series.[8] Only shared forms of life can narrow down the potential openness of rules to a single outcome. To judge the application of rules in practice requires one to be a properly socialized member of the community, a member who knows how to interpret the rules in conformity with the way others do in a shared form of life. Controversy at the cutting edge of science does exist, but the community of science is able ultimately to enforce one interpretation because scientists share a form of life.

SSK offers two related, but distinct, interpretations of Wittgenstein to "sociologize" epistemological topics. Scientific reasoning is related to forms of life by adopting either a *collective* or an *interactional* interpretation of science. The collective interpretation sees scientific work as occurring within well-defined communities that act to ensure consensus about rule-following. This view forms the basis of most core SSK work, including the methodological relativism of Collins and Pinch and the strong program of Barnes and Bloor.

The interactional interpretation denies the significance of all larger contexts (methodologies, disciplines, institutions, research programs, etc.) by focusing upon very local, emergent, interactional orders and the work that goes into maintaining them through mundane interaction. This approach involves the application of Harold Garfinkel's ethnomethodology to science by Michael Lynch and Eric Livingston, though the laboratory ethnography of Latour and Woolgar and the actor-network theory of Law, Callon, and Latour incorporates elements of this approach (see chapter 5).

7. Collins, "An Empirical Relativist Programme," p. 98. Compare Wittgenstein, *Philosophical Investigations*, § 121: "Philosophy may in no way interfere with the actual use of language; it can in the end only describe it. For it cannot give it any foundation either. It leaves everything as it is." Collins does criticize science when distortions of the self-regulated, "core set" of a scientific community exist. See the discussion of violations of the core set in chapter 6.

8. Wittgenstein, *Philosophical Investigations*, § 185; Feyerabend, "Wittgenstein's Philosophical Investigations," pp. 470–71.

The collective interpretation maintains that scientific communities are real, cohesive entities, while the interactional interpretation argues that such apparently cohesive communities are essentially constructs we credit through a process of ongoing face-to-face meaning-making wherever such claims are made. Whichever interpretation is adopted, outside critique is blocked.

For the collective Wittgensteinians, outside critics of a form of knowledge can only be followers of *different* forms of life—they cannot engage that knowledge on fair terms. Thus, in Shapin and Schaffer's influential account of the early Royal Society, Thomas Hobbes was excluded from the Royal Society because he could not accept their rules of the game requiring experimentalists to set aside debate about underlying natural causes for agreement on shared facts. Hobbes's only hope for prevailing in his dispute with Boyle about whether air-pumps created vacuums would have been to succeed in setting up his own community with like-minded natural philosophers, perhaps through the intervention of his former tutee King Charles II.[9] Participants in different forms of life see the world in incommensurable ways, which is why the sociologist must invoke social causes to explain how one would displace another.

For Bloor, a form of life is "a pattern of socially sustained boundaries," the "substratum of conventional behaviour" that underlies rule use. "Competing usages" imply "rival groups."[10] For Collins, a form of life is a social group for which intersubjective agreement in rule use obtains: "This leads one to expect groups to be able to communicate readily within themselves because of their members' common ways of going on, but equally we would expect difficulty in communication between culturally diverse groups."[11] Once one has adopted a form of life, it comes complete with its own rules for correctly viewing the world. In this sense, group membership is the crucial criterion that determines one's view of the natural world.

For the interactional Wittgensteinian, outside critics of a form of knowledge can only proceed by falsely imposing their interpretive meaning on the ongoing, interactionally negotiated cognitive orders associated with practical scientific activity. It is larger frameworks of meaning that seem to allow criticism to proceed, but they do so improperly by imposing meaning that does not fit the cognitive order negotiated in local settings. The meaning imposed could be appeal to larger social or cognitive contexts that are held to apply. Such larger frameworks obtain no purchase on the life world of scientists unless and until scientists interactively reference these debates. Even in such a

---

9. Shapin and Schaffer, *Leviathan and the Air-Pump*. On assessing the implied counterfactual statements in this argument, see Lynch, "King's Evidence" and chapter 6 below.

10. Bloor, *Wittgenstein*, pp. 140, 37, 48.

11. Collins, *Changing Order*, p. 15. The fact that machines are not socialized into our forms of life explains why they cannot think like us. See Collins, *Artificial Experts*, pp. 17–18, 41, 181–82, 211; and Collins and Kusch, *The Shape of Actions*, pp. 1, 196–97.

circumstance, the meaning of such referencing is given entirely by the work that is done with them by local participants.

Scientists might invoke the philosophy of Karl Popper or mention the gender of another participant, but the ethnomethodologist insists that we must pay attention to what work such use does for them in specified contexts considered separately. It does not mean that Popper's philosophy, or underlying gender dynamics, have had an impact on sense-making from the outside. In short, unless and until scientists invoke some social variable, intellectual tradition, historical context, or macro-structure in observable activity, it does not exist. And when they do invoke them, they remain aspects of a local, interactive moment of negotiating meaning. They do not persist. They do not form social or cognitive structures, contexts, or traditions subsisting when the moment has gone away. They do not constrain the actors behind their backs. They do not serve up lessons on how other people in other contexts face similar issues in accounting for gender or other sociological phenomena.[12]

If a critical approach to the history of science like that proposed by Lakatos and Feyerabend is to get off the ground, we must be able to think about how criticism is possible. Lakatos and Feyerabend rejected any unitary account of scientific method, but they did not argue that methodological commitments have no effects or that outsiders could not evaluate and criticize the methods employed by scientists. Lakatos envisioned writing a book, *The Changing Logic of Scientific Discovery*, that would show the second-order impact that changing methodological doctrines had on the practice of science, which presumes that one can sort out the one from the other.

Feyerabend declared himself "against method," but that reflected his belief that even capacious reconstructions of the rules for good science like Lakatos's Methodology of Scientific Research Programs (MSRP) failed to capture all of the desirable moves within the history of science. Feyerabend argued that if we insisted on finding a methodological rule for all of science, then "there is only *one* principle that can be defended under *all* circumstances and in all stages of human development. It is the principle: *anything goes*." The slogan "is *not* the one and only 'principle' of a new methodology," but

---

12. Garfinkel, *Studies in Ethnomethodology*, ch. 5, presents an analysis of how a transsexual (who later turned out to have been born intersexed) passed as a woman, but this did not lead to any generalizations or comparisons with the practical work of performing one's gender by other transsexuals, or by men and women in general. As Bologh, "The Promise and Failure of Ethnomethodology from a Feminist Perspective," p. 200, puts it, "Garfinkel's ethnomethodological study does not include any such reflection on the social conditions or form of life presupposed in the doing of gender. Hence it lacks a political dimension. The study does not become the basis for reflection on how we are all 'doing gender,' creating reality, producing our gendered world in and through our own practices, and what this means: the practical purposes and consequences for women and men of this form of life and the possibility that it could be otherwise. Nor does Garfinkel reflect on his own doing of masculinity or that of his colleagues in relation to Agnes's display of femininity."

the only way rationalists committed to universal standards would be able to understand him.[13]

For any method that generates positive results, there is an opposite principle that also does so in *some* times and *some* places. This is why he spoke of the effectiveness of *counterinduction* as a strategy to force an *external* criticism of entrenched theories and associated natural interpretations.[14] He believed that Galileo's violations of the accepted rules of Aristotelian science made possible a productive transition in the history of science that rectified the limitations of Aristotelian observation and reasoning. He did not believe that Galileo's criticisms were a simple mistake, flouting Wittgenstein's interpretation of rules as habits or customs. Galileo introduced a new observation language that was incommensurable with the scholastic one but that actually facilitated criticism of an entrenched perspective that otherwise would have remained moribund.

Incommensurable perspectives depend upon the introduction of a new, core theory that carries with it ontological assumptions that call into question the observation language taken for granted by the accepted theory. It is not the social uses of the theory that determine the meaning, such that an atypical use could be policed and eliminated by the social group. Instead, scientific theories themselves provide the context for establishing the meaning of observational claims. In fact, a scientific concept's meaning could change without its use changing at all, as a new theoretical context provided a new meaning for force when Newton's mechanics replaced impetus theory, despite both using the term "force" in the same contexts.[15] Inventing a new theory makes possible criticism that would be otherwise be blocked by the reigning form of life.[16]

Note that this argument states in a general, philosophical way that the production of theoretical diversity and the introduction of distinct standpoints creates a desirable pluralism in science that is capable of dislodging entrenched perspectives, precisely the claim of methodological reformers from the Royal Society to standpoint epistemologists. Unlike stronger versions of standpoint epistemologies, Feyerabend saw such moves as a fallible tool, to be dispensed

---

13. Feyerabend, *Against Method* (1978), p. 28; Feyerabend, *Science in a Free Society*, p. 39.
14. Feyerabend, *Against Method* (1978), pp. 29–33.
15. Oberheim, *Feyerabend's Philosophy*, p. 68. Compare Fuller, *Social Epistemology*, pp. 83–84, ch. 6, on the emergence of incommensurability from unnoticed meaning change in science.
16. Feyerabend, "Consolations for the Specialist." Oberheim, *Feyerabend's Philosophy*, pp. 46–48, 57, traces Feyerabend's "top-down," contextual theory of meaning and its core consequence of incommensurability to his 1951 dissertation, in a critique of the "bottom-up," logical positivist idea that theory-free protocol sentences could form the basis for science. The argument was developed for publication in "An Attempt at a Realistic Interpretation of Experience" (1958), incorporated into his first published articulation of incommensurability in "Explanation, Reduction, and Empiricism" (1962), and is a continuing thread through all stages of his philosophical development.

with in turn.[17] Thus, while Feyerabend rejected any kind of conception of cumulative progress, or unitary conception of scientific reference, he was favorable to a dialectical account of progress, where the need to overcome stagnation and conceptual conservatism was paramount. Such progress is impossible if we, in effect, demand that critics go off and form their own clubs because they cannot be admitted to the existing ones. The continuing influence of Wittgenstein's later philosophy on STS has hampered our ability to think through the issue of the suppressed potential of science and technology.

THE MYTH OF METHOD

Historians and sociologists have shown that methodological doctrines in science have played a different role than philosophers of science have supposed. Rather than being determinate procedures that generate or justify research results, "methods" are themselves objects of contention. On the one hand, methodological narratives draw upon contextually embedded research practices to form a post hoc reconstruction of the research process. Consequently, debates about methods are at the center of conflicts between competing research programs without providing a neutral point of reference for evaluating science.

On the other hand, methodological discourse has been mobilized for a variety of purposes, such as to consolidate institutional structures or to legitimize the use of scientific results and techniques.[18] Consequently, focusing on the uses to which methodological doctrines have been put (following the Wittgensteinian dictum that meaning follows use) allows the historian to examine issues at the interface between the cognitive content of science and the social organization of science and its place in a wider culture. Despite this "symptomatic" significance, scientific method is no longer understood as an external guide to action, as rules for conducting science, or as normative criteria for judging science.

In a sense, verbal descriptions of method are too formal and abstract to be organically integrated into forms of life. Thus, Shapin and Schaffer "treat controversies over scientific method as disputes over different patterns of doing things and of organizing men to practical ends." They declare that "methodology will not be treated solely as a set of formal statements about how to produce knowledge, and not at all as a determinant of intellectual practice,"

17. Sprat, *History of the Royal Society*; Keller, *A Feeling for the Organism*; Bschir, "Feyerabend and Popper on Theory Proliferation and Anomaly Import"; D'Agostino, "Verballed? Incommensurability 50 Years On."

18. Schuster and Yeo, *The Politics and Rhetoric of Scientific Method*; Webster, "The Origins of the Royal Society"; Wood, "Methodology and Apologetics."

thereby dissolving method into patterns of behavior while leaving merely verbal formulas impotent to shape practice."[19]

Methodological doctrines have been employed traditionally as external, normative standards for judging, or even generating, action. Francis Bacon laid out rules for induction that would shape the generation of facts and inference to underlying forms or causes on their basis. What role do his "merely verbal formulas" play in shaping his scientific practice and that of his followers? Can he be said to have followed his own method? If others follow his method, will they get the same results?

Historically, it can be argued that members of communities in which there is agreement about what method to follow will retrospectively connect their findings to that method, much as Mulkay argued that Merton's norms were used as an after-the-fact way to talk about what scientists had done in practice, a practice in fact unconstrained by Merton's norms. By contrast, when scientists disagree about what methods to follow, they will argue over which method is best and connect their findings to their preferred method to try to vindicate the method itself.[20] When individuals within a community or in different communities appeal to the same "merely verbal formulas," is there anything that can be said about how those formulas constrain or enable particular outcomes? If the Royal Society claims to be following Bacon's method, is there anything more than a nominal connection with what Bacon said or did? Would the Royal Society's practice look the same if they had used a different methodological text as their point of reference?

When sociologists declare their objects of study to be embedded in coherent forms of life or wedded to locally contingent orders of interaction, socially enforced rule-following puts walls around communities, both spatially and temporally. Hobbes is excluded from the Royal Society, judged incompetent not just a critic, as he cannot perform what is required of Royal Society members to be recognized as a competent member of the group. But so too is Bacon excluded from the group because he worked in a different time and a different context. When he discusses matters of fact, it means something entirely different than when Robert Boyle does so. When the other Royal Society fellows try to follow Bacon's method, they do so not because they follow the chain of reasoning laid out by Bacon's words but because they follow the rules of their group. The rules are socially enforced behaviors, not merely verbal instructions that actually direct them.

Peter Winch's work provides the classic articulation of a Wittgenstein basis for sociology as socially enforced rule-following. Winch's claims that "the analysis of meaningful behaviour must allot a central role to the notion

19. Shapin and Schaffer, *Leviathan and the Air-Pump*, pp. 15, 14.
20. Schuster, *Descartes-Agonistes*, p. 296, footnote 54.

of a rule; that all behaviour which is meaningful (therefore all specifically human behaviour) is *ipso facto* rule-governed."[21] The interpretation of rules is determined by the "form of life" that holds to them. As a result, any effort to criticize the historical actors or ethnographic subjects under investigation from an external standpoint is beside the point, akin to criticizing cricket players for not following the rules of baseball.[22] Indeed, it is even inappropriate to contrast how participants interpreted the rules of their game with the manifest meaning of the rules as read by an external observer. Rule-following is defined in practice since only a form of life can narrow down the range of possible extensions of rules to new contexts from the logical possibilities.

In the interdisciplinary field of Science and Technology Studies, an emphasis on forms of life has broadened our understanding of scientific work by representing the ongoing construction of scientific knowledge in concrete contexts, but at the cost of conceiving methodological narratives as entirely mythical, serving merely as post hoc rationalizations of locally embedded scientific work.[23] Methodological doctrines, even when subscribed to by scientists under study, are treated as highly misleading accounts of actual scientific practice.

The irony is that the principle of interpretive charity employed by sociologists in reconstructing forms of life in terms recognizable to the actors studied breaks down when it comes to crediting these actors' own beliefs that they followed a particular scientific method in carrying out their work. And yet criticism of forms of life from the "outside," traditionally proceeding by the claim that actors failed to properly follow the rules of scientific method, are rejected as illegitimate. Not only is the idea of a single scientific method rejected for a proliferation of local methods, but even the possibility of criticizing scientists for failing to follow their own method is foreclosed by a social account of rule-following borrowed from Wittgenstein. Only other participants in a shared form of life can decide that a rule was not followed; bare examination of the rule itself cannot decide. Social (read local) rule-following trumps epistemological (read universal, outside, or post hoc) rule-following. The Wittgensteinian roots of STS are the key reason why

---

21. Winch, *The Idea of a Social Science*, pp. 51–52. In fact, habituation to identifiable behavioral practices does not imply that tacit rules are shared and generate the behaviors. See Turner, "Practices as the New Fundamental Social Formation in the Knowledge Society," p. 53.

22. For the unfortunate comparison of science to baseball, complete with a defense of science studies as a field of its own with no implications for the conduct of science itself, see Fish, "Professor Sokal's Bad Joke."

23. Barthes, *Mythologies*; Latour and Woolgar, *Laboratory Life*; Schuster, "Methodologies as Mythic Structures." See the discussion of Schuster's later work below for a development of Barthes's approach that shows specifically how methodological commitments shape practice even when they do not generate research in line with formal methodological rules.

criticism of the products and processes of science becomes inconceivable within the field.

Thus, despite the often noisy accusation from some scientists that STS inappropriately challenges scientific authority, the upshot of most core work in the field suggests the reverse: science is a field of locally maintained cognitive communities where outside criticism is unwarranted and the goal of sociological study is ethnologically thick reconstruction of scientific practice. Ironically, epistemological accounts of the success of science, philosophical criticisms of scientific practice, and explicitly cultural, ethical, or political challenges to scientific authority are equally short-circuited on this account.

Focusing on methodological discourse as anything other than a rationalization is often seen as an attempt to reintroduce a traditional, philosophical focus on the efficacy of method.[24] The introduction of a social or community-based conception of knowledge has been accompanied by a rejection of outside political or cultural critique.[25] The policing of the right to speak based upon membership in increasingly fragmented social groups within contemporary multicultural identity politics and many interpretations of anthropological relativism is arguably connected to this same members-only rule.[26]

By linking rule-following, and by implication all forms of inference, to cohesive "forms of life," Wittgensteinian sociologists have opened up such topics to sociological scrutiny only at the cost of construing critics either as deviant or as outsiders belonging to a separate form of life. Consequently, methodological discourse looks like bad faith rationalizations of arbitrary rules. Not only is epistemological critique by professional philosophers disallowed but so too is cultural or political criticism of the products and processes of science. At most, scientists are chided for presenting an unrealistic image of their work to the public, which creates unrealistic expectations and unfair criticisms.[27] The fact that scientists have often resisted this friendly representation of the import of work in STS testifies to the strong attachment to some image of scientific method and its virtues.

The linking of rationality and rule-following to a collective or interactional "form of life" has provided the license for reinterpreting epistemological

---

24. See, for example, Richards and Schuster, "The Feminine Method as Myth and Accounting Resource," for a critique of Evelyn Fox Keller's focus on the alternative methodological commitments of Barbara McClintock.

25. Rorty, *Contingency, Irony, and Solidarity*; Fish, *Doing What Comes Naturally*; Lyotard, *The Differend*.

26. See Hollis and Lukes, *Rationality and Relativism*, for the bringing together of debates about anthropological relativism and Wittgensteinian sociology, especially in the contribution by Barnes and Bloor, "Relativism, Rationalism and the Sociology of Knowledge." The theme is carried over in the debate over multiculturalism in Joppke and Lukes, *Multicultural Questions*.

27. Collins and Pinch, *The Golem*.

topics in a sociological idiom.[28] It does so by equating social structure with cognitive structure, seeing knowledge as the shared beliefs of a community and social organization as simply the existence of different knowledge communities.[29] I argue that the concept of a form of life applied to scientific life has a number of difficulties.

First, the concept introduces a sociological element only at the cost of assuming that social life is consensual at its core and that the limits of consensus define the limits of community. As consensus breaks down, communities must reestablish consensus or split apart. Criticism is failure to follow the rules and indicates a pupil who is still learning or an outsider with no claim to speak.

Second, the concept does not explain scientific change, since how forms change over time and how competing forms of life interact with other forms cannot be explained without abandoning an analysis of rule-following. Indeed, Michael Lynch criticizes Bloor for being insufficiently Wittgensteinian in appealing to individual, cognitive variability as the engine for social change, as variant interpretations form the potential basis for new ways of going on. Lynch rejects that as a "private" interpretation of a rule disallowed under Wittgenstein's philosophy. His own view focuses on the achievement of interactional order in local contexts and leaves aside consideration of larger scientific development.

Third, the idea that rule-following is part and parcel of membership in a collective makes it difficult to thematize the role of methodological discourse. Scientists often engage in efforts to make their practice conform with their methodological commitments. Their efforts to do so shape their ongoing work in important ways, even if the results look different from what they (or we) might expect. The dynamic quality of the resultant practice is one example of the importance of context-transforming and context-transcending discourses in the conduct of social life, a finding that pulls us away from the localist and holist reading of social forms to a more ambiguous and relational picture of social life.[30] Scientists—and the rest of us—transform knowledge claims and practices by subjecting them to criticism based upon standards and procedures held to be reliable bases for normative guidance.

---

28. See Bloor, *Wittgenstein*, ch. 9 ("The Heirs to the Subject That Used to Be Called Philosophy") and Collins, *Changing Order*, ch. 6 ("The Scientist in the Network: A Sociological Resolution of the Problem of Inductive Inference").

29. Collins, *Changing Order*, ch. 6; Collins and Kusch, *The Shape of Actions*, p. 11. For criticisms see Fuchs, *Against Essentialism*.

30. See, for example, the multi-pass, rather than single-pass, reading of science, where science continually reworks its results and methods in Nickles, "Justification and Experiment" and "Reconstructing Science." Pickering, *The Mangle of Practice*, p. 4, develops a similar account of "practice as the work of cultural extension and transformation in time." See also Brown, "The Functional Complexity of Scientific Evidence," pp. 70–71.

COLLECTIVE WITTGENSTEINIANS

Early arguments by David Bloor and Harry Collins that scientific knowledge could be subject to sociological analysis drew upon the later work of Wittgenstein to suggest a sociological resolution of the Duhem-Quine thesis that theory selection was underdetermined by evidence: social closure provides what the evidence alone cannot.[31]

The core of these sociological readings is an identification of inference with the socially enforced linguistic behavior of an identifiable community. The idea here is that the social collective enforces acceptable interpretations to provide limits on the application of concepts, which are held to be open-ended in principle. Barnes, Bloor, and John Henry argue explicitly that consensus defines the limits of community. It is the socially enforced character of interpretation that checks the proliferation of interpretations that the open-ended character of observation reports, concepts, or beliefs would otherwise allow. Thus, in admitting that a "snap shot" view of beliefs in a particular community "could indeed yield up results in which the isolated deviant or critic would later turn out to be right," they insist that it is a community that ultimately determines what it is to be right:

> The ultimate correctness of the consensual view shows itself . . . when we ask: What would this "turning out to be right" amount to? The answer is that it would be a new consensus.[32]

If the deviant is unable to sway the community to a new consensus, he is dismissed as incompetent or finds solace in another community, as the critic of R. A. Millikan's oil drop experiment Felix Ehrenhaft was to find in being retrospectively judged a poor experimentalist by the experimental physics community but finding favor among Machian, positivist philosophers.[33] The deviant who fails as a revolutionary joins a *different* community.

Bloor notes that rules are actor's categories, not analyst's categories that could be understood and evaluated by outside observers. Social institutions can be considered "self-referring and self-creating," "a species of self-fulfilling prophecy," and "a collective pattern of self-referring activity."

---

31. Knorr-Cetina and Mulkay, "Introduction," and other reviews of the field point to the Duhem-Quine thesis as opening up the possibility of a sociological explanation of the content of science. This appeal to empirical underdetermination is ironic in that it seems to maintain a residual asymmetry: the social steps in to fill what evidence (conceived as if it were nonsocial) cannot. See Fuller, *Philosophy, Rhetoric, and the End of Knowledge*, pp. 148–49, 321.

32. Barnes, Bloor, and Henry, *Scientific Knowledge*, p. 14.

33. Ibid., pp. 36–38. Feyerabend, *Conquest of Abundance*, pp. 161–66, describes attending experimental lectures by Ehrenhaft, which taught him the powerlessness of experiment to displace entrenched theories or worldviews. See Oberheim, *Feyerabend's Philosophy*, p. 121.

They actively create meanings that cannot be judged by outsiders because it is ultimately the consensus of the group that determines whether the application of a rule is correct or not. Members of the group learn these rules purely by ostensive training, correcting misunderstandings and confirming correct applications. Since the experience of any pupil is finite, that means the extension to new cases is not determined in advance, though it does create dispositions to carry forward the rule in certain ways. Despite the existence of underlying psychological dispositions internalized through social learning, Bloor maintains that speech, belief, and actions are all *internally* related, not connected to each *externally*, as separate entities that could encourage us to talk about how the one influences the other. Thus, there is not, on the one hand, a rule and, on the other, its application; they are the same thing.[34]

To see the difficulties of this approach, consider objections to the parallel notion of a speech community that have been raised within sociolinguistics. Mary Louise Pratt has drawn on Benedict Anderson's observation that modern nations are "imagined communities" in order to suggest that the same situation obtains with the communities identified by linguists. Linguists tend to identify a single set of rules shared by a speech community and identify certain forms of speech within that community as privileged and against which all other speech acts are judged either as normatively legitimate for that speech community or as deviant according to those rules. Hence, internal divisions within speech communities either are not studied or are identified as subcommunities with their own distinctive rules.

What such an approach ignores is the relationship—often one of hierarchy, domination, or conflict— between such defined subcommunities. Moreover, in analyzing particular institutionally located language games, the analysis reproduces the structures of authority existing in particular linguistic exchanges, for example, by classifying as "normal" those patient-doctor interactions in which the patient agrees to the recommended course of action (exactly what Collins would recommend). Analysis of the communicational norms of this language game treats resistance as failure to internalize the rules of the game rather than as involving disagreement and conflict over appropriate norms.[35] The boundaries of interpretive communities in science may be similarly "imagined" by participants and analysts alike, and reifying such boundaries can obscure the true nature of cognitive agreement or disagreement.

---

34. Bloor, *Wittgenstein, Rules and Institutions*, p. 33; Kusch, "Rule-Scepticism and the Sociology of Scientific Knowledge," pp. 577–78, 587. Bloor develops the self-referential view first articulated by Barnes, "Social Life as Bootstrapped Induction." Dear, "Narratives, Anecdotes, and Experiments," develops a similar, self-referential approach where actions are understood as fully constituted by the actor's accounts of their actions.

35. Pratt, "Linguistic Utopias"; Anderson, *Imagined Communities*, pp. 6–7.

In Wittgenstein's focus on teacher-student interactions, the focus is on the student's ability to adopt appropriate rule-following behavior. Thus, Collins's Wittgensteinian-inspired account, while allowing interpretative flexibility and conflict over research findings to take place, confines this conflict to a "core set" of researchers.[36] In a three-stage model, initial interpretative flexibility is then limited by the mechanisms within the core set that lead to closure. Only in the third stage is there an attempt to connect the dispute to the "wider social and political structure."[37] In most cases, "distance lends enchantment," so that expert knowledge is detached from the private core-set negotiation and a methodological propriety is lent to such claims as wider communities accept them without awareness of the radical contingency of their production.[38] As with the model of Barnes, Bloor, and Henry, Wittgensteinian emphasis on the flexibility and openness pertaining to concept use is limited by community-enforced consensus. Collins's model has been criticized for consigning rivals to the third stage of controversy.[39] Indeed, despite rejecting epistemological analysis, Collins's model provides a defense of de facto expertise without providing for the possibility of independent debate over methods within a larger community:

> Rather than taking a philosophical model of science as a standard, or an ideal model of the trajectory of good research, or a model of conduct within the research community, one may ask "*who* should have control over protocols and interpretations?"[40]

This allows Collins to formulate a sociological measure of "'distorted' interpretative license," which is only possible given that the limits of core sets are taken to be unproblematic:

> This is not an attempt to get an "is" from an "ought," because the argument starts with an ought. The ought is simply the common-sense idea that *only* those who are reasonably scientifically qualified and/or experienced should be allowed to contribute to core-set work, and that *all* those who are specially well

---

36. Collins, *Changing Order*, pp. 142–45.
37. Collins, "Stages in the Empirical Programme of Relativism," p. 7.
38. Collins, *Changing Order*, p. 145. This picture is similar in many ways to the account found in Putnam, "The Meaning of 'Meaning,'" with the exception that Putnam stresses the fixing of reference by experts while Collins stresses the radical flexibility of reference at this stage.
39. Chubin, "Collins's Programme and the Hardest Possible Case." It is true that rivals exist within core sets, yet Chubin's point is that this does not include all potential rivals but only those who mutually recognize each other as members.
40. Collins, "Public Experiments and Displays of Virtuosity," p. 740. Compare Nelson, *Who Knows*, where an argument that it is ultimately a community "who knows" is taken to provide license for wider examination of the value assumptions of technical knowledge. A rejection of the fact-value distinction leads to the conclusion that "the autonomy of science is a myth and a pernicious one" (p. 137), and the development of a feminist empiricist approach to science critique on that basis.

> qualified or experienced in the area in dispute should be allowed to contribute if they wish.[41]

Pathologies in science amount to distortions of the core set, whether through "over-restrictiveness" (Lysenko affair, military testing) or an "over-extension of core-set boundaries" that would allow outsiders to illegitimately comment on technical research.[42]

There is much to appreciate about this attempt to salvage a residual basis for evaluation within a Wittgensteinian perspective, as I will consider in chapter 6. Criticism from the sociologist, however, is confined to making sure that the relevant social groups are properly populated. His position still maintains deference to the judgment of the properly constituted scientific community. Here, Collins uses a collective interpretation of Wittgenstein to defend a view of science as an autonomous community, subject to distortion by political interference, a view Polanyi used to counter a different kind of social conception of science associated with the British radical scientists who wished to plan science for public benefit.[43]

Scientific communities, however, are not preexisting and self-contained. They are generated through sociological processes that are not themselves reducible to group membership. Their boundaries are defined by observers, often in ways that are contested and fought over. Their boundaries are also fluid and involve changing relationships with other groups. The former idea ultimately draws on Benedict Anderson's claim that communities are "imagined," that is to say, constructed and defined in ways that reflect historical and political processes shaping the process of in-group and out-group classifications with which people identify. The modern nation-state constructs a version of a national community through mass media propaganda in a way that reinforces the power of the state.[44]

The latter idea draws on the concept of hegemony, particularly as developed by the Italian Marxist Antonio Gramsci, who focused on coalitions of different groups who consent to submit to common leadership. Combining elements of both consent and coercion, Gramsci envisioned that hegemony underlay the dominance of the capitalist state, as well as the social movement of distinct classes that would overthrow it. Hegemony emphasized that

---

41. Collins, "Public Experiments and Displays of Virtuosity," p. 740.

42. Ibid., pp. 740, 741.

43. Polanyi, *Personal Knowledge*; Nye, *Michael Polanyi and His Generation*, ch. 6. On the clear divisions between Polanyi and the leftist scientists, see the criticisms of Nye in Turner, "Polanyi Defanged," pp. 950–52.

44. Anderson, *Imagined Communities*, pp. 37–46. Recent work in STS develops the concept of imaginaries based in part on Anderson's work. See Jasanoff, "Future Imperfect," pp. 6–7; Jasanoff and Kim, *Dreamscapes of Modernity*; McNeil et al., "Conceptualizing Imaginaries of Science, Technology, and Society."

the interests of any particular class were fluid, contradictory, and changeable and that groups would have to make sacrifices to best pursue their long-term interests.[45] (Gramsci's own attempts to define what those interests should be and how they should be pursued represents a kind of political method talk very much like scientific method talk that likewise seeks to transform unreflective practice.)

Later, in theories of "autonomist" Marxism, the point was taken further in showing how the internal structure of groups could reflect wider effects of political economy, as workers are divided into differences of caste, race, waged or unwaged status, or gender that play different roles in social reproduction. The unwaged labor of housewives was crucial to a system of waged, male labor under capitalism, for instance, which meant that potential challenges to the system could come not just through strikes but via campaigns seeking "wages for housework."[46] Such a possibility is inconceivable when social identities are treated as distinct, self-contained, and separate from larger frameworks of social power.[47]

Aspects of these approaches that counter Wittgenstein's static view of social groups are evident in much STS work, especially with the growing infusion of work in political science and its focus on power. Jasanoff developed an emphasis on the coproduction of science and society where the drawing of boundaries was a key component shaping how science gets interpreted in practice, including who gets to count as an expert in policy forums. On this view, scientific communities are not simply given and self-regulating but contested by participants in policy-relevant science.[48] Michael and Birke, applying actor-network theory to the concept of core sets, note that core sets themselves are constituted by nontechnical conflicts; they do not preexist nor can they be delineated apart from political processes.[49] In other words, any definition of the core set itself is "imagined" or "constructed" through social processes from Collins's third stage of controversy, a theme developed in chapter 6. Hamlin argues that there can be rival core sets that compete for recognition such that "the 'leave it to the core set' approach will beg the question." When one group is able to exploit political and institutional support, they may displace competing approaches as a kind of "core-set imperialism."[50]

---

45. Gramsci, *Prison Notebooks*.

46. Vishmidt, "Permanent Reproductive Crisis." On autonomist Marxism, see Cleaver, *Reading Capital Politically*.

47. This is the theme of Žižek, *The Courage of Hopelessness*, p. 231.

48. Jasanoff, "Contested Boundaries in Policy-Relevant Science"; Jasanoff, *States of Knowledge*.

49. Michael and Birke, "Enrolling the Core Set."

50. Hamlin, "Third Wave Science Studies," pp. 171, 172. See also Rip, "Constructing Expertise," pp. 419–22.

The concept of hegemony broadens our view of social life beyond consensus. Gottweis challenges the idea that consensus among established actors is the only possible aim in showing how the controversy over genetic engineering in Germany led to the reconstitution of boundaries and identities themselves. Attempts by state and industry actors to impose a policy consensus through a 1990 law regulating the use of genetic engineering and defining the bounds of acceptable dissent were resisted by activist appeals to the democratic value of heterogeneity and dissensus over uniformity and consensus. Gottweis denies "that we can simply assume the 'existence' of actors or structures in politics" or that "stable boundaries between sectors such as politics, the economy, and science" exist prior to an active process of constructing them.[51] Institutions do have established rules, but they are never simply applied according to the consensus of stable groups; they are articulated actively in a contested political process. What this means is that existing identities or social groups are defined by their interrelationships with other groups and by meanings more or less imposed by hegemonic actors, what Gramsci calls a historical bloc, which can break down during an organic crisis.[52]

In this approach, hegemony is a never completely successful effort to establish a uniformity across diverse discursive contexts. As in the concept of interpretive flexibility, rules or discursive forms are never simply "applied" to a new realm; they must be articulated anew in each extension. Yet in contrast to Wittgensteinian-inspired consensual models like that of Collins, the process of extension is subject to conflict at all levels and from alternative, counterhegemonic discursive forms.[53] By contrast, once a competing interpretation fails to gain currency in a community, as in Bloor, Barnes, and Henry's analysis of Ehrenhaft or Shapin and Schaffer's analysis of Hobbes, the counterview is excluded and forgotten. Scientific change is driven by two elements, conceptual variability (among individuals for Barnes, Bloor, and Henry; within core-set negotiations for Collins) and its acceptance or rejection by the community. In short, society is modeled on Wittgenstein's teacher-student interaction.

In Gottweis's application of hegemony to science, the conflict between different social "imaginaries" not only can lead to mutual adjustment of the boundaries between discursive communities but also can reconstitute core identities themselves, questioning the well-defined character of forms of life to which knowledge is relativized. Larger dialogical or conflictual rela-

---

51. Gottweis, *Governing Molecules*, p. 4.

52. Gottweis, "German Politics of Genetic Engineering and Its Deconstruction," p. 198; Gottweis, *Governing Molecules*, p. 264.

53. Laclau and Moufee, *Hegemony and Socialist Strategy*; Laclau, *New Reflections on the Revolution of Our Time*; Connolly, *Identity\Difference*; Keohane, "Central Problems in the Philosophy of the Social Sciences after Postmodernism."

tion between communities or subcommunities cannot merely be relegated to "public relations" (legitimation or boundary maintenance) apart from ongoing rule-following internal to a community. This in turn calls into question approaches that would sharply separate different levels of methodological discourse: those "internal" to ongoing practice, explicit post hoc rationalizations, and general methodological prescriptions.[54]

INTERACTIONAL WITTGENSTEINIANS

Ethnomethodological approaches draw on an alternative reading of Wittgenstein that denies the reality of larger "contexts" for explaining scientific negotiation over the application of concepts. Michael Lynch has argued that the collective Wittgensteinian's interpretation of Wittgenstein errs by providing a skeptical interpretation of rule-following. Like Saul Kripke, sociologists such as Bloor interpret Wittgenstein to have provided a skeptical account of how rules determine action, according to Lynch. Such an account conceives of rules as "external" to action, to be related by a "causal" theory of underlying sociological and psychological dispositions. Unlike Kripke, Bloor explicitly invokes sociological categories to establish the link between rules and action that epistemology has failed to provide.[55]

Lynch begins from an alternative, non-skeptical reading of Wittgenstein. Rather than portraying rule-following as a "problem" to be solved by finding a means of determination of action that will supplement the inadequacy of rules, Lynch denies that there is a meaningful separation between a rule and its extension.[56] This interpretation of internal rule-following denies the separation between explicitly formulated rules and the actions they are held to determine. If they are not analytically separate entities, then they cannot be connected together by sociological theories, like Barnes and Bloor's self-referential theory of institutions. Lynch's approach disavows any attention

54. Schuster, "Methodologies as Mythic Structures."

55. Lynch, *Scientific Practice and Ordinary Action*, p. 162; Kripke, *Wittgenstein on Rules and Private Language*. For the debate about the interpretation of Wittgenstein, see Lynch, "Extending Wittgenstein"; Bloor, "Left and Right Wittgensteinians"; and Lynch, "From the 'Will to Theory' to the Discursive Collage." Bloor, *Wittgenstein, Rules and Institutions*, pp. 7–8, 68; Kusch, "Rule-Scepticism and the Sociology of Scientific Knowledge," p. 578; and Bloor, "Institutions and Rule-Skepticism," pp. 596–97, deny that Bloor follows Kripke's skeptical interpretation of Wittgenstein or that rules and action are externally related.

56. Lynch, *Scientific Practice and Ordinary Action*, pp. 172–73. Here, Lynch draws upon Baker and Hacker, *Skepticism, Rules and Language*, who argue against the "community view" interpretation on this basis. This might appear strange to those who see ethnomethodology as part of the discipline of sociology. Lynch does seek to use ethnomethodology to elucidate the practical production of social order, yet this is accompanied by a critique of explanatory social theory and appeals to social context (see Lynch, *Scientific Practice and Ordinary Action*, p. 172, n. 42).

to macroscientific change as a topic of investigation, instead opting for an analysis of locally accomplished productions of meaning.[57]

Barnes and Bloor see a new form of life as potentially emergent with each new use of a term. Any apparent misunderstanding or misapplication of a rule could be the basis for a new collective understanding. In the normal course of events, however, the dominant extension of a rule is accepted by all within a community and those who persist in alternative formulations are described as hostage to misunderstanding, even irrational. Using Wittgenstein's number series example, a student can typically be expected to continue a number series (2-4-6 . . .) in one way, a socially enforced rule (8). In principle, however, a different answer could be the basis for a new rule, if it gained the support of the community.

Where Bloor sees all "misunderstanding" of a rule as a potential "alternative" interpretation privately held by the student (and, in principle, capable of wider adoption), Lynch argues that such merely logical possible alternatives are not relevant to an account of *situated* rule-following. In situated rule-following, deviant continuations of the number series are not the basis for potential alternative rules but simply a misunderstanding in that context. To think otherwise is to imagine standing apart from the actual context of rule-following and picturing different, imaginary continuations that have no social reality. Lynch's methodological prescriptions for the sociologist essentially enforce the members only rule:

> The initial characterization of the action as a "misunderstanding" of the rule only makes sense from a standpoint that is already situated in (i.e., internal to) the "accepted institution of arithmetic," so that there is no comparable standpoint from which to characterize what the pupil is doing as a "competing understanding."[58]

If deviant understandings develop, they do not depend upon an individual's private (psychological) interpretation of the rule, following Wittgenstein's argument against private languages.

Language is a social accomplishment; one cannot think in a language not shared by others. A deviant student cannot have an alternative language "in

57. Lynch, *Scientific Practice and Ordinary Action*, pp. 299–311; Livingston, *The Ethnomethodological Foundations of Mathematics*, pp. x–xi, 1, 16; Doing, *Velvet Revolution at the Synchrotron*, pp. 25–26.

58. Lynch, *Scientific Practice and Ordinary Action*, p. 177. Garver and Lee, *Derrida and Wittgenstein*, ch. 5, contrast Derrida unfavorably with Wittgenstein on this point. In contrast to Wittgenstein's anti-skeptical conventionalism, Derrida proceeds by analytically separating a text from its context and then showing its indeterminacy of meaning, which is only possible given this separation from the context in which meaning obtains. Yet arguably philosophy and history alike make possible dialogical engagement with the past precisely by various forms of decontextualization, which reframe the meaning of a text. See LaCapra, *Rethinking Intellectual History*.

their head" ready for wider adoption. So Bloor's appeal to individual (presocial) cognitive variability feeding social change is blocked. As a result, the meaningfulness of considering merely logically possible alternative inferences, as in the usual understanding of the Duhem-Quine thesis, is called into question. This leads to a non-skeptical account of the import of SSK that refuses to conclude that SSK research has established an antirealist thesis about science (since we cannot just imagine logical alternatives to situated, scientific theories), while at the same time refusing to endorse the usual conception of realism (since accepted theories do not apply to all social settings).[59]

Thus, rather than treating the topic of indexicality—broadly addressed to the contextual references necessary to make sense of language use—as a "methodological horror" to be resolved through more careful analytical philosophy or as a reason for a thoroughly antirealist and skeptical position, Lynch suggests that indexicality in and of itself is not something that needs to be avoided or overcome.[60] Instead, the mere existence of indexicality is not taken to imply a less certain or precise claim in the context of situated productions of meaning. This view differs from the standard attitude toward indexicals within SSK. The step-by-step removal of indexical references to truth claims in scientific work is the key finding of Bruno Latour and Steve Woolgar's *Laboratory Life*.[61]

Trevor Pinch addresses the same point in terms of the degree of "externality" attached to observational claims depending upon how much instrumental and theoretical mediation is taken for granted.[62] Facts of nature are stated without reference to conditions of discovery. Scientific controversies can lead to changes in the externality of observational claims, depending upon

59. Compare Arthur Fine's similar, phenomenologically inspired rejection of realism and antirealism in Fine, *The Shaky Game*, chs. 7–8; Fine, "Unnatural Attitudes"; and the discussion in chapter 5.

60. Lynch, *Scientific Practice and Ordinary Action*, pp. 18–22, traces Garfinkel's extension of Bar-Hillel's focus on the problem of computer translation of indexical expressions, such as pronouns, deictic expressions (*here*, *there*, *this*, etc.), and similar contextually dependent expressions, to a general contextual dependence of all meaningful expressions. See Bar-Hillel, "Indexical Expressions" and Garfinkel, *Studies in Ethnomethodology*, ch. 1. For a treatment of the problem in analytical philosophy, see, for example, Perry, "The Problem of the Essential Indexical." For indexicals as a "methodological horror" leading to a skeptical and antirealist epistemology, see Woolgar, *Science*, ch. 2. Putnam, "The Meaning of 'Meaning,'" p. 234, extends the notion of indexicality within philosophy beyond obviously "token-reflexive" terms like *here* or *now* to terms like *water*, which are understood as referring to "water around here." The upshot is to argue against the position that the intension (meaning) of a concept determines its extension (reference). Such a "descriptivist" position is shared by Carnap, Kuhn, and Kuhn-influenced positions within SSK. See Boyd, "Constructivism, Realism, and Philosophical Method," pp. 164–69. For criticisms of the original "baptism" of referring terms in the causal theory of reference, see Fuller, *Social Epistemology*, ch. 3. For an "anti-descriptivist" position, where reference is fixed only retroactively as an effect of the (changing) signifying process, see Žižek, *The Sublime Object of Ideology*, ch. 3. Žižek incorporates postmodernism's emphasis on discourse, while denying that signification ("intension") exhausts reference ("extension").

61. Lynch, *Scientific Practice and Ordinary Action*, pp. 101, 194–95; Latour and Woolgar, *Laboratory Life*, pp. 174–83.

62. Pinch, "Toward an Analysis of Scientific Observation."

the level of critical scrutiny to which the claim is subjected. The experimenter must assess the risk of challenge versus the significance of a claim in deciding what level of externality an observational claim should have.

Lynch's point is that even following the removal of the indexical elements in such claims as they become established, unquestioned truths, the context of the use of such claims will always involve indexical components in their conditions of use (call this the thesis of essential indexicality). Even facts of nature are indexical in practice. (Thus, this is more than the claim that a "black-boxed" truth could be reopened to reveal its indexical construction.)

Lynch's approach would be compatible with the view that hegemony is an open process and that the identity of any community depends upon how social and political processes construct them in the first place were it not for ethnomethodology's scrupulous and exclusive focus on interaction. The ethnomethodologist imposes rules on the appropriate use of language for sociologists, requiring that they not use reference to contexts of meaning transcending the immediate, interactional environment. This is a requirement that no other practice holds itself to, as appeals to wider frames of reference are routine parts of interactional sense-making. From the fact that scientists have to work out the meaning of appeal to scientific method in practical interaction does not mean that analysts should avoid reference to anything outside of local interaction—or in fact, that they could avoid such reference if they tried.[63] The result is that attention to sociological contexts shaping interaction are rejected. For example, Lynch rejects feminist treatments that (according to Lynch) depend upon stable categories such as gender. In fact, recent feminist treatments have questioned the stability of categories of gender quite extensively while noting that the situated, practical details of coalition building can create links between struggles evident in distinct, local contexts.[64]

The critique Lynch levels is more general, however, and depends upon a rejection of the scientific pretensions of the social sciences and a limitation on the appeals to social context that can be made. On the first point, Lynch is most explicit in rejecting Bloor's "external" approach to rule-following:

---

63. Ethnomethodology suffers from the paradoxes of anti-essentialism in philosophy and social theory in general. Thus, Fuss, *Essentially Speaking*, suggests that all seemingly "essentialist" moves cannot be eliminated without the anti-essentialism position itself becoming tacitly essentialist, perhaps in a more misleading—because hidden—fashion. Thus, if one wants to develop an anti-essentialist position, one cannot "act as if essentialism has an essence" (p. 21), for instance, by treating all accounts of the "natural" in feminist discussion as representing that which is unchangeable (and hence, for anti-essentialist feminism, suspect and reactionary) and reducing them everywhere to a production of the social.

64. Lynch, *Scientific Practice and Ordinary Action*, p. 267, n. 5; Lynch, "Going Full Circle in the Sociology of Knowledge"; Haraway, "A Manifesto for Cyborgs"; Fuss, *Essentially Speaking*, pp. 36–37.

> It is misleading to ask "how we attach meaning" to the sign, since the question implies that each of us separately accomplishes what is already established by the sign's use in the language game. This way of setting up the problem is like violently wresting a cell from a living body and then inspecting the now-dead cell to see how life would have been attached to it.[65]

What is noteworthy in this formulation is the more general, Wittgensteinian basis of this objection: meaning must be understood in the context of use, not by decomposing its components. After all, ethnomethodology and SSK show the importance of intervening and reworking the everyday life world in practical scientific work. However, a similar move is denied to the social sciences (or, more to the point, any kind of macro-level sociopolitical criticism). In the debate about how to relate micro and macro perspectives in sociology, ethnomethodologists are eliminative micro theorists, rejecting the reality of macro-structures or social and historical context.

Aside from a rejection of the recontextualizing work of the social sciences and the naïve claim that ethnomethodological investigation is somehow able to avoid such recontextualization,[66] Lynch also relies upon arguments in ethnomethodology and conversational analysis against the invocation of social "contexts" to "connect" the micro and macro realms of society. Schegloff's argument against this formulation of a micro-macro "problem" is cited by Lynch in order to undermine appeals to stable, empirically investigable contexts apart from the temporal production of meanings in situated action. Invocation of such "contexts" is crucial to both historical explanation and political critique and its blocking by ethnomethodological restrictions is key to the complaint that ethnomethodology has a quietist political function in the contemporary academic world.[67]

Schegloff argues that research examining tape recordings of ordinary speech in careful detail (conversational analysis, or CA) creates significant

65. Lynch, "From the 'Will to Theory' to the Discursive Collage," p. 289.

66. See Bloor, "The Living Foundations of Mathematics," for an argument that despite ethnomethodologists' "refusal to theorize," they nevertheless subscribe to a very strong theoretical claim, which Bloor dubs "the locality thesis" (p. 351). Similarly, Bloor, "Left and Right Wittgensteinians," p. 276, notes that the difference between his position and ethnomethodology cannot be that the one constructs rules for understanding rules and the other does not: "If this means trying to achieve ideally objective, nonindexical formulations (about the use of indexical formulations), then it is indeed impossible. But the sociologist of knowledge, no less than the ethnomethodologist, is aware of the fact that his own cognitive processes are as indexical and context-bound as those under study. We are doomed to speak indexically about indexicality, as about everything else." Bloor's point here would tend to undermine any principled distinction between ethnomethodological accounts and broader appeals to context.

67. Lynch, *Scientific Practice and Ordinary Action*, pp. 32, 241, 258; Bauman, "On the Philosophical Status of Ethnomethodology," pp. 17–21; Lynch and Fuhrman, "Ethnomethodology as Technocratic Ideology"; Gleeson and Erben, "Meaning in Context," pp. 474–76.

difficulties for attempts to "link" such "micro" research to traditional "macro"-oriented sociological topics:

> One of our most insistent and recurrent findings is the so-called local character of the organization of interaction (that is, its turn-by-turn, sequence-by-sequence, episode-sensitive character), and this is one basis for the problems that arise in attempting to relate its analysis to so-called macro.[68]

Such findings are held to challenge three varieties of possible links with the macro realm: (1) cultural variation in the organization of micro interactions; (2) variation within micro interactions as the result of class, ethnicity, or gender differences; or (3) relating micro interactions to larger contexts such as institutional or organizational contexts.

In the first instance, Schegloff considers variation in rules governing turn-taking in conversations and reports broad uniformity in conversational "repair" work whereby situationally produced order is maintained. Schegloff then argues that what variation exists is not significant in challenging the conversational rules discovered by analyzing conversations in the West. Speakers of Tuvaluan, a South Pacific language, do not attempt to guess at what word another speaker intends when that speaker has trouble finishing a sentence, perhaps because the concept of "intention" is weaker in South Pacific societies. Speakers of Quiche in Guatemala do not interrupt their speech stream (that is, use a "cutoff" or "stop") to indicate that they will correct something they have said, since such stops are linguistically meaningful (i.e., phonemic). Likewise, they do not use brief sound stretches that are similarly meaningful, but they do use longer sound stretches to indicate self-repair.[69] Yet, other conversational repair practices are the same, and what variation exists can only be detected against the recognizability of the formal structures in the first place.

Schegloff has shown that the formal regularities identified in Western societies by a particular research tradition can pick out closely similar regularities in other cultures, but that alone does not show that a different set of regularities could not be picked out, either based upon a different research tradition or where initial discovery involved study of (or was conducted by) non-Western societies. The significance of such variation as exists has been shown to be minimal only with respect to the project of identifying formal regularities. Such variations could be quite significant when considering the social effects of such variation on social life and institutions.

Failing investigation into whether fewer (quick) options to initiate self-repair have effects for the cognitive life of the Quiche or whether the reduced

68. Schegloff, "Between Micro and Macro," p. 209.
69. Ibid., pp. 211–13.

place of intention among speakers of Tuvaluan has effects on political organization or property rights, for instance, we cannot conclude that such variation is insignificant but merely insignificant as it affects conversational turn-taking rules. This is an important point, since the claim for the limited value of research questions and methods outside a broadly ethnomethodological framework depend upon this kind of tacit privileging of ethnomethodological findings as a normative standard. This is precisely the kind of approach Feyerabend targeted for criticism, an overadapted theoretical perspective that determines its own observation language to be the only one possible.

This privileging of ethnomethodological findings as a normative standard for judging competing claims and methodologies can be seen in Schegloff's criticisms of the second mode of connecting micro and macro research. Here, the attempt is made to examine how variation in the attributes of individual speakers affects turn-taking rules.

Schegloff addresses himself in particular to the body of research suggesting that men tend to interrupt women much more than the reverse. Here, Schegloff calls into question whether the identification of participants in terms of particular attributes—in this case, gender—is meaningful to the participants themselves:

> One [problem] concerns the need in this type of analytic enterprise to show that characterizations the investigator makes of the participants are grounded in the participants' own orientations in the interaction. This is not at all clear (except, perhaps, statistically) for the characterization of the participants in gender terms in this research tradition (or in class, ethnic, or other such terms in cognate research traditions).[70]

Once again, a criterion valued by ethnomethodologists and conversational analysts—acceptability of analysis to those studied—is unproblematically treated as a natural desideratum for all social research enterprises whatever. Indeed, precisely the most obvious "scientific" means for establishing the relevance of gender as an operative characteristic affecting ongoing interaction—statistical evidence—is dismissed almost as an afterthought.

This is similar to Lynch's objection to applying in the social realm the type of natural scientific investigative strategies that remove objects from their lived context for greater understanding. This is doubly ironic since conversational analysis does identify regularities in a manner that Lynch himself has identified as scientistic.[71] In both cases, however, there is a shared commitment to a kind of radical empiricism that seeks to avoid theory, or at least what it sees as theory. In effect, Schegloff denies that it is a meaningful question to

70. Ibid., p. 215.
71. Lynch, *Scientific Practice and Ordinary Action*, ch. 6.

ask why men are more likely than women to interrupt others in conversational turn-taking except to point to the individual, locally produced reasons for it in any particular instance, reasons that are explicable by formally identifiable features and require no appeal to gender or other personal attributes.[72]

The third type of link between micro and macro domains that Schegloff considers is the invocation of "context," especially institutional contexts of various sorts: legal, bureaucratic, medical, and the like. Schegloff notes that there is a problem of multiple descriptions: a given action may fall under more than one context so that it is not obvious what context is operative at any particular moment.[73] Schegloff does not grapple with this problem so much as use the existence of the problem to deny the legitimacy of invoking such contexts in explanation, except insofar as participants in interaction themselves invoke various contexts:

> It is not, then, that some context independently selected as relevant affects the interaction in some way. Rather, in an interaction's moment-to-moment development, the parties, singly and together, select and display in their conduct which of the indefinitely many aspects of context they are making relevant, or are invoking, for the immediate moment.[74]

The problem of determining which contexts to invoke is "solved" only by making a substantive, undefended assumption at the outset—namely, that relevant contexts will be invoked by the participants, just as was assumed when considering the effect of class, gender, or ethnicity in micro interactions.[75]

Moreover, Schegloff relies upon CA protocols to discover such participant-invoked contexts, rather than pre- or post-interaction questioning or appeal to institutional facts of which the participants would be aware. Thus, Schegloff rejects studying how doctors tend to behave. Instead, we are urged to specify what is involved in "doing being doctor."[76] Yet clearly not just anyone can learn forms of interaction establishing themselves as doctors in routine interaction and actually become doctors and operate on patients. Moreover, the fact that doctors are known to be doctors—by administrators, nurses, patients, and other doctors—is a necessary condition for such interactional performances to work. Such a condition may not be sufficient, for

72. Schegloff, "Between Micro and Macro," p. 216.
73. Ibid., p. 218.
74. Ibid., p. 219.
75. Schegloff might object that this is not an assumption but the outcome of a body of empirical research. However, from the fact (if it is a fact) that moment-to-moment interaction can be modeled in this fashion, it does not follow that "nonvisible" contexts are not operative, nor that interaction constitutes the only meaningful social phenomenon.
76. Schegloff, "Between Micro and Macro," p. 220.

instance, if a doctor acts in ways that undermine his or her role. Yet this does not imply that anyone can be a doctor if he or she learns such interactional skills of behaving in the appropriate way in routine interactions.[77]

Schegloff concludes that the focus on interaction obviates the need to link micro and macro realms, since whatever macro links one might appeal to either will be situationally invoked in the interactional context itself or will fail to be so invoked and hence held to be nonexplanatory. The limitations of such a strategy can be seen more clearly in Schegloff's concluding example: the attempt to reorganize the forms of interaction regulating Ronald Reagan's press conferences in the early years of his presidency. Schegloff analyzes the press conferences as a variant of two-party exchange systems, albeit with one party constituted by a number of people, the press corps.[78] Rather than having the president call on the next questioner after responding to the previous question, the changed procedure involved providing a random, preselected order to the president. This would avoid requiring the president to choose from shouting reporters upon completing an answer.

The change was not very successful, however, since instead of scanning the room for the next questioner, the president returned his gaze to the previous questioner, inviting undesired follow-up questions. Moreover, following a particularly pointed question, the president did quickly avert his gaze to consult his list, presumably to avoid a follow-up question. This turned out to be a much more blatant evasion than scanning for the next questioner had been, and in the end, the system was scrapped.[79]

The analysis is very revealing in illustrating how identifiable forms structure interaction in ways that can be changed but not entirely in line with some preplanned objective. Yet the explanation is hardly complete. What is missing is the "context" that might explain why such a change was attempted and might link such changes with other activities engaged in by the president, his staff, and those carrying out his directives, and other activities around the world. For instance, one possible explanation for such an attempted change might involve the effort to gain greater compliance from the press corps and increased "spin control" for political reasons. Consider, for the sake of argument, a "political" criticism that the new system for press conferences was implemented to avoid sustained questions about presidential policies, such as covert and possibly illegal activities to support military intervention in

77. Schegloff's strictures entail a kind of willed ignorance of anything outside whatever interaction is currently being studied. This is another way of saying that sociologists would have to avoid even the kinds of interactional invocations of context that Schegloff maintains that actors themselves engage in.

78. Schegloff, "Between Micro and Macro," p. 222.

79. Ibid., pp. 224–26.

Central America.[80] It would not be possible to object that some reporters did make such a larger "context" evident, since the change could have been tested and failed without anyone invoking the "right" context for its attempted implementation. Nor would simply telling a number of different stories of local interaction suffice (staff meetings, bank transfers, arms shipments, military training, etc.), since the point is the relationship between particular local contexts, which is why some sociologists seek to build macro-structures out of the agglomeration of interactional contexts.[81]

The promise of a strict focus on interaction to avoid the problems of individuating relevant "contexts" leaves us even more impoverished in understanding the place of methodological discourse. In some ways, the thesis of essential indexicality and a focus on interaction helps improve our understanding of how actors actively construct meanings in local contexts, but without an understanding of the *relationships* between different contexts of interaction and different scales of social life, we cannot understand the forces shaping social life. By rejecting contexts outside local, temporally produced interactional orders, understanding the role of methodological discourse across spatial and temporal settings becomes more difficult. Lynch promotes the study of "epistopics," whereby traditional topics of philosophical interest—such as discovery, observation, and replication—will be used to pick out widely varying sites to study. The goal of such study will be to undermine any notion of "coherence" linking separate examples of the epistopic so that research questions about whether something called "methodological discourse" could have determinate "effects" would be rejected based on a prior commitment to a methodological doctrine, in this case ethnomethodology itself.[82]

Methodological discourse would be in the same predicament as rule-following in Lynch's "internal" reading: to construe any "method" as separate from its application in concrete use would commit the same error as separating the formulation of a number-series rule from its extension. Lynch notes that while this interpretation "may seem puzzling given the fact that formulations of rules are commonly set down on paper and posted on walls, and they are often recited separately from any acts that do or do not follow them,"[83]

80. Chomsky, *Necessary Illusions*, ch. 9. It is ironic that Lynch, *Scientific Practice and Ordinary Action*, pp. 304–305, invokes Noam Chomsky's appeal to "normal science," which Chomsky offers as a defense of his comparative method in order to fend off complaints about the methodological adequacy of his analysis of U.S. press coverage of international events. Lynch borrows this idea to promote a non-scientistic deemphasis on methodology, yet Lynch (and Schegloff) argue against the kind of comparative and generalizing approach Chomsky actually employs.

81. See Knorr-Cetina, "Introduction"; Callon and Latour, "Unscrewing the Big Leviathan"; and the discussion in chapter 5. The limits of the focus on interaction that are shaped by political power can be seen in Lynch and Bogen, *The Spectacle of History*.

82. Lynch, *Scientific Practice and Ordinary Action*, p. 282.

83. Ibid., p. 173.

it is a mistake to treat such separate formulations as intelligible apart from particular acts, invoking Wittgenstein's appeal to "quiet agreement," which

> refers . . . to the practical adherence that supports a rule's intelligibility, that is, the order of concerted activities already in place when a rule is formulated, notably violated, disregarded, or evidently followed. The statement of a rule or order is a constituent part of such activities, and there is no way to contain or determine those activities in even the most elaborate version of the "naked" statement.[84]

For Lynch, the same "naked" statements about method, appearing in different local contexts, do not indicate that a common method shaped diverse practical activities. The problem here is that a Wittgensteinian worry that a propositional rule will be taken to "contain or determine" its application results in avoiding the consideration of how "naked" statements move from place to place and take on new meanings in new situations.[85] While such a "statement" may take on quite different practical meanings, it does not follow that the circulation itself cannot be studied for its effects or constraints on its variable use considered. Naked statements about methodology or theory are not just Rorschach tests to have meanings attributed to them entirely unconstrained by what went before. The historical dimension of human life would be thoroughly eviscerated on this interpretation of Wittgenstein's strictures.

THE RELATIONS OF SCIENCE

After having considered both collectivist and interactional interpretations of Wittgenstein, we can return to the point that identifiable forms of life are not self-contained, but are often established and maintained by reference to outside groups. On this view, a dialectical approach obtains to the extent that individual identities are internally related to each other and to the whole, which in turn reshapes the identities in turn.[86] In looking at how a reasonably well-bounded community comes about, it is not sufficient to focus only on that group, since the group will likely be defined in part by differentiation from, and opposition to, existing groups. In looking at the formation of the Royal Society, attention to this process shows that surrounding (and sometimes overlapping) groups may constrain the manner in which members come to

84. Ibid.
85. See Lynch's argument against Latour's semiotic approach in ibid., pp. 290–92.
86. Quadrio, "Hegel's Relational Organicism." Descombes, *The Institutions of Meaning*, pp. 201–2, shows that whether objects are understood as internally or externally related to each other (that is, as lacking or possessing independent reality), is relative to a description, which can vary depending upon the particular theory of changes of state applied based on the pragmatic goals of explanation.

account for themselves.[87] Likewise, no clear criterion exists for demarcating different forms of life; such forms were frequently overlapping (many Royal Society fellows were university professors or physicians).

Moreover, there is no reason to believe that communities are cohesive in the sense of depending upon uniformly shared rules or understandings. We have seen that the limits of consensus define the limits of community for the strong and relativist programs. At best, the undetected or temporary existence of deviant understandings may provide conceptual variability leading to a new consensus, but the ongoing coordination of competing or contrasting perspectives within the *same* community is not expected. The recognition of such a conflict would be an "autogarfinkel" in Collins's terms, pointing to the need to resolve the scientific controversy to restore community cohesion.[88] That disagreement underlying scientific controversies is equated with Garfinkel's breaching experiments (where refusing participation in routine social understandings led to social breakdown) demonstrates just how much cognitive disagreement is equated with social breakdown in this view.

In fact, it would be more realistic to suggest that apparently coherent forms of life are temporary stabilizations of various ongoing, socially situated scientific activities. Ongoing meaning shifts may be disguised by apparent agreement.[89] Shared methodological commitments may nevertheless constrain scientific activity in similar ways, even while a variety of different contexts shape the use of such method in particular settings. Consequently, social cohesion does not imply cognitive consensus. If a relatively coherent community exists, specific social causes should be identified, typically involving explicit efforts to establish and maintain a shared community structure. As such efforts "drift apart," the associated social meanings can be expected to change.

Consider a classic case of methodological reform of the sciences: the Baconianism of the early Royal Society of London. The statutes of the early Royal Society call for its institutional life to be regulated by ideas drawn from Francis Bacon's methodological writings. The statutes call for society meetings to record observed matters of fact apart from speculation about causes.[90] The polemical writings of the Royal Society simultaneously defend its approach to natural philosophy from accusations of atheism and religious enthusiasm and attack the limitations of natural science practiced in university settings (even though many significant members had university positions and the

87. Lynch, *Solomon's Child*, ch. 5; Hannaway, *The Chemists and the Word*.

88. Collins, "An Empirical Relativist Programme in the Sociology of Scientific Knowledge," pp. 94–95.

89. Fuller, *Social Epistemology*, chs. 5–6. The historian can often trace the origins of such divergences in meaning leading up to explicit conflicts. See Roberts, "Setting the Table."

90. Shapin and Schaffer, *Leviathan and the Air-Pump*.

Royal Society itself originated in earlier meetings at Gresham College and Oxford University).[91]

Nevertheless, many historians of science have seen this Baconian rhetoric as window dressing providing a surface coherence to a group with varied interests. One reason for this claim is that Baconian empiricism has fallen out of fashion among philosophers and historians of science. Those Royal Society fellows who clung most obviously to the tenets of Bacon's empiricism were the virtuosi who collected strange facts seemingly without regard to their significance for knowledge.[92] The real significance of the early Royal Society was judged by reference to larger narratives of method, whether emphasizing systematic experimental inquiries or mathematical natural philosophy. In short, the early society's "significant" work was understood in reference to the later emergence of Newtonian experimental and theoretical science; all else was judged reactionary or as merely institutional cover for varied interests.

Shapin and Schaffer's work on the Royal Society does revive something of the significance of the Baconian view through a collective Wittgensteinian account of the Royal Society by examining what gets to count as legitimate facts of nature (Boyle's operational descriptions of the experimental vacuum, for instance) and what gets rejected as outside the legitimate bounds of the experimental community (Hobbes's causal explanations of the leaky air-pump). Critics are judged to speak for a different form of life, while "method" can only refer to the taken-for-granted behavior of a well-defined community. They deny that method can act "as a determinant of intellectual practice."[93]

In effect, there are two levels here: the level of practice and the level of post hoc methodological redescription. The first follows taken-for-granted ways of behaving, which means it is a product of the internalization of a particular kind of social life and not a set of decontextualized rules that can be followed. The second is a way of talking about that practice after one has already generated one's results, its own form of practice, no doubt, but one that has no connection with the first level of practice itself, except to draw on bits and pieces of that work to tell a story to give it a methodological propriety. Like Mulkay's approach to norms discussed in chapter 1, ordinary practice is unconstrained by methods or norms, but a second-order practice pretends that they are to convince outsiders—or oneself—that the practice was carried out in a proper way to generate knowledge deserving of support.

But these levels are not as separated as this meta-methodological account pretends and the effects of actors attempting to implement methodological

91. Lynch, *Solomon's Child*, ch. 5.
92. Hunter, *Science and Society in Restoration England*, p. 67.
93. Shapin and Schaffer, *Leviathan and the Air-Pump*, p. 14.

prescriptions goes deeper than post hoc legitimation. My own research on the early Royal Society began from an examination of the seriousness with which its fellows tried to implement Bacon's previously articulated philosophical system for reforming knowledge in terms appropriate to their setting and interests. It does this by examining Bacon's *Novum Organum*, the key methodological text influencing early Royal Society fellows, identifying three related but distinct interpretations of Bacon's much-cited injunction to study things themselves, rather than the empty words of university scholastics. First, Bacon talked of things themselves as objects that we passively observe (specular objects), picked up within the Royal Society in its constant emphasis on the need for shared witnessing without imposing dogmas. Second, Bacon considered things themselves to refer to manipulated objects of art (manual objects), where our ability to actively manipulate nature testified to our understanding of nature's causes. Third, Bacon conceived things themselves as underlying forms that could be recombined in new ways like letters of the alphabet to produce phenomena at will (generative objects). These varied metaphors for the same concept (focusing upon things themselves) supported three visions of objectivity broader than stereotyped pictures of Baconian empiricism: empiricism (knowing is seeing), constructivism (knowing is doing), and theoreticism (knowing is uncovering hidden structures).[94]

This analysis begins from the existence within Bacon's writings of semantic variability in the articulation of his methodological system. Interpretative flexibility is built into the system through the suggestive metaphors that articulate the meaning of Bacon's injunction to focus upon things themselves as well as the role these ideas play in Bacon's program for an inductive ascent from facts to causes to build useful knowledge. Without noticing that Bacon's program was broader than later stereotypes suggest, it is difficult to recognize how early Royal Society fellows saw their work as building upon Bacon, since Hooke's experimentalism or William Petty's theoretical political economy does not obviously fit the picture of idle empiricism. Interpretive flexibility within Bacon's own thinking and interpretive flexibility in the further development of Bacon's system by later users are key to this approach, but the variability of ways Bacon could be used was not unlimited. Early fellows did not have to take Bacon's methodological system seriously, but given that they did, it is possible to trace similar articulations of the injunction to focus upon things themselves in their own work, shaped, however, by the different situation of the Royal Society and the interests of its members.

The earlier English Baconianism of the Interregnum period revealed a greater interest in social and political reform than the Royal Society's studied

94. Lynch, *Solomon's Child*, ch. 1.

avoidance of such controversial topics,[95] more suitable to establishing a royal charter and avoiding blame for destabilizing the nation. Consequently, the larger political and religious context must be studied to understand why Baconians shifted the semantic associations of Baconian "reform," even where fellows did not explicitly invoke such larger contexts in their writings. At the same time, the hidden continuities before and after the Restoration found in the work of many fellows must be determined by close examination of their work. Thus, William Petty continued his attention to social and political reform issues, although now articulated in terms addressed to the magistrate's need to keep the subject in "peace and plenty" to ensure political order. Finally, no shared, tacit agreement on ideological issues characterized their work, as their explicit commitment to Bacon's method is used to "translate" different interests in terms allowing cooperation, as with the communication between the class-conscious gentleman John Evelyn and John Beale, a reformer rooted in the Puritan Hartlib circle's tapping of craft knowledge.[96]

The picture that emerges does not fit an image of a united form of life agreed upon ways of proceeding, since the tensions in their interpretation of Bacon ultimately pulled natural philosophers in different directions, as virtuosi, experimental philosophers, and mathematical natural philosophers increasingly went their separate ways by the turn of the century. Early fellows were explicitly engaged in building an institution exhibiting Bacon's new way, and their efforts helped ensure the coherence in outlook in the products produced in the first two decades of the institution's life. Explicitly licensed books carried out parts of Bacon's program through contributions to natural history, horticulture, instrumentation, language reform, cultural criticism, statistical epidemiology, and political economy, while the *Philosophical Transactions* carried the work of the Royal Society meetings to a larger international audience.[97] Understanding of Bacon's method is crucial to this history, but the picture of method's results varied significantly depending upon the individuals and fields involved. Moreover, the longer the period of time we examine, the more diffuse the influence of method becomes as meanings drift, recombine, and change over time.[98] To sort these changes out, we must appeal to a variety of different social and intellectual contexts applicable in different moments of the method's application. Neither coherent forms of life nor purely local, interactional orders will work.

John Schuster has developed a similar analysis of method as "mythic speech" as threaded into the level of practice in a treatment of that other great methodologist of the seventeenth century, Rene Descartes. For Schus-

95. Webster, *The Great Instauration*.
96. Lynch, *Solomon's Child*, ch. 2.
97. Ibid., pp. 20–33.
98. Schuster, "The Scientific Revolution."

ter, "'myth' is not a colloquial term of abuse," but an application of Roland Barthes's semiotic approach to myth and ideology. For Barthes, ideological beliefs are "second-order signs" that add social meanings supporting consumer culture or bourgeois values in modern society to ordinary objects presented in popular media. These myths naturalize modern social values by effacing the complexities of history and politics to *naturalize* current social arrangements. By naturalizing theoretical and axiological aims, Schuster argues that the grand methodologies of the seventeenth century (and select forerunners like Aristotle) are "perhaps the first example of a characteristically modern Western myth." Methodologies are "discourses so structured that they necessarily lie about their own powers and capabilities in the interest of turning *culture* (how the natural sciences are actually practiced) into *nature* (a simple outgrowth of human rationality and nature's amenability to it)."[99]

Although Descartes's formal rules of method cannot in fact generate the corpuscular-mechanical theories he constructs, as he believed, they are worked into the practice that constructs them in ways that support that illusion of efficacious method. The result is that this practice looks different than if Descartes had simply decided to construct corpuscular-mechanical models without attempting to constrain them by the application of proper method. Three interacting levels are identified here.

First, before explaining anything, Descartes worked out his basic metaphysical ideas, which would "constrain and condition the formulation of specific corpuscular-mechanical models." These models were constrained not by "clear, consistent, rule-bound procedural glossing," but by a kind of "on the spot interpretation and 'negotiation,' if only with himself!" Second, Descartes selected, weighed, and applied empirical evidence he perceived to be relevant in an active process that constructivist sociologists and historians have described well as subject to interpretive flexibility. Third, Descartes takes the methodological constraints and evidence collected, keeping them in view while constructing specific corpuscular-mechanical models. Descartes, and those who attempted to follow his method, shaped their work at the level of social practice in their forms of life, while carrying out a "'voicing over' or monitoring of practice [which] could be institutionalized in pedagogy to entrench both the method in question and the theories embedded in the routines so glossed."[100]

---

99. Schuster, *Descartes-Agonistes*, pp. 270–71. See also Rip, "Constructing Expertise," p. 430, n. 9, who argues that "the methodologies of science can be seen as socio-epistemic strategies, . . . part of productive practices and interactions, and then reflected upon, by practitioners and by philosophers."

100. Ibid., pp. 276, 279. See also Law, "STS as Method," pp. 31, 39, for the claim that STS focuses on "the study of method in practice," which includes examining "how methods are staged" to be convincing in particular contexts.

In effect, there are not distinct realms of unreflective practice and post hoc rationalizations, but intertwined practices incorporating lesser or greater reflexivity.[101] To return the point to the kind of interdisciplinary science of science originally envisioned by the strong program, we can look at human behavior as the product of distinct, but interacting levels of practice, spanning from our biological nature, in-born and learned psychological mechanisms, and institutions of social and cultural transmission. Any piece of knowledge transmission can be expected to have resulted from the combined effect of analytically distinct levels in any particular lineage, which is why historians of science need to incorporate knowledge of biography, scientific traditions and institutions, social and political context, and the long-term and short-term history of science relevant to the case studied.[102] Applied to the transmission of methodologically justified knowledge, this means that competent actors are capable of not only responding to their direct, visible, "local" environments but are also processing and transforming that environment via non-visible contexts and commitments that may be more or less explicitly formulated and shape behavior in turn.[103]

Consider a case where outside critics of orthodox science question the methods employed by practicing scientists and, in the process, alter the products of scientific practice as well. The criticism of AIDS drug-testing protocols by AIDS activists analyzed by Stephen Epstein is a contemporary case of science criticism conjoining methodological and political components, with similar findings regarding the variability of application of a seemingly straightforward methodological criticism. Epstein examines the impact of lay AIDS activists on scientific research on AIDS. For instance, AIDS activists challenged clinical trials for AIDS drugs for their reliance upon double-blind protocols, arguing that the use of control groups for research into a fatal disease was not only ethically objectionable but also epistemologically problematic. In particular, the use of control groups ensured that patients would hedge their bets by using outside drugs in addition to tested drugs, thereby introducing unknown contaminating factors to the research results. Instead, they proposed that scientists work with patients to test drugs in combination with additional drugs. In challenging existing scientific consensus, AIDS activists reopened a latent scientific debate between proponents of "fastidious" and "pragmatic" protocols. Orthodox, "fastidious" protocols sought out "clean" data and "pure" subjects in order to control for extraneous factors

101. Schön, *The Reflective Practitioner*. See the discussion of ethical reflection by scientists and engineers in chapter 6, which likewise sees an interaction of practice and reflection that can have real effects, but not as the simple application of an ethical methodology.

102. Campbell, "Variation and Selective Retention in Sociocultural Systems"; Hull, *Science as a Process*; Lynch, "Cultural Evolution and Social Epistemology."

103. Odling-Smee, Laland, and Feldman, *Niche Construction*.

beyond that under examination but at the potential cost of "representativeness" in real world terms. "Pragmatic" protocols are more comfortable with variability in subject populations and more concerned to study diverse populations to ensure that approved drugs did not evidence unforeseen side effects in populations unstudied in clinical trials.[104]

What is interesting about this case is that AIDS activists employed their status as outsiders to science and as representatives of the affected communities to argue that they were in a position to object to existing scientific procedures in a way that would not occur to most scientists. They did so by becoming "lay experts," studying the scientific fields criticized intensively and achieving some level of acceptance among researchers as a result. Nevertheless, the implications of such a practice of criticism were not straightforward, as some "lay experts" changed their position as it became apparent that no ready cure would be found, shifting back to support of fastidious protocols. Like the case of Baconianism in the early Royal Society, criticism of existing knowledge could incorporate both epistemic and ethical components and could result in unexpected outcomes as criticism evolved. Nevertheless, no shared form of life encoding taken-for-granted rules encompassed scientists and AIDS activists. To the extent that dialogue was achieved, it required an active process of reconsidering taken-for-granted methods and practices. Outside critics can have a legitimate effect upon scientific debates, bringing points of view excluded from mainstream debate—but only if critics show a willingness to learn and to adapt their criticisms to new circumstances.

AGAINST MONISM

A tacit assumption that the history of science has been directed by proper method has often been used to assess the causal significance of various figures in the history of science. To the extent that historical actors could be positioned within a larger narrative of the unfolding of progressive method, their influence upon the history of science could be assessed. With the rejection of such larger narratives by proponents of a sociological approach, different assumptions about causes in the history of science had to be worked out. One way to overcome the privileging of a larger, methodologically directed narrative of the progress of science is to examine how methodological programs were implemented in particular historical settings, noticing how different views about proper method and legitimate boundaries of scientific communities are contested and resolved.

104. Epstein, "The Construction of Lay Expertise"; Epstein, *Impure Science*.

In other words, scientific development is subject to conflict and power. Not everyone agrees with dominant perspectives, and they seek to support alternatives they favor. At the same time, dominant perspectives are dominant not just because they are voluntarily accepted by most scientists but because some approaches have more social power, drawing resources from the larger society and controlling their organization and use.[105] The relative hegemony of scientific knowledge can be studied to see to what extent minority positions challenge dominant views and the extent to which they try to mobilize a different conception of scientific leadership.

The intellectual form of Bacon's philosophy and its distinctive, metaphorical deformations in three directions constrain its use. These three components in turn get detached and reworked, contributing to long-lasting, if diffuse, traditions of interpreting and interrogating knowledge. When analyzing a development of these ideas, a variety of analytic "contexts" responding to different identities, reference groups, processes, and cognitive traditions shape individual applications of these ideas in practice. No scientist operates in a single form of life.[106] Nor do scientists define themselves only by groups that they belong to but also by outside reference groups.

But no purely contextual studies will be satisfactory either. There is a need to trace the ephemeral but real impact of methodological ideas in reshaping scientific practice, an idea that should not be a surprise to social theorists who realize that actors actively construct folk sociologies of their world and intervene on their basis. Anthony Giddens see this "double hermeneutic" aspect of our social life as key to how macro and micro are interrelated, as actors' conceptions of social structure help shape that structure itself. While Giddens denies that something similar applies to the natural sciences, as electrons and planets cannot appropriate our theories of them for their own purposes or reshape their behavior in kind, there is a sense in which scientific and lay accounts of scientific phenomena do something similar.[107] While taking on board a recognition of the knowledgeable agency of actors, borrowed from ethnomethodology and interpretive social science, Giddens notes that the

105. Collins, *Conflict Sociology*; Mann, *Sources of Social Power*.

106. Clarke, "A Social Worlds Research Adventure," pp. 18–22; Fujimura, "Crafting Science," pp. 168, 170–72. In Collins, *Changing Order*, pp. 145–52, his argument for sociological closure depends upon the existence of distinct, but overlapping, forms of life. The fact that they are overlapping simultaneously provides novelty and a shared basis for ultimate agreement within the core set. This is a way to ensure that Wittgenstein's "monism" (see below) gets adapted to a recognition of interpretive flexibility that sits uneasily with it. Bloor, *Wittgenstein: Rules and Institutions*, pp. 9–22, uses "meaning finitism," the open-ended character of the extension of concepts subject to conflict, as a similar way to extend the closed and self-referential concept of language games to account for change over time.

107. Lynch, "What Does the Double Hermeneutic Explain/Justify?"; Giddens, *New Rules of Sociological Method*, pp. 11–13, 170.

unintended consequences of their actions recursively shapes the structural possibilities of their action.[108]

We can also see this interaction between levels of contextual behavior in the way research programs direct practice. The burden of Lakatos's particular theory of research programs was to show how methods shape practice via positive and negative heuristics that tell the scientist where to look for constructive solutions to scientific problems within the parameters of a particular research program. This means that being trained in a particular research program can help direct change in scientific practice in a certain direction, while other contextual aspects of the scientist's work can also shape it. Moreover, such programs eventually break down as a productive guide to practice and are replaced with different programs.

In some ways, Feyerabend was even more attuned to the way that methodology could be a tool for generating new solutions that forced a change that was not part of the natural trajectory of the scientific program at a particular moment in time. Feyerabend shared the later Wittgenstein's rejection of realism for nominalism, seeing facts as conventions rather than pictures of reality. However, he rejected the view that single paradigms drive out competitors in practice or that they should do so. For Feyerabend, it is not social contexts of use that define the meaning of scientific concepts but the theoretical context that gives different meanings to different observation statements by defining different ontologies of the world. Theories are not just any kind of social rule-following and they manifestly are capable of having a meaning that transcends socially enforced rule-following.

Indeed, the whole reason that incommensurable observation languages are introduced, for Feyerabend, is to force conceptual change that is blocked by the dependence of observed facts on their theoretical contexts of meaning. Observation based upon "non-instantial" or general, scientific theories are not the same as everyday experience, which can be generally shared among all humans and provide no special obstacles to intersubjective communication. Theories, however, become so intertwined with the observations that they generate that they are, in effect, overadapted to their particular niche. So Feyerabend argues that incommensurable perspectives can do what the Wittgensteinians rule out of court: provide a means for outsiders to begin to subject core-set members to withering criticism.[109]

---

108. Giddens, *Studies in Social and Political Theory*, ch. 2; Kim, "Critical Theory Criticized," pp. 30–31.

109. Feyerabend, *Science in a Free Society*, pp. 65–69; Lynch, "Between Kin Selection and Cultural Relativism," pp. 306–7. Feyerabend's point about the theoretical holism of general physical theories traces back to Duhem, *Aim and Structure of Physical Theory*, p. 185. See Descombes, *The Institutions of Meaning*, pp. 104–5.

It is helpful to compare Collins and Feyerabend on this point to see the difference this makes for the possibility of criticism by outsiders. For Collins, even if one gets lucky and bets on the right long-shot contender to replace orthodoxy, one gets no credit because it is entirely up to core set members to regulate the view of nature we all must accept.

> It is the community of scientists who decide on what we are all going to accept as the truth of the matter; we, as lay-individuals, have no technical grounds to decide that [dissenting scientists] are right before the scientific community has done its work, however slowly they go about it.[110]

By contrast, Feyerabend rejects those who want to assimilate his view to Wittgenstein's, precisely because he opposes the rejection of criticism by appeal to a members-only view of scientific meaning.

When Herbert Feigl criticized Feyerabend's claim that there was no core of observational facts immune from theoretical contamination, he drew a comparison between Feyerabend and Wittgenstein on this point. Feyerabend rejected this comparison:

> I cannot understand why Feigl should find "neo-Wittgensteinian traces in my reasoning. . . . "Wittgensteinian" analyses are characterized by the fact that they are restricted to a single "language game"; they are monistic. Alternatives are brought in, not in order to arrive at a better theory through a criticism of the existing one, but rather in order to get better insight into the existing theory. And getting better insight into the existing theory (the existing "language game") means revealing its hidden strength, i.e., its capability to deal with the problems that have arisen and to remove the impression that a revision might be needed. A Wittgensteinian uses alternatives with a dogmatic, or a conservative, purpose. Knowing where the argument is going to lead, he does not really take them seriously, although he may admit that they possess an important therapeutic function. Our own point of view is radically different. Alternatives are taken seriously. They are all candidates for a better theory in the future. Their function is critical and progressive, that is they are used with the purpose of finding the weakness of the customary ideas. The discussion of such weakness does not have an exclusively dramatic purpose, viz., to let the accepted point of view shine forth the brighter once the philosophical fog has been removed. A greater difference can hardly be imagined.[111]

Where Feyerabend is concerned to promote alternatives to spur science to greater progress, Collins is more concerned that lay people may be misled because they lack tacit knowledge of whom to trust.

---

110. Collins, *Are We All Scientific Experts Now?*, p. 92.

111. Feyerabend, "Problems of Empiricism," p. 227, footnote 20, quoted in Oberheim, *Feyerabend's Philosophy*, pp. 70–71.

For Collins, this is like betting on the horses without knowledge that would increase one's odds of winning. Internet science critique represents a kind of social problem akin to addiction:

> No non-specialist should be able to get credit for aligning themselves, on technical grounds, with a maverick scientist who may, one day, turn out to be right. There are a huge number of mavericks and to align oneself with any one of them is nearly always simply a matter of taking a gamble with very, very, long odds. It makes no sense for ordinary people to take such a gamble. It makes very little sense for a specialist to take such a gamble unless they have some privileged source of information—either something to do with the science itself or the fact that the maverick scientists' work is being treated unfairly. But what specialists tend to know about mavericks, that it is hard for ordinary people to know, is the history of how their work has been treated in the scientific community.

For example, Collins points out that Joseph Weber lost credibility for his rejected claim to have measured gravitational waves in the 1970s so he was ignored when he returned with new claims in the 1990s. He was rejected without being read and legitimately so for Collins. But his claims would appear to be convincing to those outsiders who looked at this primary source publication.[112]

By contrast, Feyerabend argues that knowledgeable critics can use the generation of incommensurable perspectives to force consideration of anomalies shelved by the dominant perspective, what Hoyningen-Huene has called the anomaly importation thesis. In the terms we have discussed in chapter 3, the generation of an alternative approach that can predict and corroborate facts not recognized or explained by the dominant perspective forces a confrontation between the two perspectives, at least if one does not use empiricist philosophy of science to rule it out of court. The idea here is that the use of more than one theory to test theories *can allow a scientist to change his or her mind, rather than waiting for the Planck effect to weed out proponents of the dominant view. And it is precisely what Collins's epistemological regress, based in a Wittgenstein interpretation of science, does not allow you to do.*

The experimenter's regress ultimately roots closure of scientific debate in social causes that determine which side's view becomes the established one. Wittgenstein's sleight of hand does the work of conceptual change by reducing it to collective judgments of appropriate rule-following. Without social closure, one cannot have a definitive interpretation of a rule, for Wittgenstein:

> This was our paradox: no course of action could be determined by a rule, because every course of action can be made out to accord with the rule. The answer was: if everything can be made out to accord with the rule, then it can also

112. Collins, *Are We All Scientific Experts Now?*, pp. 85–93; Collins, *Gravity's Ghost*, pp. 8–11.

> be made out to conflict with it. And so there would be neither accord nor conflict here. It can be seen that there is a misunderstanding here from the mere fact that in the course of our argument we give one interpretation after another; as if each one contented us at least for a moment, until we thought of yet another standing behind it. What this shows is that there is a way of grasping a rule which is not an interpretation, but which is exhibited in what we call "obeying the rule" and "going against it" in actual cases.[113]

Carrying this over to controversy on the research frontier, Collins argues that one cannot know whether an experiment has been correctly done or not if it is not already established socially what is the correct interpretation of an experiment.

Like Wittgenstein's rules, this is the case because "usually, successful practice of an experimental skill is evident in a successful outcome to an experiment, but where the detection of a novel phenomenon is in question, it is not clear what should count as a 'successful outcome'—detection or non detection of the phenomenon."[114] Thus, scientists could readily credit the production of a TEA (transversely excited atmospheric pressure CO_2) laser because its workings were well understood and its functioning easily demonstrated through vaporizing concrete or the like. By contrast, Joseph Weber's attempts to detect gravitational radiation were disbelieved because he claimed to have found more radiation than cosmological theories predicted, so his experimental competence was doubted. For such a novel phenomenon, if scientists believe in its existence then they will credit the experimenter demonstrating it with having competently performed the experiment; if they do not, they will not. In this case, "the experimenters' regress can only be avoided by finding some other means of defining the quality of an experiment; a criterion must be found which is independent of the output of the experiment itself."[115]

That independent criterion that cannot be found in the outcome of the experiment itself can only be found in the taken-for-granted rules of the scientists' forms of life. Where novelty is introduced by a scientist, it can only gain a footing against the background of a greater degree of sociological order among the overlapping forms of life in which he or she participates. That is, the new interpretation "must be such as to seem to preserve most existing institutions; the new account must not appear to be completely unreasonable." The members of the core set, as "delegates" from different disciplines or schools, ultimately make this determination based on their judgments of plausibility, their interests, and the influence of material support, in the process jettisoning those who will not go along with the core set. These dissenters

113. Wittgenstein, *Philosophical Investigations*, § 201.
114. Collins, "Son of Seven Sexes," p. 34.
115. Collins, *Changing Order*, pp. 81, 84, 146.

are seen as "scientific non-persons . . . who, because of some individual pathology, are unable to accept the 'truth.'"[116]

This changing order is then returned to stability, as the accepted consensus becomes as taken for granted as testing the TEA laser against concrete. Collins calls this "the sociological resolution of the problem of inductive inference," connnecting this to a well-defined Wittgensteinian form of life:

> We perceive regularity and order because any perception of irregularity in an institutionalized rule is translated by ourselves and others as a fault in the perceiver or in some other part of the train of perception. Thus we arrange our perceptions into boxes. . . . If we do not care to do this, then others will not be able to communicate with us; they will treat us as though we spoke a private language.[117]

If you do not behave as a member, you will be excluded from the institution. Your attempts at communication will be rejected as pathological if they do not conform to the socially enforced rules, "the equivalent of a rude noise or complete nonsense."[118]

For Feyerabend, by contrast, comparison of two theories, each constituting distinct, taken-for-granted observation languages, actually facilitates comparison of the two under what Oberheim calls the pluralistic test model.[119] Feyerabend held his approach to be superior to Kuhn's defense of the dominance of single paradigms because comparison of incommensurable alternatives can make you change your mind. Changing one's mind is crucial for scientific change, especially scientific revolutions, which advance progress on Feyerabend's dialectical point of view.

It is the construction of an alternative point of view that suggests an alternative framework for interpreting scientific evidence that makes possible a change in mind by the scientist. As Heit and Oberheim remind us,

> It is important to keep in mind that for Feyerabend incommensurability was not the same as incomparability; rather, it merely indicates the lack of a common standard. Incommensurable theories cannot be internally related to one another; they can be compared only from a certain point of view, and this point of view should not be automatically viewed as superior, since its standards of comparison and evaluation are themselves always part of a worldview.[120]

---

116. Ibid., pp. 151, 148, 152.
117. Ibid., p. 147.
118. Ibid., p. 148.
119. Oberheim, *Feyerabend's Philosophy*, pp. 227–34. Oberheim (pp. 212–13) points out that a pluralistic test model was never a universal prescription for Feyerabend, as scientists in different contexts could legitimately ignore conflicting evidence to maintain a commitment to a given theory or could use incommensurable theories to compare alternatives.
120. Heit and Oberheim, "Paul Feyerabend, an Historical Philosopher of Nature," p. xix.

Feyerabend made this point to Kuhn in letters written to him before *The Structure of Scientific Revolutions* went to press, while both were at the University of California, Berkeley. These letters help clarify the view of incommensurability that Feyerabend had already developed in the draft of his soon-to-be-published "Explanation, Reduction, and Empiricism" (1962), which developed his contextual theory of meaning published in his "An Attempt at a Realistic Interpretation of Experience" (1958), based on material drawn from his 1951 dissertation criticizing the basic sentences of the logical positivists. Feyerabend's position at this time can be seen as a radicalized version of Popper's claim that new theories in science contradict older theories, as Kepler's laws cannot be deduced from Newton's. The view is radicalized because Feyerabend did not argue that incommensurable theories logically contradicted each other but that they could not be logically related to each other at all, though scientists could compare how the choice of a perspective offered practical benefits in generating new insights.[121]

The construction of an incommensurable alternative makes possible tests that would otherwise never have been considered by the dominant perspective, as Feyerabend explained to Kuhn using the case of Brownian motion. The key objection he had to Kuhn's notion of normal science and the dominance of single paradigms is that without the forcible intervention of an outside perspective, dominant paradigms would not break down of their own accord. According to Feyerabend, classical thermodynamics would never have been able to discover Brownian motion, with its construal of heat as based upon the motion of particles rather than the flow of energy, as it would continue to measure energy flow in a way that would rule out Brownian particles:

> Consider the physical situation: in order to discover in a direct way that the Brownian particle moves by absorbing heat from the surrounding medium one would have to be able to measure the heat disappearing from the medium, the changes of kinetic energy of the particle, the resistance overcome in the fluid (viscosity) and one would have to show that the heat lost by the medium balanced the gain of kinetic energy of the particle. It is quite obvious that the measurements involved in this procedure cannot be carried out as the thermometer would be itself involved in the fluctuations of temperature to be measured. Elaboration of the paradigm of phenomenological thermodynamics therefore will never be able to discover the anomaly presented by the existence of the Brownian particle. But this anomaly was discovered. It was discovered by the elaboration of an alternative account, viz. of the kinetic theory which then

121. Feyerabend's "Explanation, Reduction, and Empiricism" is simultaneously the crux of a priority dispute between Popperians and Feyerabend, and an argument that gets interpreted in two different ways by each camp, signifying the existence of what Hoyningen-Huene, Oberheim, and Andersen, "On Incommensurability," call meta-incommensurability. Compare Watkins, "Feyerabend among Popperians, 1948–1978," pp. 48–49, and Oberheim, *Feyerabend's Philosophy*, pp. 104–15.

> produced predictions that could be tested by experiment. Now I think that this is not just an isolated case. I think that very often anomalies can be discovered only in this way; i.e. not by further and further elaborating a given paradigm, but by elaborating an alternative paradigm and producing with its help testable predictions which, if they are confirmed, show that the first paradigm is in trouble. Your insistence upon faithfulness to one and only one paradigm is bound to result in the elimination of otherwise very important tests and it is bound in this way to reduce the empirical content of the paradigm you want to be accepted.[122]

On this way of looking at things, certain refuting facts are "hidden," which is why the construction of alternatives is necessary to promote progress, here considered in broadly falsificationist terms.[123]

At this time, Feyerabend saw the need to promote tests that would counter the tendency of theoretical systems to drive out empirical content, which the logical positivist view would only encourage. Kuhn's approach would also do so since "[t]he practice of repressing novelties, the attempt to let the theory penetrate every part of nature (and of thinking) usually leads to a situation where people are *psychologically* unable to imagine alternatives."[124] While the focus here was on theories that applied to nature as a whole, Feyerabend's later development of incommensurability pushed the point further, as not only general theories but also worldviews, art forms, perceptions, mythologies, theologies, and complex cultural practices are subject to incommensurable alternatives that would not break down of their own accord.[125] Common to both Feyerabend's early period as a kind of radicalized falsificationist and his later role as a critic of science and defender of alternative traditions of knowledge is the claim that dominant perspectives do not break down on their own but need to be pushed to do so by outside criticism.

Feyerabend saw Kuhn's *Structure of Scientific Revolutions* as an "ideology covered up as history" because it disguised his normative framework by seeing normal science as simply given by the historical facts. Kuhn essentially endorses the dogmatism of the group as desirable, as a methodological precept demonstrated by the history of science and a practice that was functional for science. He objects that Kuhn "seems to think it psychologically impossible for a scientist and, indeed, for any human being to be able to entertain various alternative hypotheses and to discuss them impassionately."[126]

---

122. Hoyningen-Huene, "Two Letters of Paul Feyerabend to Thomas S. Kühn," p. 365.
123. Feyerabend, "Problems of Empiricism," develops this argument. For criticism of this case and the development of a modified role for alternative theories in uncovering refutations, see Laymon, "Feyerabend, Brownian Motion, and the Hiddenness of Refuting Facts."
124. Hoyningen-Huene, "Two Letters of Paul Feyerabend to Thomas S. Kühn," p. 358.
125. Feyerabend, *Against Method*, p. xi–xii; Feyerabend, *The Tyranny of Science*, pp. 3–31.
126. Hoyningen-Huene, "Two Letters of Paul Feyerabend to Thomas S. Kühn," p. 356.

By contrast, Kuhn ignores the tendency for good intentions to go astray, as the pursuit of science does not necessarily comport with its initial aims but may show evidence of degeneration as an institutional practice forms. Thus, appealing to normal science may only indicate how far afield scientists have come from their initial aim and approach.[127] Explicitly introducing an alternative perspective may reform the institution by bringing it back into contact with nature after the imposition of a dogmatic perspective had reduced its empirical content.[128]

Crucial to the ongoing viability of science as an institution, then, is its permeability to new perspectives, something that philosophers of science can promote by helping force consideration of different points of view when the institution of science blocks it. Incommensurable theories can be compared by scientists who are not in fact oversocialized and overintegrated by their membership in scientific forms of life.[129] Even Kuhn allowed by 1969, in the second edition of *The Structure of Scientific Revolutions*, that shared values like accuracy, simplicity, and fruitfulness could ensure that incommensurable paradigms were compared by scientists, possibly leading to eventual agreement that one view was better than the other in these terms.[130]

For Feyerabend, scientists could be bilingual and view the old theory in terms of the new, creating an alternative basis for interpreting the evidence, though it failed to fit empiricist philosophical assumptions. Scientists, according to Feyerabend, are actually good at comparing incommensurable theories by comparing them in practice—code-switching in linguistic terms—and examining their effects, as when physicists employ separately both Newtonian and relativistic conceptions of mass. While a lack of bilingualism by scientists may cause misunderstanding and failure to communicate, only philosophers see logical incompatibility because they mistakenly believe science is a collection of statements that are invariant in meaning rather than a human practice where theoretical comparison proceeds via "an open, ambiguous, and often contested manner."[131]

D'Agostino puts the point in a slightly different way, emphasizing the way that empiricist philosophy of science obscures the practical competence of working scientists who are more flexible and pragmatic at comparing perspectives:

---

127. Feyerabend, *Against Method*, pp. 355–56.

128. Compare Hoyningen-Huene, "Paul Feyerabend and Thomas Kuhn," p. 112, and Oberheim, *Feyerabend's Philosophy*, pp. 245–53.

129. Wong, "The Oversocialized Conception of Man in Modern Sociology," pp. 184–87.

130. Kuhn, *The Structure of Scientific Revolutions*, pp. 199–200.

131. Oberheim, *Feyerabend's Philosophy*, p. 131, quoting the third edition of Feyerabend, *Against Method* (1993), p. 211.

> In particular, we will see that neither Kuhn nor Feyerabend, despite what their critics often say about them, held that incommensurability of theories implies the rational incomparability of theories. This is especially clear in the case of Feyerabend, whose argument is NOT that theories in a scientific tradition are on occasion incommensurable, but, rather, that, when the relations of theories in a tradition are represented in a particular way, they may on occasion be incommensurable according to that representation and hence incomparable if that representation is taken as providing the mechanisms of comparison. And the point of this claim is not to establish something about science, but, rather, to establish something about the representations of science which yield this result (i.e., that two theories might be incommensurable). Feyerabend in other words invokes incommensurability (according to the standards of a particular representation) as a reductio of that mode of representation. And this argument in fact depends precisely on the comparability of theories which are, according to the representation, incommensurable. Feyerabend's argument is about the ways in which we should understand progress in science and he is concerned, in particular, to establish that a historically informed approach is superior to an approach which, if applicable, is applicable only to what he calls "abstract traditions."[132]

Feyerabend's abstract traditions are self-referential and closed to external engagement and should be opened up by the construction of theories that force comparison of views that are blocked by Kuhnian normal science. Feyerabend's point was that this kind of breaking down of monistic perspectives manifestly had taken place in the history of science and that it *should* take place whenever the dogmatism of experts seemed to have blocked progress.

It is here that discussion of the revival of low-carb diets and the science addressed to it provides a salient example. The point of this example is not that low-carb diets are necessarily superior to low-fat or low-saturated fat diets. Rather, we can see how an outsider to the debate, science journalist Gary Taubes, managed to *force an argument*, where before there was neglect of the minority position. Doing so can lead to progress by unblocking development and allowing closed questions to be reopened. As we will see, a case can be made that premature closure of the debate had occurred through both self-selection by certain experts and its reinforcement by a political process of seeking dietary guidelines to be promulgated by the U.S. government. The limitation of the recognition of expertise to those recognized by their peers who employ their tacit knowledge to recognize who else can be held to possess it—the circular process evident in Collins's analysis that I have called the self-warranting view of expertise—was the main stumbling block to a fairer debate. The result may have been that experts subjected the public to

132. D'Agostino, "Verballed? Incommensurability 50 Years On," p. 517.

a dangerous dietary experiment that encouraged the consumption of excess carbohydrates as a substitute for dietary fat.

CARBS, FATS, AND THE GREAT DEBATE OVER DIETARY HEALTH

In his popular 2008 book, *Good Calories, Bad Calories: Challenging the Conventional Wisdom on Diet, Weight Control, and Disease*, science journalist Gary Taubes made a case that certain minority views within science were being neglected and that these lines of research were uncovering new insights not recognized by the dominant approach. While not averse to pointing out anomalies in the dominant approach, Taubes spent much of his time demonstrating that these neglected approaches were fruitful research programs. In Lakatosian terms, they were progressive to the extent that they generated novel, confirmed predictions.

Moreover, like Feyerabend's take on Galileo, Taubes made the case that this research needed additional support to carry out promising lines of research that had been marginalized by the dominant approach. In effect, minority approaches needed breathing space to present their case and required explicit financial support on the promissory note of what they had been able to accomplish so far. An alternative research program existed that focused on carbohydrates' effects on insulin but was neglected compared to the dominant program focused on saturated fat as the cause of disease, the diet-heart hypothesis. The key idea is that the alternative research was needed to make such anomalies salient by showing that an alternative existed that explained the anomalies as a consequence of its own development as a research program, rather than anomaly mongering.

Notice that this hits at the core of the disagreement between Kuhn and Feyerabend. For Kuhn, normal science was usually a *functional* component of science in that it efficiently advanced knowledge except where the dominant paradigm had begun to break down and a new perspective was needed. Unless a revolution was in the offing—which scientists themselves could handle—the diet-heart hypothesis should be accepted as authoritative, and working out its puzzles should proceed without interference or foundational criticism. For Feyerabend, the goal was to bring about revolutions quicker, so normal science impeded that whereas the proliferation of alternatives accelerated the process. In this sense, normal science was dysfunctional for science and represented a possible breakdown in the institutional mechanisms that in the past were functional for science.[133]

133. Hoyningen-Heune, "Paul Feyerabend and Thomas Kuhn," pp. 111–12.

In that case, we can conclude that if "alternative paradigms are needed in order to accelerate the development that in the end leads to scientific revolutions, then there must exist *scientific* institutions which are capable of dealing with polarizations that may arise in the scientific community and which prevent these polarizations to become 'incompatible modes of community life.' A theory of such institutions is as yet still missing."[134] Rather than developing a model that traces how scientific institutions vary in their ability to promote puzzle-solving progress and revolutionary changes, and how these differences would reverberate through the history of science, constructivist sociologists have developed their own version of a cryptically normative defense of autonomous science.

It is likely that Feyerabend overstated the functional importance of scientific revolutions vis-à-vis "normal science." While non-revolutionary science is never truly monistic, as Lakatos and Feyerabend have argued, simply maximizing revolutions could well prevent the efficient improvement of knowledge and suffer from its own functional deficits. What is clear is that post-functionalist sociology of science and post-positivist philosophy of science neglect the question of what institutional features of science are conducive to the growth and progress of knowledge. The development of an alternative model "for dealing with polarizations that may arise" is needed (see the discussion of Merton in chapter 7).

Much like Kuhn's defense of normal science as normally functional for science, Collins and Evans's model of expertise tries to deal with such polarizations through a downward discriminating model, as we have seen. Experts know more than laypeople, even lay activists, while within the core set, those who really know better also know whom to ignore within the core set. Collins and Evans challenge the concept of lay expertise and its role in science criticism by outsiders like Taubes. Central to their approach is a taxonomy of kinds of expertise that distinguishes between the contributory expertise of the practicing scientist and the merely interactional competence of the competent lay critic. Interactional expertise is its own kind of specialized expertise to be sure, and all contributory experts—who conduct research in a field—also possess interactional expertise that allows them to engage their colleagues in other sub-specialities in a shared language. However, journalistic critics like Taubes, and participant-observing sociologists like Collins, are interactional experts but not contributory experts.

This is a key point emphasized by Collins and explains the limit beyond which outsiders cannot go. Science criticism based upon interactional exper-

134. Hoyningen-Huene, "Two Letters of Paul Feyerabend to Thomas S. Kühn," p. 375. Feyerabend's reference to "incompatible modes of community life" quotes Kuhn's manuscript of *The Structure of Scientific Revolutions*, p. 94.

tise is not appropriate for Collins. Launching an outside critique based on extensive research and engagement with primary source literature, even with close interaction with contributory experts, cannot overcome the self-warranting nature of expert communities. When Taubes appeals to the research program of nutritionist John Yudkin or the observations of clinical physicians George Campbell or Thomas Latimore Cleave on sugar, he cannot know that they are experts that deserve—or still deserve—credit.[135] Contributory experts in the field can exercise downward discrimination based upon their tacit knowledge of who are trustworthy experts in the field. Outsiders like Taubes cannot do so, because their discrimination would be "upward" and would require tacit knowledge of who is doing good work in the field, something that itself is absorbed from long experience in the field.

Key to their defense of self-warranting communities is the assumption that knowledge is the sort of thing that is possessed by individual members of a scientific community, but not easily shared except by a process of apprenticeship designed to transfer tacit knowledge. Such a view makes criticism impossible for those not sharing the font of tacit knowledge at issue (or at least not possessing a different set of tacit knowledge of their own that is relevant to core problems in the field). Another approach, however, considers knowledge to be the kind of thing that involves representations of some kind *detached* from bodies or brains.

In viewing knowledge as primarily tacit knowledge, Collins and Evans put their focus on who possesses knowledge. Knowledge is a kind of social phlogiston that can flow between contiguous bodies given enough time and exposure. Indeed, they consider just how much of a body one needs in order to know anything at all in this sense in a side dispute with phenomenologist Hubert Dreyfus's view that full embodiment is necessary. Likewise, Collins has argued that machines can't know in the way that humans can, so in this sense their view is a fully humanistic picture of knowledge, where knowledge is a particular possession of humans that cannot be readily replicated by artificial intelligence.[136]

By contrast, a dialectical approach would see knowledge in the first instance as disembodied and socially distributed—in minds, databases, books, journals, institutional routines, interactional practices, and the like. Such products are produced by a certain kind of human engagement with the world; however, the products of a scientist's labor become objectified in the course of scientific work and alienated through the division of labor and the introduction of labor-saving and labor-enhancing instruments as well (see

135. Taubes, *Good Calories, Bad Calories*, pp. 108–21.
136. Collins, *Artificial Experts*.

chapter 5). On this view, there might not be any individuals who know, who can encapsulate and articulate the knowledge gained in a domain.[137]

Knowledge even in a narrow specialty may exceed what any one individual can possess. Extracting what knowledge we possess as a species may not be as easy as asking the relevant experts, since experts may have all kinds of blind spots and may have to rely upon all manner of constructed automata that factor into any practical use of knowledge. On this view, knowledge would tend to be alienated even from the contributions of those who helped construct it. As is clear from our discussion so far, this view follows Hegel rather than Polanyi as an inspiration. Now, it is enough to notice that such a view carries with it a very different idea of what normative evaluation of expert knowledge entails than the tacit-based picture of knowledge, with its asymmetry of downward and upward discrimination. Here we consider the role of lay critics in challenging the diet-heart hypothesis connecting (saturated) fat consumption to heart disease.

It is not that we are all scientific experts, spontaneously able to criticize and judge the experts. However, outside critics can call attention to neglected research traditions and evaluate fruitful directions free of the limitations—and the advantages—of the tacit judgments of core experts. After all, tacit judgments of peers who deserve attention can also be described as stereotypes and prejudice. Tacit knowledge may give scientists a deep understanding lacking to those without their experience, but it also, at the same time, can be "an obstinate addiction, to the forms of some private life," as Thomas Sprat put it in the epigraph to this chapter.[138] Core experts have manifestly dismissed minority positions in nutritional science, as well as the critique put together by Taubes, but that, by itself, is not sufficient grounds for rejecting their alternative views when a promising track record and a potential program for further research can be spelled out.

The public dispute about Gary Taubes's recommendations have been extensive, among scientists, journalists, and bloggers. He has been accused of rejecting a diet—lowering fat to lose weight and prevent heart disease—for lack of evidence, then endorsing a low-carb diet that is equally unproven by the evidence. Similarly, he has been accused of mining the scientific literature in a way that is skewed by confirmation-bias: he endorses studies that support his point of view and ignores or rejects those that do not. Moreover, he has been accused of focusing on a causal mechanism that is simplistic and reductionist in identifying carbohydrate consumption with insulin rise and fat

---

137. Bartley, *Wittgenstein*, p. 179, summarizes this point as the recognition that "we never know what we are talking about." See also Bartley, *Unfathomed Knowledge, Unmeasured Wealth*.

138. Sprat, *History of the Royal Society*, pp. 676–67.

storage in turn. Considered as an example of the dangers of "citizen science," as Collins describes it, he could be a clear example of its dangers.

Considered another way, Taubes can be seen as someone who has developed interactional expertise on the subject, with a special focus on minority views that are only beginning to gain a hearing in science. In this case, he is reviving minority positions that were never completely vanquished from science, much as the AIDS activists studied by Epstein did with regard to clinical trial design. Indeed, in response to charges of lack of evenhandedness, he argues that his purpose is to make the case for the neglected view:

> The point of writing the [initial] article was precisely to note that some equally respectable scientists question this dogma, and then to explain why. I defined the state of the argument as having undergone "a subtle shift in the scientific consensus" over the past five years: "It used to be that even considering the possibility of the alternative hypothesis, let alone researching it, was tantamount to quackery by association. Now a small but growing minority of establishment researchers have come to take seriously what the low-carb diet doctors have said all along.[139]

In this sense, Taubes could be the spokesperson for a minority position in the core set. The idea that the core set's final consensus is only the result of core-set negotiations by experts who recognize each other is problematic. In this case, a political process of establishing dietary goals influenced the experts selected, via hearings initiated by Senator George McGovern and leading to the publication of *Dietary Goals for the United States* in 1977.[140] Moreover, this skewed process exploited a situation where scientists taking different approaches failed to recognize each other as part of a common core set. Often, in scientific controversies, there are dueling core sets that compete for influence, in this case taking their case to the public in an ongoing war of position.[141]

Taubes's case begins with a discussion of the ebb and flow of research programs on obesity over the past two centuries. He discusses the rise of "banting," the low-carb craze of the nineteenth century, based upon the weight

139. Taubes, "An Exercise in Vitriol Rather Than Sound Journalism."

140. Taubes, *Good Calories, Bad Calories*, pp. 9–10.

141. Hamlin, "Third Wave Science Studies," pp. 171–72. See the counteroffensive against Taubes by Fumento, "Big Fat Fake," who defended orthodoxy in *Reason* magazine, a venue for criticizing pseudoscience. Fumento accused Taubes of ignoring scientific studies conflicting with his thesis and distorting the testimony of scientists he interviewed. Fumento's ire is directed primarily at a subsidiary issue in the controversy, whether low-carb diets are superior to weight loss compared with *isocaloric* low-fat diets. Taubes, *Good Calories, Bad Calories*, chs. 16–17, argues that calories are not equivalent in their effects on the body, with high-carb diets preferentially leading to fat storage, while Fumento argues that diets are only effective if they lower calorie consumption whatever the source. Some low-carb proponents adopt a middle position: cutting calories are what matters but low-carb diets are more satiating so calorie deficits are more likely to be maintained.

loss of William Banting who avoided sweets and starch. His 1863 pamphlet, *Letter on Corpulence, Addressed to the Public*, details the advice of his doctor, William Harvey, who had attended a lecture of the physiologist Claude Bernard identifying the accumulation of glucose in the blood as the cause of diabetes. Taubes argues that reducing starch and sugar continued to be widely advocated in medicine and popular culture until a shift toward targeting fat in general, or saturated fat specifically, by the 1970s.

The shift was facilitated by identification of saturated fat as the cause of heart disease in Ancel Keys's work in the 1950s, facilitated by a metaphor of saturated fat as "clogging arteries" like grease in plumbing pipes. The unhealthiness of saturated fat led investigators to focus on limiting calories, which further led to a focus on fat, which has more calories per gram than carbohydrates or protein. The shift away from low-carb diets was a by-product of new ideas about heart disease, not obesity. Moreover, when Keys and other researchers first identified blood cholesterol levels as a risk factor for heart disease, they did not yet have reliable laboratory tests to differentiate its components, including HDL and LDL cholesterol ("good" and "bad" cholesterol, respectively) and triglycerides. When the role that glucose played in influencing unhealthy levels of these blood lipids emerged along with an understanding of insulin resistance, focus did not shift back to carbs per se but refined carbs and the need for fiber.[142]

As historian of science Anne La Berge shows, the new emphasis on saturated fat was judged authoritative as the result of a questionable reliance upon one camp of researchers on the McGovern committee. The result was that after 1980, correlation between dietary fat and blood cholesterol levels led to the demonization of fat as an "overarching ideology" while the obesity epidemic was in full swing. Her identification of low-fat diets as part of an overarching ideology reflects the extent to which the diet-heart hypothesis was reinforced by particular U.S. cultural trends such as an established dieting culture among American women and mass media promotion of low-fat diets. Calorie counting was preferred in women's magazines, which amounted to a low-fat approach as dieters knew that there were nine calories per fat gram compared to four calories per carbohydrate or protein gram. Personal control of bodily form had been achievable by a quantitative focus on calorie consumption and weighing oneself on newly available scales before scientists began to investigate the topic in earnest. Insurance companies compiled statistics of healthy weights by the mid-twentieth century, believed to be applicable over the course of a lifetime. Scientists took from the broader culture

142. Taubes, *Good Calories, Bad Calories*, pp. ix–xxi, 122–35; Olson, "Discovery of the Lipoproteins, Their Role in Fat Transport and Their Significance as Risk Factors"; Lustig, "Prophecy and Propaganda," pp. xiii–xiv.

a set of assumptions about how behavioral change would, or should, lead to weight loss.[143]

In 1957, the American Heart Association (AHA) took one side in the scientific debate, endorsing Keys's claim that a low-fat diet would lower the incidence of heart disease. The advice, targeted to those with certain risk factors for heart disease, was gradually extended—first, to those who were overweight and then to all. The AHA advocated the replacement of animal fats by vegetable oils and margarine in 1961, supported by vegetable oil trade groups, but the low-fat mania did not take off until after the publication of the McGovern report in 1977. Staffers for McGovern were unaware of scientists opposed to Keys and the low-fat view, so the composition of the committee was skewed.[144]

The drive for low-fat was reinforced by political movements for vegetarianism launched in particular by Frances Moore Lappé's 1971 blockbuster, *Diet for a Small Planet*, which had displaced earlier health movements looking to revive traditional diets reliant on unprocessed meat and plant foods or that sought to recover hunter-gatherer ways of eating.[145] Portrayal of one side in a still-open, scientific controversy as scientific truth by the U.S. government shifted the presumption of evidence thereafter and brought about a radical, population-level experiment on the American public, as consumption of vegetable oils increasingly replaced animal fats. The American Medical Association registered its objections at the time, essentially appealing to the precautionary principle urging restraint in conditions of scientific uncertainty (see chapter 6), arguing that "there is a potential for harmful effects for a radical long term dietary change as would occur through adoption of the proposed national goals."[146]

By contrast, early research programs identifying sugar as a risk factor for heart disease and obesity date to the 1950s. At that time, physiologists John Yudkin and Ancel Keys put forth competing hypotheses on the causes of coronary heart disease. Yudkin argued that dietary sugars increased the risk of heart disease, while Keys put the blame on fats. Much of the emerging

143. La Berge, "How the Ideology of Low Fat Conquered America," pp. 139–43. See also Yager, *The Hundred Year Diet*.

144. La Berge, "How the Ideology of Low Fat Conquered America," pp. 145–47; Taubes, *Good Calories, Bad Calories*, pp. 45–47.

145. Lappé, *Diet for a Small Planet*; Taubes, *Good Calories, Bad Calories*, ch. 3; Price, *Nutrition and Physical Degeneration*; Audette, *Neanderthin*; Stefansson, *The Fat of the Land*; Shepard, "The Corvidean Millennium," pp. 339–41. The irony is that the scientific case for a "Paleo diet" had emerged at the same time, although initially even Paleo proponents emphasized lean meats rather than ones high in saturated fat and cholesterol. See Eaton and Konner, "Paleolithic Nutrition"; Cordain et al., "Plant-Animal Subsistence Ratios and Macronutrient Energy Estimations in Worldwide Hunter-Gatherer Diets"; Kuipers et al., "Estimated Macronutrient and Fatty Acid Intakes from an East African Paleolithic Diet"; Taubes, *Good Calories, Bad Calories*, p. 69.

146. Quoted in Taubes, *Good Calories, Bad Calories*, p. 47.

research on sugar was actively suppressed by the sugar industry in a case parallel to that of tobacco companies suppressing research linking cigarettes and cancer. The Sugar Research Foundation (SRF), founded in 1943 and later renamed the Sugar Association, sought to challenge research emerging by the 1960s that sugar could elevate cholesterol levels. The SRF convinced Fredrick Stare, chair of the Harvard University Nutrition Department, to oversee a review of the literature, titled "Carbohydrates and Cholesterol Metabolism," authored by D. Mark Hegsted and Robert McGandy and published in the *New England Journal of Medicine* in 1967. Internal documents show that the SRF was explicitly concerned to counter studies of sugar's role in elevating serum cholesterol levels and that Hegsted and McGandy were paid $6,500 by SRF for the publication. The published review did not disclose industry funding and questioned the data and interpretations of the studies carried out by Yudkin and his colleagues.[147]

Yudkin settled a libel suit against the World Sugar Research Organization when their *Quarterly Bulletin* referred to his book as a work of "science fiction."[148] Yudkin details both industry-sponsored attempts to edit or censor his work, via the industry-funded British Nutrition Foundation, and efforts by core-set researchers to exclude him from conferences and to reject his argument as emotional, lacking evidence, and even fraudulent.[149] Keys criticized Yudkin's hypothesis as "tendentious," "propaganda," and a "mountain of nonsense," seeing his treatment of evidence as selective and "flimsy," the same objection that critics had made of Keys's famous correlational Seven Countries Study, which had provided the (selective) basis for identifying saturated fat as the cause of heart disease.[150] Robert Lustig's work on the role of fructose in metabolic syndrome has faced similar criticism.[151]

Taubes's role in goading the medical establishment to better research low-carb diets shows the extent to which a self-taught "interactional expert" can help revive neglected but promising research programs like the carbohydrate-insulin hypothesis. His broadsides in the *New York Times Magazine* and *Science* elicited positive and negative reactions from researchers, bringing neglected perspectives back into debate. In September 2012, Taubes established

147. Kearns, Schmidt, and Glantz, "Sugar Industry and Coronary Heart Disease Research," pp. E1–E4.

148. Yudkin, *Pure, White, and Deadly*, pp. 189–92.

149. Ibid., pp. 192–212.

150. Keys, "Sucrose in the Diet and Coronary Heart Disease," pp. 193, 200; Keys, "Coronary Heart Disease—the Global Picture," p. 175; Keys, *Coronary Heart Disease in Seven Countries*; Taubes, *Good Calories, Bad Calories*, pp. 119–21, 31–33; Yudkin, *Pure, White, and Deadly*, p. 210.

151. Lustig, Schmidt, and Brindis, "The Toxic Truth about Sugar"; Khan and Sievenpiper, "Controversies about Sugars"; Khan and Sievenpiper, "Metabolic Improvement with Fructose Restriction"; Lustig, "Response to 'Metabolic Improvement with Fructose Restriction'"; Teicholz, *The Big Fat Surprise*, ch. 2; Blackburn et al., "Review of *Big Fat Surprise* Should Have Questioned Author's Claims"; Teicholz, "Response to Critique of Review of the *Big Fat Surprise*."

the Nutrition Science Initiative with researcher Peter Attia to raise money to carry out long-term, randomized clinical trials comparing low-fat and low-carb diets. The thought was that they would create a "Manhattan project for nutrition" that would carry out costly, but definitive studies that would allow nutrition science to make fundamental progress. The foundation raised $40 million from the Laura and John Arnold Foundation (with money John made as an Enron trader) and sought to crowdsource more. They brought on board researchers independent of Taubes's view to establish credibility and carried out a pilot study that led to disagreements about methodology and academic freedom between the researchers and Taubes. The organization fell apart and research results failed to establish Taubes's holy grail: that low-carb diets would have a metabolic advantage over isocaloric low-fat diets. They did produce research supporting low-carb proponents' claims that the diet could lead to weight loss without clear evidence of coronary heart disease risk.[152] Additional research was funded directly by the Arnold Foundation, which also sought to fund a variety of science reformers concerned about a potential crisis in reproducibility and prevalence of fraud in science (see chapter 7).[153]

It was not until 2003 that long-term studies of low-carb diets first started coming in, suggesting that low-carb diets worked about as well as low-fat diets for losing weight, and often improved HDL and triglyceride levels. Studies emerging from NuSi and the Arnold Foundation showed similar results.[154] Given the rough equivalence in the effects of low-carb and low-fat diets on weight loss, low-carb proponents began to highlight evidence that shows that low-fat diets can improve risk factors for heart disease only if weight is reduced. By contrast, low-carb diets seemed to improve risk factors even in the absence of weight loss, suggesting that carbohydrates' effects on insulin are dangerous.[155] This line of evidence counters the claim that has been made frequently since the 1970s that the Atkins diet was "dangerous," whatever its effectiveness as a weight loss tool. Atkins's own death became politicized with his wife insisting that he had slipped on ice, while leaked autopsy records showed he had had a heart attack and was obese.[156] There remains continuing controversy about how to interpret recent studies, but it is fair to

152. Molteni, "The Struggles of a $40 Million Nutrition Science Crusade"; Hall et al., "Energy Expenditure and Body Composition Changes after an Isocaloric Ketogenic Diet in Overweight and Obese Men"; Guyenet, "Nutrition Science Initiative (Nusi) in Retrospect."

153. Apple, "John Arnold Made a Fortune at Enron."

154. La Berge, "How the Ideology of Low Fat Conquered America," pp. 164–65, 171–72; Hu et al., "Effects of Low-Carbohydrate Diets versus Low-Fat Diets on Metabolic Risk Factors"; Gardner et al. "Effect of Low-Fat vs Low-Carbohydrate Diet on 12-Month Weight Loss in Overweight Adults and the Association with Genotype Pattern or Insulin Secretion."

155. Feinman and Volek, "Low Carbohydrate Diets Improve Atherogenic Dyslipidemia Even in the Absence of Weight Loss"; Krauss at al., "Separate Effects of Reduced Carbohydrate Intake and Weight Loss on Atherogenic Dyslipidemia."

156. Yager, *The Hundred Year Diet*, pp. 197–201.

say that the low-carb hypothesis has been revived as a competing research program.[157] The emphasis on the impact of sugars on insulin, in particular, has received much attention even for scientists who do not advocate a low-carb diet as a solution. On the other hand, there had been a move to lower saturated fats among some low-carb diets, such as the South Beach diet, an approach followed by Atkins's dieticians, who advised doctors directly; this approach was not, however, included in a 2002 update to his popular book. Currently, the Atkins diet emphasizes that saturated fat is only harmful in combination with a high-carbohydrate diet.[158]

Other researchers focus on factors distinct from both the impact of saturated fat and from Taubes's focus on insulin, including the role of polyunsaturated fatty acids (PUFAs) in vegetable oils or other aspects of the endocrine and neurological systems that control weight or that reset the "set point" that the body maintains against dietary changes.[159] In general, there has been a shift away from a focus on behavior as a cause—eating less and exercising more to lose weight—to behavior as an effect. For instance, eating more and exercising less is now seen as an *effect* of leptin resistance rather than an independent cause of weight gain.[160] This is especially significant because voluntary compliance with *any* diet is limited, particularly over the long term. The extent to which U.S. dietary culture over the past century or more has reinforced researchers' assumptions that successful weight loss depended upon will power and conscious behavioral change neglected this question of the causes of non-compliance despite the onslaught of expert and governmental propaganda.[161]

Finally, efforts continue to put dietary science in an evolutionary framework, where cultural mismatch between our evolved characteristics in ancient

157. Seid and Rosenbaum, "Low Carbohydrate and Low-Fat Diets."

158. Yager, *The Hundred Year Diet*, p. 200; "The Saturated Fat Paradox," https://www.atkins.com/how-it-works/library/articles/the-saturated-fat-paradox.

159. Muhlhausler and Ailhaud, "Omega-6 Polyunsaturated Fatty Acids and the Early Origins of Obesity"; Yeo and Heisler, "Unraveling the Brain Regulation of Appetite"; Schwartz et al., "Obesity Pathogenesis: An Endocrine Society Scientific Statement"; Rosenbaum et al., "Low Dose Leptin Administration Reverses Effects of Sustained Weight-Reduction on Energy Expenditure and Circulating Concentrations of Thyroid Hormones."

160. Jung and Kim, "Molecular Mechanisms of Central Leptin Resistance in Obesity." In *Good Calories, Bad Calories*, pp. xxii–xxiii, Taubes mentions leptin once, as part of the complex "mechanisms of weight regulation and disease" that can be set aside when considering their "fundamental *causes*," which need not be complex. For Taubes, "if refined carbohydrates and sugars are indeed the reasons why we fatten—through their effect on insulin and insulin's effect on fat accumulation—and if our goal is to prevent or remedy the disorder, the salient question is why any deeper explanation, at the moment, is necessary." Taubes's focus on the idea that carbs drive insulin levels which drive fat storage is as limited in the long run as the idea that saturated fats "clog" arteries. See Hall, Guyenet, and Leibel, "The Carbohydrate-Insulin Model of Obesity Is Difficult to Reconcile with Current Evidence."

161. La Berge, "How the Ideology of Low Fat Conquered America," pp. 168–72.

environments and current obesogenic environments is key.[162] The ready availability of highly palatable food exploits the brain's reward circuits, altering homeostatic mechanisms in ways that cause the body to defend greater body fat as a set point.[163] Tellingly, the diet change that is most likely to be highlighted by those emphasizing that the contemporary food system exploits our physiology with exaggerated, "supernormal" stimuli is the combined presence of fat and starch in the same meal.[164]

The process by which fat consumption and poor health outcomes like heart disease and obesity were linked reveals a number of instances in which the social historical process of consensus formation could be subject to questioning for rigging the issue of who is a member in good standing of the scientific core set. (1) Early epidemiological studies like those by Ancel Keys introduce *path dependence* to the development of the field, as preliminary epidemiological findings in his Seven Countries Study shaped the conceptual framework for future research. (2) The diet-heart hypothesis benefits from the *substitution* of laboratory measurement of serum cholesterol for health outcomes believed to be linked to it. (3) The *uneven development* of scientific research programs affects the controversy, as laboratory measures and understanding of triglycerides, component lipoproteins of cholesterol, and insulin lag. (4) The *political construction of consensus* allows a skewed representation of the core set to establish itself as the set of recognized experts and distort or delay contributions from competing programs, what Hamlin called "core-set imperialism."[165] As we saw above, Feyerabend put this point to Kuhn: "alternative paradigms are needed in order to accelerate the development that in the end leads to scientific revolutions." If so, "then there must exist *scientific* institutions which are capable of dealing with polarizations that may arise in the scientific community and which prevent these polarizations to become 'incompatible modes of community life.'"[166]

162. Ball et al., "Eating Behaviors and the Food Environment"; Giskes et al., "A Systematic Review of Environmental Factors and Obesogenic Dietary Intakes among Adults"; Gluckman and Hanson, *Mismatch*, ch. 1; Cordain et al., "Origins and Evolution of the Western Diet."

163. Guyenet and Schwartz, "Regulation of Food Intake, Energy Balance, and Body Fat Mass."

164. Barrett, *Waistland*, ch. 3; Barrett, *Supernormal Stimuli*; Gremillion, *Ancestral Appetites*, pp. 146–47.

165. Hamlin, "Third Wave Science Studies," pp. 171, 172.

166. Hoyningen-Huene, "Two Letters of Paul Feyerabend to Thomas S. Kühn," p. 375.

Chapter Five

The Place of Science and the Turn to Material Practice

A Lab of One's Own for Patient Activists

Laboratories are where the relationships between recalcitrant, material reality and malleable, social constructions are forged. If early SSK extended sociological explanation to science via Wittgenstein's concept of a form of life, the social dimension of science was understood to ensure that scientific communities shared certain tacit assumptions about how to proceed. The influence of Wittgenstein on the strong program and the empirical program of relativism (EMPOR), in particular, placed the emphasis on how communities shape the interpretation of scientific concepts by socially enforced rule-following. This represented a distinctive conception of what constituted anything that was "social," emphasizing communally shared beliefs. If knowledge was inherently social, then belief-formation had to be understood by reference to the consensual rules of the game within any community.

With the introduction of a semiotic perspective in Latour and Woolgar's *Laboratory Life* (1979), and its development into actor-network theory (ANT), a different conception of the social emerged that sees the social dimensions of science developing from the play of signifiers and signified that make up any meaningful unit of communication (a sign). Signifiers are material things (objects, letters, sounds) that signify—have meaning—by virtue of their connection to the signified, the concept associated with the material signifier. As applied to the laboratory setting, this approach sees the point of experiments to be to construct inscriptions—material traces—that can be used to represent natural processes, eventually resulting in scientific publications. Scientific laboratories are ultimately places where a host of complex practices are carried out to produce meaningful and convincing scientific statements as output.

The process uses laboratory equipment understood as "inscription devices" that "transform pieces of matter into written documents."[1] Statements initially are accompanied by "modalities" that modify statements by contextualizing the putative claim about the world, qualifying their validity or specifying their probability.[2] The ultimate goal is to produce textbook statements that are free of modalities and are accepted as unqualified statements of fact about the natural world. Actor-network theory extended the network of material activities designed to support scientific representations beyond the laboratory walls.[3]

Key to this approach is the assembling of networks of social and material objects in ways that support scientific representations, which means that there is no realm of the "social" set off from the material objects of the natural world. To be sure, material objects manipulated through laboratory activities have been processed through scientific work and no longer represent natural entities in their unmodified state. This means that science is "constructed" in a literal sense, like manufactured objects are constructed in factories. It also means that science is not the production of an autonomous "social" realm, and this led Latour and Woolgar to change the subtitle of their work from "The Social Construction of Scientific Facts" to "The Construction of Scientific Facts" in the second edition of *Laboratory Life* (1986).[4] With a shift from shared, communal beliefs to practical activities within a laboratory setting designed to mobilize human and non-human entities, the focus became the fine-grained activities of scientists at the work bench (even if Latour and Woolgar made the point primarily by the thought experiment of a naïve observer ignorant of scientific culture confronting laboratory activities).[5]

Increasingly, the field shifted from speaking of communities, forms of life, or the "experimental life" to discussion in terms of *practice*, with the implication that scholars were getting more realistic and specific about what

1. Latour and Woolgar, *Laboratory Life*, p. 51.

2. Ibid., pp. 77, 90, n. 19. See also Akrich and Latour, "A Summary of a Convenient Vocabulary for the Semiotics of Human and Nonhuman Assemblies," p. 259, for the claim that "the key aspect of the semiotics of machines is its ability to move from signs to things and back." The semiotic method is borrowed from Greimas, *The Social Sciences*. Ethnomethodology has developed a similar point about "modalities," referring to the essential "indexicality" of statements, where the meaning of statements depends upon the context in which the statement is embedded. However, ethnomethodology rejects any claim that general statements of fact escape indexicality or that indexicality represents a skeptical challenge. See Garfinkel, *Studies in Ethnomethodology*, pp. 4–10; Lynch, *Scientific Practice and Ordinary Action*, pp. 17–22, 101; vom Lehn and Dingwall, *Harold Garfinkel*, pp. 95–98. Compare also Pinch, "Toward an Analysis of Scientific Observation," p. 9, on reducing the degree of "externality" of observation claims.

3. Latour, *Pasteurization of France*; Latour, *Science in Action*; Latour, *Reassembling the Social*.

4. Latour and Woolgar, *Laboratory Life*, p. 281; Latour, "Technology Is Society Made Durable," pp. 129–30.

5. Doing, "Give Me a Laboratory."

scientists actually did in the course of their work.[6] Key to this rhetoric of specificity is an accompanying register whereby conceptions of science that were idealized constructs were to be replaced by *material* accounts.[7] Latour spoke of finding the "missing mass" that explained how social order could be maintained, as weak social norms are supplemented by delegated non-human entities that constrain choice by engineering design, leading to a distribution of distinct competences between human and non-human entities.[8] By implication, an emphasis on practice invoked materialism over idealism, in the fraught language of philosophy.

By the 1990s, sociology of scientific knowledge (SSK) and sympathetic philosophical and historical approaches began to speak of the need to understand "practice," "material practice," or "material culture" with the understanding that these terms delimited something localized and specific in time and space and more material than the shared "ideas" of a scientific research program.[9] The implication was that the emerging field of Science and Technology Studies (STS) was more "materialist" than past idealized or idealist treatments of science as disembodied knowledge.

Latour, Michel Callon, John Law, and Arie Rip developed an actor-network approach that built upon and extended the materialist semiotics of Latour and Woolgar's *Laboratory Life*, showing that the material world became locally linked to scientific actors in the effort to extend scientific representations. "Local" human and non-human actors seek to enroll each other to extend the sociotechnical networks that make knowledge a material force extending beyond laboratory walls to work in the broader world. In effect, knowledge itself is understood as a distributed material entity, not just a set of shared

6. Shapin and Schaffer, *Leviathan and the Air-Pump*; Pickering, *Mangle of Practice*; Galison, *Image and Logic*. Lynch, "From Normative to Descriptive and Back," pp. 115–21, discusses the "practice turn" in STS in the 1990s. Soler, Zwart, Lynch, and Israel-Jost, *Science after the Practice Turn*, p. i, summarize the development as follows: "In the 1980s, philosophical, historical and social studies of science underwent a change which later evolved into a turn to practice. Analysts of science were asked to pay attention to scientific practices in meticulous detail and along multiple dimensions, including the material, social and psychological."

7. On the significance of specifically *materialist* conceptions of semiotics, see Lenoir, "Was That Last Turn a Right Turn?," pp. 120–25; Haraway, "The Promises of Monsters," pp. 298–99, 304–5.

8. Latour, "Where Are the Missing Masses?," pp. 225–32. See also Winner, "Do Artifacts Have Politics?"; Baird, *Thing Knowledge*.

9. Pickering, *Science as Practice and Culture*; Schatzki, Knorr-Cetina, and von Savigny, *The Practice Turn in Contemporary Theory*; Soler, Zwart, Lynch, and Israel-Jost, *Science after the Practice Turn*. The concept of practices has deeper roots in twentieth-century philosophy, including the work of Wittgenstein, Heidegger, Kuhn, and Polanyi, among others, such that practices are "the vanishing point of twentieth-century philosophy" (Turner, *Social Theory of Practices*, p. 1; see also Schatzki, *Social Practices*, pp. 12–13 for the link with Wittgenstein). My focus in this chapter is on the concept of concrete, local, and material practice that went beyond an earlier focus on tacit knowledge and forms of life. This approach has its ultimate roots in the Young Hegelian development of a theory of ideology as a forerunner of the sociology of knowledge. McLellan, *The Young Hegelians and Karl Marx*, pp. 9–10, traces priority here to August von Cieszowski's rejection of Hegelian speculative philosophy for a philosophy focused on practical action and his coining of the term "praxis" in 1838.

beliefs.[10] Science and technology gain "reality" as a heterogeneous assembly of human and non-human entities are made to work together across time and space, assembling society as it builds knowledge or technological artifacts.[11]

EXPERIMENTAL LIFE AS HISTORICAL MATERIALISM: FROM FORMS OF LIFE TO PRACTICES

The term "practice," and the trend to replace abstract treatments of culture with a concrete, material focus, recapitulates the early history of the sociology of knowledge in the efforts by Karl Marx to come to terms with the "Young Hegelians." The "Young" or "Left" Hegelians sought to subject religion to criticism in a way parallel to SSK's efforts to do the same for science. For Ludwig Feuerbach, for example, the religious believer inverted and projected his weaknesses onto the concept of an all-powerful God. In so doing, he turned the project of philosophy on its head, which was understood to provide the authoritative basis for science and ethics alike, instead seeing philosophical beliefs as reflecting social reality in distorted form.[12]

This "transformative method" common to the Young Hegelians was the core of a theory of ideology, which involved inverting Hegel's idealist philosophy for an account of philosophy as reflecting and disguising its social causes.[13] Where Hegel's philosophy of spirit announced the coming of social harmony and rationality with the rise of the Prussian state that would overcome brute, sensual needs and reconcile parochial interests, his followers found the Prussian state, and their place in it, inadequate for such a philosophical project.

Mah shows that the Young Hegelians transitioned from belief in the sovereignty of philosophy at the end of the 1830s to belief that philosophy was an ideology that "implicitly registered the social, political, and intellectual dislocation that accompanied the uneven consolidation of "modern" market society and popular politics in Prussia."[14] Marx and Engels transformed this

---

10. Callon and Law, "On Interests and Their Transformation"; Callon, "Some Elements of a Sociology of Translation"; Callon, Law, and Rip, *Mapping the Dynamics of Science and Technology*; Law, *Power, Action and Belief.*

11. Latour, *Aramis*. See also Pickering, *The Mangle of Practice*, pp. 14–23; Pickering, "Living in the Material World." The "coproduction" of society and knowledge became an important organizing concept for STS work drawing on history and political science. See Jasanoff, *States of Knowledge*.

12. Feuerbach, *Essence of Christianity*, pp. xxxiv–xxxv, 9–14, 27–28; McLellan, *The Young Hegelians and Karl Marx*, pp. 112–16; Henderson, *The Life of Friedrich Engels*, I, pp. 14–15; Livergood, *Activity in Marx's Philosophy*, pp. 4–5.

13. Kainz, *Hegel's Philosophy of Right*, pp. 4–5; Avineri, *The Social and Political Thought of Karl Marx*; Rosen, *Bruno Bauer and Karl Marx*, pp. 201–11; Harvey, *Feuerbach and the Interpretation of Religion*, p. 10, ch. 1.

14. Mah, *The End of Philosophy, the Origin of "Ideology*," p. 3.

tacit theory of ideology into a formal theory that the dominant ideas in any society reflected the interests of the ruling class in *The German Ideology*, an unpublished manuscript written in 1846.

Marx developed his concept of "sensuous practice" in a critical appraisal of Feuerbach's approach, undermining the viability of a correspondence theory of truth reaching across a gulf between human concept and natural object. In the process, he developed a distinctive "historical materialism" that tied scientific work (and all other representations produced in each society) to their basis in the social relations of production. It was the emphasis on material practice that facilitated the development of his account of capitalism as a relational system, first presented in lectures to workers in Brussels in December 1847:

> In the process of production, human beings work not only upon nature, but also upon one another. They produce only by working together in a specified manner and reciprocally exchanging their activities. In order to produce, they enter into definite connections and relations to one another, and only within these social connections and relations does their influence on nature operate, *i.e.*, does production take place.[15]

As in the efforts by Marx to understand how the process of production emerged at the interface between humanity and nature in "sensuous human activity," the turn to practice in SSK likewise sought to transform abstract talk of nature and society into concrete, spatiotemporally delineated practices.[16]

Marx inverted Hegel precisely by abandoning the Hegelian belief in the capacity of abstract thought to overcome the limitations of the atomized and purely sensual empiricist subject, as the rise of impoverished landless former peasants made a mockery of Hegel's belief in the emergence of a rational Prussian state that would restore social harmony. Thus, it is not Marx's engagement with Greek atomism that marks his transition to materialism, but his attempt to criticize the distorting abstraction of both Hegel and the Young Hegelians.[17]

---

15. Marx, *Wage-Labour and Capital*, p. 28. For the application of historical materialism to science, see Murray, *Marx's Theory of Scientific Knowledge*, pp. 74–78; Lynch and Fuhrman, "Recovering and Expanding the Normative."

16. Rockmore, "Is Marx a Materialist?," pp. 63, 75, has argued that Marx is a materialist only in a philosophically "non-standard sense of the term as concerns the focus on concrete, social problems" rather than the ontological primacy of matter over ideas, arguing instead that Marx "bases his theories on the concrete or real social context, from which they emerge and to which they return." However, Marx's emphasis on "sensuous human activity" links the two, as it is the practical engagement with a material world that explains dynamic human cognition. See Mah, *The End of Philosophy, the Origin of "Ideology,"* pp. 41–42, ch. 8.

17. See the discussion of Marx's dissertation on Democritus and Epicurus's natural philosophies under Young Hegelian Bruno Bauer in Mins, "Marx's Doctoral Dissertation," pp. 158–59, and Browning, "Marx's Doctoral Dissertation." Livergood, *Activity in Marx's Philosophy*, pp. 3–11,

Marx did not use the expression "historical materialism," but he did refer to a "materialist method" in *The German Ideology*, which begins from "real individuals, their activity and the material conditions under which they live, both those which they find already existing and those produced by their activity." Such an approach is subject to empirical confirmation and grounds history on the "physical organization of these individuals and their consequent relation to the rest of nature," which is in turn transformed "when they begin to *produce* their means of subsistence . . . indirectly producing their actual material life."[18] The key to understanding this turn to practice is that it does not begin from an elaborate theory of the economic base and superstructure of modern society but from a concrete focus on material realities that rejected essentialist concepts of human nature like Feuerbach's species-being, which had previously influenced Marx's thinking, most notably in the 1844 *Economic and Philosophical Manuscripts*.

Marx's decisive break with "the Hegelian preoccupation with abstract thinking" came only with *The German Ideology*, with its focus on concrete human practices divorced from teleological statements of human essence or purpose. Marx's materialist move was facilitated by his experiences with the workers' movement in Paris and his collaboration with Friedrich Engels, who had already drafted his classic sociological treatment of the working class in Manchester.[19] Marx's focus on practice, like those in STS focusing on material practice a century and a half later, sought to go beyond a negative critique of philosophical abstraction. As Mah puts it,

> What was still required to lay Hegelian idealism to rest was a positive alternative: a fully worked out understanding of the evolution of human sensuousness. Such a demonstration needed to be antiphilosophical in two senses: it had to renounce the use of ideal philosophical categories and arguments, and it had to show how and why ideal processes were in principle necessarily derived from sensuous or material processes. Marx had to explain in an unphilosophical manner why philosophy should be seen as the efflux of human sensuousness.[20]

---

argues that it is the notion of *activity* that Marx incorporates into his materialist system by drawing upon Epicurus's concept of the declination of atoms in preference to Democritus's atomism where the principle of motion is external to the object. As such, Livergood argues that Marx provides an idealist core to his materialism by emphasizing the self-moved nature of human action. McLellan, *The Young Hegelians and Karl Marx*, pp. 9–10, identifies August von Cieszowski's 1838 concept of praxis as significant here in combining theory and practical action. Browning, however, shows that at the time of his dissertation, Marx fully accepted Hegelian idealism, which he applies wholesale to the discussion of ancient atomism (pp. 131–33).

18. Marx, *German Ideology*, p. 61.

19. Mah, *The End of Philosophy, the Origin of "Ideology,"* pp. 210, 194; Engels, *The Condition of the Working Class in England*; Hollander, *Friedrich Engels and Marxian Political Economy*, pp. 43–58. For Marx as a theorist against abstraction, see Sayer, *The Violence of Abstraction?*

20. Mah, *The End of Philosophy, the Origin of "Ideology,"* p. 210.

Several aspects of Marx's focus on "living human individuals" or "real premises" can be identified, but they all begin from a concrete focus on empirically observable human beings collectively producing solutions to their basic material needs for food, shelter, and reproduction. Each solution to a set of needs invariably introduces new needs, which in turn alter the solutions in kind.[21]

With the emergence of consciousness, initially focused on the immediate natural environment, greater social complexity and division of labor becomes possible. The separation of material and mental labor are the preconditions for seeing consciousness as separate from the natural world rather than rooted in it—as it is in fact. As the division of labor develops, distinct roles and interests emerge that are nonetheless interdependent in practice. The state emerges as the supposed guarantor of an interest common to all, a view that is in fact an ideology insofar as the state actually represents the disguised private interests of a distinct class.[22] Marx's awareness of how private material interests took control of the state dated to his journalistic treatment of how forest owners denied peasants traditional rights to gather wood, using the existence of "the wood thief to purloin the *state itself*."[23]

After the Prussian state suppressed the newspaper he edited, the *Rheinische Zeitung*, the point was further underscored by his experience with working men's associations in his time in Paris and his admiration for the revolt of the Silesian weavers.[24] Whereas the other Young Hegelians struggled with how their critique of ideology could itself escape ideology, Marx's identification with the industrial proletariat led him to reject philosophical foundations and abstract debate for a focus on the concrete situation of a group who experienced exploitation and who could transform that situation through revolutionary practice. It was this rejection of philosophical abstraction and his endorsement of material practice that led him to a critical, rather than the neutral, concept of ideology that ultimately prevailed in the sociology of knowledge and Marxism itself.[25]

Whereas Lenin thought of ideology as a tool where intellectuals wielded theory on behalf of competing classes to shape their interests, Marx's focus on practice began with an attempt to develop and facilitate means for a revolutionary imperative that he identified already among the new laboring classes. In that sense, it was the identification of a latent tendency in proletarian practice itself that provided the basis for predictions of revolution. When the Communist League, with its manifesto authored by Marx and Engels,

---

21. Ibid., p. 211.
22. Ibid., pp. 214–15.
23. Ibid., pp. 184–85; Marx, *The German Ideology*, pp. 77–78, 92–93, 113–14.
24. Mah, *The End of Philosophy, the Origin of "Ideology*," p. 200.
25. Ibid., pp. 226–27. Lynch, "Ideology and the Sociology of Scientific Knowledge," pp. 199–202.

was suppressed and its members split between competing factions, Marx and Engels adjusted their account of that practice to reflect a longer gestation for revolution that responded to the new realities on the ground.[26] On the other hand, when the Paris Commune showed how an alternative to the modern state emerged in revolutionary practice, Marx sought to build upon that basis.[27] Common to both moves was a rejection of philosophical foundations for material and political practices as a ground for theory.

The move toward practice in STS shares this contempt for philosophical foundations of knowledge and its call for an embedded, sociological account of practice, even if it likewise rejected Marx's critical stance. In *Laboratory Life*, Latour and Woolgar present the equivalent of a historical materialism applied to science insofar as they focus on the production and reproduction of scientific practice. Laboratories are shown to allow for the development of sophisticated practices for producing representations held to be true for all, though they in fact represent the disguised interests of select technical workers at privileged sites of scientific production. Laboratory practices are designed to produce the conditions for ensuring the acceptance of scientific claims by processing and transforming natural inputs through technologically mediated practices carried out in particular locations. Scientific reproduction also takes place as new generations of scientists are trained in the relevant ways of representing and transforming natural phenomena. The production and reproduction of scientific practice can be expected to transform itself over time.

Both Marx and Latour argued that the objects of science are not brute, natural things themselves but objects made and remade by complex, sociomaterial practices. Thus, Thyrotropin Releasing Factor (TRF) is not a hormone lying ready to find in "nature" but an isolated and processed product of specialized laboratory practices that in turn facilitated new, hybrid laboratory research programs.[28] The complex division of labor and the ongoing transformation of the material conditions of labor that Marx saw as crucial for understanding capitalism were also applied by Marx to the production and reproduction of scientific practice. Already in the *Economic and Philosophical Manuscripts*, he argued that within science, the "material of my activity [is] given to me as a social product."[29] In *The German Ideology*, Marx and En-

26. Marx, *The Cologne Communist Trial*. See also Hollander, *Friedrich Engels and Marxian Political Economy*, pp. 21–22, on Marx and Engels's endorsement of reform as they observed capitalism evolve, making them the "first revisionists." Blackburn, "Marxism," p. 34, argued that they supported advancing proletarian interests via "all possible legal means," including workers' parties under bourgeois democracy, while nonetheless anticipating "a future test of force" between the working class and the state.

27. Marx, "The Civil War in France."

28. Latour and Woolgar, *Laboratory Life*, pp. 106–7, 112–19, 124–29.

29. Marx, "Economic and Philosophical Manuscripts of 1844," p. 298.

gels argue that even "'pure' natural science is provided with an aim, as with its material, only through trade and industry, through the sensuous activity of men." Human activity transforms natural processes in the laboratory and remakes the world outside the laboratory to such a degree that an undisturbed nature hardly exists at all and the nature we encounter reflects "this unceasing sensuous labor and creation."[30]

Despite Latour's opposition to Marxism, his approach represents a similar inversion of self-contained idealist conceptions of the social in SSK. Similar to Marx, Latour sought to provide material explanations for seemingly abstract thought that situated it within complex practices at the interface between humans and the world.[31] Laboratory practices, and those practices "extending" the laboratory to allow laboratory knowledge to apply to a wider world, so transform brute, unmodified nature as to bring into being entirely new hybrid objects.[32] Latour and Woolgar rejected a dichotomy between minds and nature and called for a moratorium on cognitive explanations of scientific knowledge, with its focus on "ideas of reason intrinsic to the mind."[33]

PHILOSOPHY OF SCIENCE AGAINST IDEALIST CONSTRUCTIVISM

When one speaks of the local and material basis of the production of *knowledge*, philosophers instinctively reverse the associations emerging from Marx's historical materialism and the current sociological focus on practice, suggesting that a "reduction" of knowledge to the peculiarities of local social and material causes is inappropriately idealist, reducing the world to our thought about it. The view that every culture has its own view of the natural world is understood as an anthropological update of Protagoras's claim that man is the measure of all things.

On this view, social constructivism commits the "epistemic fallacy" in relativizing natural knowledge to cultures, thereby treating the way the world

30. Marx, *German Ideology*, p. 91; Lynch and Fuhrman, "Recovering and Expanding the Normative," p. 242; Murray, *Marx's Theory of Scientific Knowledge*, pp. 75–77.

31. Latour and Woolgar, *Laboratory Life*, p. 277; Berreby, "And Now, Overcoming All Binary Opposition"; Kofman, "Bruno Latour, the Post-Truth Philosopher, Mounts a Defense of Science." Compare Feyerabend, "Art as a Product of Nature as a Work of Art," p. 88, for an argument reducing the separation of the arts and sciences by showing on the one hand "that, like rocks and flowers, works of art are products of nature" and, on the other hand, that the natural world "is an artifact, constructed by generations of scientists-artisans from a partly yielding, partly resisting material of unknown properties."

32. Latour, "Give Me a Laboratory and I Will Raise the World"; Callon and Latour, "Unscrewing the Big Leviathan"; Latour, *We Have Never Been Modern*.

33. Latour and Woolgar, *Laboratory Life*, pp. 280–81, 258, n. 2.

works (ontology) as dependent on the way we think it works (epistemology).[34] However, SSK proponents would see philosophers as committing an "ontological" fallacy, by treating the way we believe the world works as dependent upon the way the world actually works. This view comes out quite clearly in the strong program, where the recognition of a recalcitrant reality "out there" provides no comfort in believing that our representations of that reality reflect that reality itself.[35]

Since the world in itself is inaccessible to us directly, and we require communities and specialized equipment to see the world in a particular way, SSK theorists believe that talk of correspondence to reality *simpliciter* is idealist, treating one special way of representing the world as if it was the only one. For Barnes and Bloor, the strong program admits that *all* scientific representations are constrained by the natural world, which doesn't provide a basis for distinguishing between those views that are true or false, rational or irrational, or successful or unsuccessful. In a sense, the natural world drops away as unknowable in any unmediated sense, while sociological variables, including the way that communities organize their inquiry into nature differently, are invoked to explain scientific belief. This is why in the other dominant program of early SSK, the empirical program of relativism (EMPOR), nature is seen to play "a small or non-existent role in the construction of scientific knowledge."[36] Whether nature is omnipresent but unable to differentiate between competing representations on its own or nature if seen as playing no role for "methodological" reasons, invoking a greater approximation to underlying truth in successful science is rejected.

Post-Kuhnian philosophy of science has attempted to square the circle by continuing to reject constructivist approaches as idealist, while also, in a sense, reducing science to the products of a local reference, mediated by a professional community. Building on the realist semantics Putnam and Kripke set up to counter Kuhn's talk of incommensurable worlds, modern scientific realism has tried to reclaim the materialist high ground by conceding the constitutively social dimension of science to Kuhn and the social constructivists, but seeing it as affording a causal—rather than deductive—

34. Bhaskar, *A Realist Theory of Science*. Idealism is specifically invoked by Calvert-Minor, "Blurry Humanism," pp. 148–49, in criticizing SSK. Calvert-Minor argues as well that the attempts by Barnes and Bloor to distinguish their position from "methodological idealists" fails since their position is a "realism emptied of all epistemic value" due to their symmetrical treatment of scientific theories. By contrast, Fuchs, *The Professional Quest for Truth*, p. 2, argues that "[r]ealist epistemologies have systematized the worship of science as the model and metaphor for the sacred values of truth, impartiality, and objectivity. The eminently religious character of this worship is evident in the dualistic structure of epistemological realism," evident in its internalism and asymmetry between truth and falsehood.

35. Barnes, Bloor, and Henry, *Scientific Knowledge*.

36. Collins, "Stages in the Empirical Programme of Relativism," p. 3.

connection to underlying reality. When we refer to an object as gold, the accuracy of our reference is held to depend upon a causal connection between an initial reference to gold and our own reference, mediated by a continuity of reference that is not disrupted by changing conceptions of what constitutes gold. In the race with sociology to profane science by associating it with real places, people, and things, modern philosophy of science has nonetheless sought to preserve the higher "spiritual" place of knowledge on the new basis of a spatiotemporally limited "baptism" linking an underlying ontological reality with a community invested with the power of apostolic succession.[37]

Constructivists, by contrast, held on to the localist thesis that scientific knowledge reflected its construction in specific settings without guaranteeing that different communities would come to see the same light. In a sense, constructivists remained the Protestants to realism's Catholicism, by conceding that scientific truth is subordinate to political reality, the cognitive authoritarianism we saw wedded to skepticism in chapter 1. Indeed, the localist thesis gets connected to the view that one must concede the scientific community's considered judgment as there is no higher standard. The struggle to define the post-Kuhnian world among philosophers and sociologists consequently evidences a curious combination of materialism and idealism. Philosophers and sociologists compete to be more "materialist" than each other, with Putnam drawing upon Lenin's historical materialist polemic against Ernst Mach's sensation-based empiricism and Latour eliminating mind for an agonistic materialist semiology.[38]

WHO'S TO BLAME FOR ALL THE FUSS?

The squabble between philosophers and sociologists might have continued as a minor intellectual dispute had it not been for the Sokal hoax. In 1996, physicist Alan Sokal succeeded in publishing a spoof of postmodern treatment of quantum physics in the journal *Social Text*, a ploy that led to a front-page story in the *New York Times*. This helped launch the so-called Science Wars as a number of scientists and other critics charged that Science and Technology Studies, seen as an application of postmodernism in the humanities to the study of science, were part of a no-nothing movement to undermine the authority and status of science. Biologist Paul Gross and mathematician Alan Levitt's 1994 *Higher Superstition: The Academic Left and Its Quarrels with*

37. Fuller, *Social Epistemology*, ch. 3.

38. Putnam, "The Meaning of 'Meaning'"; Putnam, "Explanation and Reference"; Latour and Woolgar, *Laboratory Life*, p. 280. In his later work, Putnam was critical of metaphysical realism. In Putnam, *Naturalism, Realism, and Normativity*, p. 118, he would refer to "Lenin's (disastrously incompetent) polemical book against Machian positivism."

Science had inspired Sokal and controversy even erupted (in early online bulletin boards) over the relative guilt science studies researchers should feel about the possible influence of their field on the Unabomber.[39]

The FBI had even sent investigators to the 1994 joint meeting of the Society for Social Studies of Science (4S), the History of Science Society (HSS), and Philosophy of Science Association (PSA) in New Orleans that year based upon letters accompanying some of the bombs that seemed to indicate a knowledge of the history of science. This perception was reinforced later by the Unabomber's manifesto, sent to media outlets in June 1995 and eventually published, which the FBI took to show "that the bomber is immersed in the most radical interpretations of the history of science."[40] At that 1994 meeting, they had subpoenaed the membership roster of 4S and HSS, and sounded out the 4S council about possible suspects. It became clear that the FBI believed the Unabomber was, at the least, a fellow traveler of the history and social study of science.[41]

The Unabomber, as it turned out, was, like Levitt, a Harvard-educated mathematician critical of academic leftism. He had also been subject to abusive psychological experiments by Henry Murray, a psychologist and former consultant to the Office of Strategic Services (OSS), forerunner to the CIA. The experiments were designed to test how gifted students responded to "vehement, sweeping, and personally abusive" attacks, so arguably unethical scientific experiments shaped his murderous impulses.[42] Be that as it may, the accusation that science studies was connected to an irrationalist plot against science and modern civilization had become more widely known than the field itself by the time notice of Alan Sokal's hoax was published in the *New York Times*.[43]

The controversy resulting from the "science wars" had been percolating for some time, perhaps dating to the derailing of the appointment of Bruno Latour to the Institute for Advanced Study by resident scientists and mathematicians in 1990.[44] With Gross and Levitt's attack on the field and Sokal's hoax, opposition to the field by some scientists grew. In defending the field from outsiders, debate became intense about who should take the blame for opening the field to criticism, despite the fact that many in the field espoused what they called "radical" approaches to the study of science. Aren't radicals supposed to be shouted down by the establishment? Don't they wear the

39. Sokal, "Transgressing the Boundaries"; Scott, "Postmodern Gravity Deconstructed, Slyly"; Gross and Levitt, *Higher Superstition*.

40. Broad, "Esoteric Wedge of Academia Is Roiled by Hunt for Bomber." See also Gottlieb, "Pattern Emerges in Bomber's Tract."

41. Shrum, "We Were the Unabomber"; Restivo, "4S, the FBI, and Anarchy."

42. David Kaczynski, *Every Last Tie*, pp. 11–12. See also Chase, *Harvard and the Unabomber*.

43. Scott, "Postmodern Gravity Deconstructed, Slyly."

44. Berreby, "And Now, Overcoming All Binary Oppositions."

label like a badge of honor? In fact, many preferred to scapegoat particular approaches within the field in an effort to appear respectable and accepted by natural scientists and other academics. The conditions for the emergence of full-fledged ideological critique of science would probably have required the disestablishment of STS in the academy, paralleling the marginalization and oppression of the Young Hegelians in Prussia. In the event, a more modest rhetoric and successful disciplinization did take place. Exactly which form of radical STS would win out as sufficiently moderate remained at issue.

EXTENDING SYMMETRY AND LABORATORY PRACTICE

Before the science wars broke out officially, the emerging field of Science and Technology Studies had already begun to suffer internal divisions. This can be seen most clearly in the debates characterizing the influential 1992 anthology *Science as Practice and Culture*, and the divergence in approaches to the categories of the "natural" and the "social" by Collins's methodological relativism and the actor-network theory of Latour, Callon, and Rip. The anthology included primarily representatives of the sociology of scientific knowledge (SSK), along with broadly sympathetic philosophers and historians. The book set its aim to challenge the idealized image of scientific knowledge as purified of social contaminations. In this context, the most heated part of the debate concerned whether Latour's move to incorporate nature back into scientific representations, albeit as local "actors" linked in networks with each other and human actors, constituted a betrayal of the fundamental discovery of SSK.

For Latour, the "symmetrical" treatment of human and non-human actors was more true to the underlying thrust of SSK, even if it moved the field away from disciplinary sociology for a new kind of field studying the non-pure, networked "associations" of previously artificially separated natural and social things. Where Collins saw Latour's move as a violation of the symmetry principle since now nature would no longer play "little or no role" in explaining scientific representations, Latour's view did provide a way for STS scholars to defend themselves against the accusation that they rejected scientific engagement with the natural world in the context of the science wars.

In *Pandora's Hope*, Latour recounts an exchange with a scientist where he is interrogated about whether he believes in reality and in science's knowledge thereof, which he answers in the affirmative, admitting to being confused by the question. He wondered who doesn't believe in reality, yet perhaps scientists could be forgiven for taking Collins seriously when he argues that nature plays "little or no role" in the production of scientific knowledge or for interpreting Pickering's claim that quarks are constructed as denying that they are discovered in an independent reality.[45] As we saw in chapter 1, Collins does have his own account of how science and science studies can each engage in separate enterprises with distinct metaphysical import on the question of natural reality. In his case, this led to a gulf between natural and sociological accounts of the world, which Latour wished to mix together. In many contributions to the Pickering volume, a new emphasis on "sensuous practice" did something similar, serving to unite concept and world via the work of transforming nature in the direct engagement of human activity with material objects. Such a view was seen as superior to separating ideas and nature analytically, followed by attempts to bridge the gap at the level of representations.

Thus, despite Latour's political opposition to Marxism,[46] his approach recapitulates Marx's efforts in the 1840s to ground human knowledge in practice, involving the *active* transformation of the material world:

> The chief defect of all hitherto existing materialism—that of Feuerbach included—is that the thing, reality, sensuousness, is conceived only in the form of the *object* or of *contemplation*, but not as *human sensuous activity, practice*, not subjectively. Hence it happened that the *active* side, in contradistinction to materialism, was developed by idealism—but only abstractly, since, of course, idealism does not know real, sensuous activity as such.[47]

Thus, Latour's discussion of "quasi-objects" and the effort through practice to add "reality" to technological objects represents a kind of semiotics-tinged historical materialism.[48]

To see how this move makes sense, recall the spirit that linked early SSK. Rejecting the view that genuine knowledge must escape or transcend the social and historical conditions under which they were produced, sociologists generalized the point by articulating a principle of symmetry, whereby even true, rational, or successful knowledge claims must be understood by reference to the same kinds of causes that would explain false, irrational, or unsuccessful science (however defined or determined). The principle of symmetry linked most dominant approaches to the field, whatever their other differences. Supported primarily by laboratory ethnographies examining the ongoing construction of meaning in a local setting, SSK aggressively promoted the idea that the success of science had local causes, in opposition to

45. Latour, *Pandora's Hope*; Pickering, *Constructing Quarks*. See Haraway, "A Giant Bumptious Litter," for her response to the questioning scientist.

46. Berreby, "And Now, Overcoming All Binary Oppositions"; Latour, *We Have Never Been Modern*.

47. Marx, "Theses on Feuerbach," p. 143.

48. Latour, *We Have Never Been Modern*; Latour, *Aramis*.

traditional philosophical stereotypes of knowledge as distinctly placeless, a view from nowhere (or everywhere).

The idea of SSK's core mission survives in the subtitle to Steven Shapin's 2010 book *Never Pure: Historical Studies of Science as if It Was Produced by People with Bodies, Situated in Time, Space, Culture, and Society, and Struggling for Credibility and Authority*.[49] So long as the goal remained undermining philosophers' conceptions of a timeless, placeless, and asocial knowledge (however stereotyped their view of philosophy might be), a certain unity of perspective obtained, uniting a group portraying themselves as rebels at the margins of a scientific culture. Unlike the 1830s, when followers of Hegel failed to gain university appointments and state support, SSK was successful in making that transformation. One wonders if a more politically engaged STS would have evolved had the discipline failed to be established in the academy.

As the field gained respectability, as testified to by the explosion of publications by noted presses and National Science Foundation training grants, this surface rhetoric of unity behind the science-as-local thesis retreated primarily to the role of consolidating the view of the field for outsiders, while internal debate shifted to the varying implications of the localist thesis for interpreting science. Pickering's anthology demonstrated these new conflicts of a disciplinizing field, even as continued general commitment to the localist thesis obscured the differences in perspectives. Thus, when the field was attacked in the science wars, the different sides could simultaneously come together as one (defending themselves against a common enemy who denied the insight that science was local), while intensifying internal debate as blame was sought for who was responsible for allowing the field to be "mischaracterized" as it was.

THE PRIVATE AND THE PUBLIC: THE ACCOUNTABILITY OF THE LABORATORY SITE

Shapin and Schaffer have shown us that the development of experimental natural philosophy in the seventeenth-century Royal Society required managing a crucial tension between the "private," situated and artificial nature of experimental manipulations and the need for "public" knowledge of the real physical world outside the confines of the laboratory. Experiment required the production and disciplining of artificial, "new" phenomena in the restricted space of

49. Shapin, *Never Pure*. See also the editorial summary of consensus in the field in Felt et al., *The Handbook of Science and Technology Studies*, pp. 1–2: "[A]ll knowledge is local and reflects the specific historical moment, cultural context as well as networks within which it is made."

the laboratory. The observation of the air-pump's behavior in several trials was unavailable to those without an air-pump, while even calibrating accounts of different air-pumps was problematic, requiring travel and negotiation between sites. As such, the "public" nature of knowledge could be called into question, a value shared both by scholastics and by mechanists like Hobbes. Indeed, the tradition of alchemical laboratories could provide no comfort for Boyle and company since its private nature resembled the discredited "enthusiast" desire for private illumination rather than publicly justifiable claims, an error widely blamed for the political disruption of the Interregnum period.[50]

Boyle's solution was ingenious but not definitive. First, he developed an experimental vocabulary aimed at allowing competing theoretical perspectives to make common progress in describing what had occurred in the experiment. Shapin and Schaffer used the polemics with Hobbes to show that there were limits to the efficacy of this strategy: Hobbes denied that the air-pump produced phenomena appropriate for deciding disputes in natural philosophy and hence refused to accept that the experimental vocabulary *referred* to anything at all. Only proper use of language based on consistent reference to conceivable material causes was deemed appropriate. Moreover, the test of the vocabulary's adequacy must involve the ability to handle commonly available empirical experiences in principle available to all rather than artificially produced "special cases," which fail to illuminate nature's general laws and fail to be accessible to outsiders.

Second, Boyle sought to solve the problem of public authority by developing social, literary, and material "technologies" to re-present experimental results to a wider community, thereby extending knowledge produced in the laboratory to a wider scientific community. Of note is the strategy of "virtual witnessing," which multiplied witnesses by presenting the experimental activity in prolix detail in order to remove the reader's doubt that the experiments occurred as they did. The authority of such accounts was further enhanced by the fact that a gentleman had vouched for them.[51] Once again Hobbes had a different account of what insured legitimacy, and he argued that the Royal Society would usurp proper authority over the representation of things leading to a "divided" polity, much as he held that the clergy had inappropriately usurped the king's authority, thereby causing civil war.

Making any direct connection between this case and modern science is problematic. Arguably, the development of a mathematical account of experiment played a more important role in developing modern approaches to physical science.[52] Boyle's rhetorical focus on facts over theories does not

50. Shapin and Schaffer, *Leviathan and the Air-Pump*.
51. Shapin, *Social History of Truth*.
52. Dear, *Discipline and Experience*; Schuster, *Descartes-Agonistes*.

actually capture his own practice very well, let alone the diversity of theoretical and methodological approaches within the scientific revolution.[53] Moreover, virtual witnessing seems to have largely disappeared from most forms of scientific journal writing to the extent that any particularities of time and place involved in experimental activity are removed, making transmission of tacit skills between laboratories that much more essential.[54] To be sure, virtual witnessing still plays an important role in legitimation for a lay public.[55] However, its temporary appearance at the birth of modern science and later replacement by more compact and transferrable forms of experimental reports would lend support to a Bachelardian account in terms of a "rupture" with common sense normalized by a philosophical account.[56]

Nevertheless, the need to deal with the tension between the locally situated nature of experiment and the global scope and justification of the knowledge produced seem endemic to modern science, at least for those sciences in which experiment plays a significant role. Logical positivism, at least in the form developed by Carnap, tended to conflate experimentation and mere observation, and saw observation as common to all in all places (or at least capable of formal reduction, one to the other), reducing science to a collection of law-like regularities available in principle to all. This converted the experimental approach into a placeless method for checking the claims of scientists, since in principle the observations underlying any scientific claim could be checked. The availability of confirming observations in principle available to all downplayed the difficulties scientists had in converting specialized manipulations of matter in lab settings into general descriptions of nature. While naïve about the difficulty outsiders would have in verifying scientific claims themselves, this did contribute to a democratic sense of accountability underlying logical positivism.

By contrast, sociological constructivists and contemporary realists in philosophy *both* emphasize the role of specialized sites and manipulations that make possible representations of the world. Contemporary realists have considered the important role that experiment plays in intervening in nature and uncovering deeper structures of reality. In other words, correct representations of nature come not from building theory statements from observation statements but from finding a true connection to the world by intervening in a way that is not available to everyday observation and that produces an esoteric language that refers approximately to underlying causal mechanisms.[57]

---

53. Schuster and Taylor, "Blind Trust," pp. 508–15.
54. Collins, *Changing Order*; Bazerman, *Shaping Written Knowledge*.
55. Gieryn and Figert, "Ingredients for a Theory of Science in Society."
56. Bachelard, *The Formation of the Scientific Mind*.
57. Harré and Madden, *Causal Powers*. Sometimes this form of realism results in an overemphasis on experimentation, as in Bhaskar, *A Realist Theory of Science*. Not all parts of science rely on

Constructivists have used opposite language to describe this process, talking of "constructing" quarks or "manufacturing" facts, rather than uncovering preexisting structures of nature. However, despite this sharp difference in how scientific representations are understood, they both agree that it is only specialized manipulations of nature available to carefully trained communities that are able to produce viable scientific representations of nature.

To a first approximation, constructivists should be seen not as classical relativists ("all positions are equal"), but rather as scientific relativists ("all scientific representations are equal" or, better, "all scientific representations are initially equal"). In Collins's program of methodological relativism, interpretive flexibility in the interpretation of experimental evidence means that different, competing representations can be expected to develop in science that have to be sociologically winnowed down before science can be said to speak with one voice. In Latour's approach, this same process is described in somewhat different language. Get yourself a laboratory, as an inscription-making device, and work to remove modalities or indexicalities from your statements, converting highly conditional statements to unquestioned facts.[58] In either case, the price of admission to get in on the game of constructing representations is a laboratory, or in the case of non-laboratory sciences, some other specialized manner of engaging natural phenomena and membership in a scientific community studying it. Both constructivists and realists believe that scientific representations are not simply the straightforward conversion of observation language into theoretical claims. The observer brings certain background assumptions and practical constraints that cannot be dispensed with, and those assumptions shape the character of the knowledge produced.

MATERIALIST CONSTRUCTIVISM AND ELITE LABORATORY SPACES

The approaches to constructivism by those emphasizing practice can be seen as more "materialist" than a sociology of shared belief. In this case, the independence from human ideas comes from technology rather than brute, unaltered nature. Latour and Woolgar point out that the process of isolating and producing substances and measures and traces is an extremely artificial

---

experimental manipulation, for example, evolutionary theory (where historical explanations in open conditions apply) and astronomy (where instruments may extend perception without manipulating the phenomenon). This is particularly problematic since Bhaskar hopes to license a Marxian social science, which uncovers real historical structures. As Collier, *Scientific Realism and Socialist Thought*, pp. 142–47, points out, the metaphor of political action and learning as "experimentation" cannot bear the full weight that Bhaskar's discussion demands.

58. Latour and Woolgar, *Laboratory Life*, pp. 154, 244–52. See also the agonistic model developed in Latour, *Science in Action*.

process (read "manufactured" rather than "found").[59] So laboratory science, at least, looks a lot different than our image of patient, objective observation of the world without interference, the standard picture of empiricism. Indeed, Daston and Galison argue that the twentieth century ideal for science is trained judgment rather than the "mechanical objectivity" of the nineteenth century, which emphasized removing subjectivity to allow nature to speak without bias, a view that itself replaced earlier notions of correspondence to natural forms.[60] In fact, mechanical objectivity persists as an ideal today, especially where highly bureaucratized settings are designed to resolve political conflicts over scientific findings.[61]

It is easy to see that this ideological function of a mechanical notion of objectivity does not actually work as advertised to overcome the role of subjectivity in expert assessment of knowledge, which is why the role of trained judgment has been ascendant in the history and philosophy of science since Polanyi.[62] The growth of large-scale laboratory science also highlights the dependence of trained judgment on specialized equipment that makes plausible the notion of constructing knowledge on the basis of artifice rather than transparent representations of unaltered nature. This involves lots of interference and lots of technology, rather than watching how nature behaves on its own, as it were. Randall Collins made this distinction explicit in contrasting idealist constructivism ("the imposition of ideas upon the world") and materialist constructivism ("the dominance of physical practices, embodied in material equipment").[63] The former is closest to Collins's EMPOR, while Callon and Latour's actor-network theory fits the latter.

Followers of Aristotle used to call for an understanding of nature's "ordinary course," not "nature vexed" through technology, which they took as the opposite of natural. It was the seventeenth-century natural philosopher Francis Bacon who explicitly reversed this scale of values. Science approached understanding of the natural world through the artificial world, through "nature vexed" rather than nature in its ordinary, non-manipulated course. So Robert Boyle builds an air-pump, a machine that pumps air and creates

59. See also Haraway, "The Promises of Monsters," pp. 296–97, where the construction of nature by technoscientific practices is compared to commodity production with the qualification that human and nonhumans co-construct nature. Both Latour and Haraway's criticisms of a purely cultural constructivist approach to nature has shaped a new emphasis on natural agency in environmental history. See Walker, "Animals and the Intimacy of History," and Nash, "Furthering the Environmental Turn."

60. Daston and Galison, *Objectivity*.

61. Porter, *Trust in Numbers*. For national cultural differences over mechanical objectivity versus trained judgment, see Jasanoff, "Acceptable Evidence."

62. Turner, *The Social Theory of Practices*; Nye, *Michael Polanyi and His Generation*.

63. Collins, *The Sociology of Philosophies*, pp. 537–38; Latour and Woolgar's emphasis on the experimental production of artificial phenomena that are the objects of scientific explanation is borrowed from Bachelard's concept of phenomenotechnique (*Laboratory Life*, pp. 64, 258, n. 1). See Rheinberger, "Gaston Bachelard and the Notion of 'Phenomenotechnique,'" pp. 320–21, 324.

vacuums, and sees that as illuminating natural laws, even though we do not ordinarily see vacuums on earth. Technologies like air-pumps are taken to be signs of how nature works, not just something peculiar and unnatural we have done with machines. The elaborate work of transforming natural inputs into highly processed outputs is a rather new idea about how to understand the world. It connects objective knowledge of the natural world to specific, historical practices that produce highly processed, artificially produced material via specialized patterns of inquiry.

As such, materialist constructivism's focus on the particular epistemic leverage that laboratory sites provide challenges how knowledge has been framed by philosophy.[64] Emphasizing the socio-historico-material situatedness of science is held to lead to an idealistic account of knowledge. Idealism is held to be bad because it reduces the world to our image of it. By contrast, emphasizing the general applicability of material laws typically comes at the cost of conceiving of science as distinctly *placeless*, as in logical positivism's reduction of science to knowledge in principle available to any individual whatsoever by reduction of experience presumed to be (structurally) uniform for all observers (so long as they are not misled by metaphysics). Indeed, logical positivism shares the characteristics of revolutionary modernism found in the classic social theorists for whom progress would result in the "annihilation of space through time," whether through the increasing displacement of feudal forms by capital, the instrumental rationalization of society, or the cumulative increase of decontextualized knowledge.[65]

This means that the movement toward a scientific approach to the study of science associated with logical positivism can be seen as part of a larger modernist movement that associated progress with the dissolving of geographically bound impediments to rationality and freedom. Geographer David Harvey noted that "[t]he reduction of space to a contingent category is implied in the notion of progress itself."[66] In his 1921 critique of Spengler's nationalism, logical positivist Otto Neurath attacked Spengler's reliance upon the concept of culture, based upon a theory of types, whereby distinct spatially bound entities such as "nation" or "culture" can get a footing, arguing that such an approach should be replaced by one identifying a difference in stage of

64. Compare Lecourt, *Marxism and Epistemology*, pp. 136–39, who builds on Bachelard's philosophy to show how the objects of science emerge from spontaneous scientific practice, and from the infrastructure of the "technological city" (p. 138), in a way that philosophical accounts of experience impede. Only a "society can send electricity along a wire." The neglect of this insight leads philosophers to miss the relationships between "the history of production techniques and that of scientific instruments, between that of scientific instruments and that of their concepts" (pp. 138, 139).

65. Harvey, *Condition of Postmodernity*, p. 205.

66. Ibid.

development.[67] His approach here represented a coming together of his logical positivist beliefs about science and his political commitment to Marxism, in an idiosyncratic blend influenced by Austro-Marxism and other schools intermediate between Bolshevik orthodoxy and reformist social democracy.

His emphasis on centralized planning in combination with his logical positivist approach to science makes it easy to look back on his approach as scientistic and authoritarian. In fact, his approach emphasized democratic control and compromise based upon freedom and a diversity of interests. The democratic component depended upon popular education and the diffusion of knowledge, as in his emphasis on a universal language for science and his visual language, Isotype, for the public representation of statistical and social knowledge. Cat judges his approach as a "humanist, welfarist, voluntarist, scientific utopianism [that] was an engineered epicurean socialism (closer in spirit to young-Marx's philosophy)."[68] He was a part of an attempt led by the Austrian Socialist Party (SDAP) in "Red Vienna" to develop a socialist culture that aspired to be more democratic than the Bolsheviks' top-down cultural policies and that supported Vienna's republican form of government as a stepping-stone to that end. Rejecting the 1930s' version of virtue-based identity politics in the form of Alfred Adler's view that a new socialist man would have to be created prior to the construction of socialist society, Neurath believed that the emergence of socialism from out of capitalist development and republican government would create the environment leading to new behavior.[69] Nonetheless, Neurath's views have been dismissed by many scholars today due to the perception that he endorsed scientistic modernism, the bête noire of postmodernist and poststructuralist thought.

In reply to theorists questioning modernist "metanarratives" of progress, reifying local identities can be as reactionary as reductive, biological accounts of race.[70] It is worth reminding ourselves of why someone like Neurath would point out the regressive function that the concept of self-contained, autonomous culture can play. Cultural explanations can substitute fluidly for racial explanations of social problems. The function of culturalist explanations can be seen in the United States when politicians blame black culture for the lack of progress made by anti-poverty programs.[71] For Neurath, the solution would be to reject ideal cultural types as provincial holdouts against social progress. The peculiarities of place are dispensable for scientific and political reasons in Neurath's materialist system. As in his arguments for a

67. Neurath, *Empiricism and Sociology*, pp. 192–93.
68. Cat, "Political Economy."
69. Gruber, *Red Vienna*, pp. 5–11, 84–85.
70. Lynch, "Between Kin Selection and Cultural Relativism."
71. See Moynihan, *Negro Family*, which can be seen as a key transition in the transformation of Cold War liberalism into neoconservatism.

unity of science, the idea was that provincial differences were temporary; unity was a project for the future where different, incommensurable values need to be compared through informed judgment.[72]

DEMOCRATIC JUSTIFICATION AND HISTORICALLY DEVELOPED THEORIES

One of the motivations for conceiving of science as distinctly "placeless" —that is, as entailing the increasing decontextualization of knowledge—is that it can be used to emphasize a concern with *democratic justification*. For the logical positivists, the program of reduction to a common observation language was motivated in part by a desire to allow metaphysics and ideology to be removed from knowledge claims by allowing (at least potentially) any citizen to verify whether the claims were true. Otto Neurath incorporated this concern into his Marxism; indeed it appears that such "materialist" concerns were one factor in his introduction of a physicalist reduction language (reducing observations to objects) in place of Carnap's earlier phenomenalist reduction language (reducing observations to sensations).[73] It was part of his opposition to metaphysics that vague appeals to either "materialism" or "realism" were to be replaced by a more precise language that avoided the metaphysical tendencies introduced by such language.[74] It is in this fashion that he hoped to avoid the idealist tendencies of Mach's positivism that Lenin had criticized in *Materialism and Empirio-Criticism* while benefiting from the anti-metaphysical aspects of Mach's approach, which served as a protection against the reintroduction of reactionary metaphysics.[75]

Whereas Putnam would later use Lenin's thoughts as a springboard for a realist semantics that would counter the idealistic tendencies of Kuhn and Carnap, Neurath hoped to avoid idealism without taking this step:

72. Cat, "Political Economy." As put by Cartwright et al., *Otto Neurath*, p. 140, Neurath argued that "[c]ommunication with members of our own communities—with whom we share many beliefs—does not differ in principle from communication with members of alien cultures."

73. Carnap, *Logical Structure of the World*, developed a phenomenalism that Neurath saw as a version of idealism. Carnap was later convinced by Neurath that a physicalist reduction language was more appropriate. Since the choice of language was taken to be a practical decision and not one of truth or falsity, Neurath prevailed upon Carnap to consider the political implications of physicalism (as well as its *intersubjective* character). As Carnap (cited in Neurath, *Empiricism and Sociology*, pp. 43–45) recalls it, although he convinced Neurath that nineteenth-century materialism contained metaphysics, Neurath maintained that unlike idealism, it was not fully metaphysical, such that its progressive core could be maintained through a commitment to physicalism. See, however, the discussion of Putnam's causal theory of reference below.

74. Neurath, *Empiricism and Sociology*, pp. 415–16.

75. Lenin, *Materialism and Empirio-Criticism*.

> But "realists" too have a strong metaphysical streak, because of their dualism [i.e., between the world and its representation]. Even with those realists who are mainly concerned with the "existence" of the physical world, we often find an incomprehensible inclination to leave a tiny door open to the problems of the "freedom of the will," and the like. Only those who stand for strict physicalism, for the study of the spatio-temporal order, are completely consistent.[76]

The emphasis on intuited "empathy" rather than behavior in explaining social life was an important target of criticism here and, in general, an emphasis on the shared physical world as the basis for science and practical life was presented as key to the critique of metaphysics. When Vienna Circle member Heinrich Neider helped convince Carnap in 1929 that the intersubjective intelligibility of science required jettisoning the reduction of scientific concepts to private experience, to be replaced by a physical observation language, Neurath was pleased. In Neider's recollection, Neurath responded "'Of course, finally!,' for with this a bridge was built towards materialism, which he valued so much as the philosophy of the workers' movement."[77]

This idea that there is not a dualism of the world and knowledge about the world, that then needs to be connected in some fashion, underlay Neurath's critical comments on Carnap's *Aufbau [Logical Structure of the World]* (1928) and set the stage for a fully naturalistic and sociological approach to science, contrary to the usual picture of logical positivism as foundationalist in orientation. Carnap himself sought to create a kind of "science of science," albeit one rooted in the formal sciences of logic and mathematics rather than the empirical sciences of psychology or sociology. Even before publication of the *Aufbau*, Neurath had urged Carnap to avoid methodological solipsism, whereby Carnap correlated scientific facts with structurally similar sensations of individuals. Despite the effort to provide a formal, structural unity to the distinct sensory experiences of individuals, this approach looked too idealist and individualistic to Neurath. Moreover, not only did phenomenal experience not guarantee intersubjective agreement—my experience may differ from yours—but it did not guarantee intrasubjective agreement, as my experience now may not be the same as my experience at another time. He advised Carnap to "[s]ay right at the start that [the] goal is an objective world which is the same for all individuals."[78]

---

76. Neurath, *Empiricism and Sociology*, p. 416. The critique of the dualism of metaphysical realism is later a theme of Hilary Putnam's internal realism position (one of his many evolving positions on the realism issue). See Putnam, *Realism with a Human Face*.

77. Quoted in Uebel, *Overcoming Logical Positivism*, p. 91. See also Cartwright et al., *Otto Neurath*, p. 152.

78. Uebel, *Overcoming Logical Positivism*, p. 73. On Neurath's critique of Carnap's methodological solipsism, see Cartwright et al., *Otto Neurath*, pp. 148–58.

Carnap would later explicitly acknowledge Neurath's efforts to see knowledge as something within the material world itself, not standing apart from it in a separate realm (as in Popper's later view of three worlds: material objects, mental images, and objective knowledge): "Neurath emphasized *from the beginning* that language phenomena are events *within* the world, not something that refers to the world from the outside."[79] Thus, speech is based on sound waves, while written language on marks on paper. Carnap objected that it was the structural characteristics of these examples, not their purely material character, that made them meaningful, which highlights the difficulties of characterizing how exactly knowledge resides within the material world. What does come across clearly is that Neurath already shared contemporary social epistemologists' aims to describe knowledge as something "in and about the world."[80]

As Thomas Uebel put it, Neurath's "point was that the justification of scientific knowledge should not, indeed could not, proceed by means of heavily idealized reconstruction which neglected the distinctive characteristics of what he deemed the real object of concern, historically developed theories."[81] It is here that Neurath develops a view closer to Marx's practice-oriented approach than Carnap's focus on the logical structure of theories. Science itself was a social activity, and any philosophy of science needed to understand how the scientific community produced knowledge in practice, not via a rational reconstruction of a hypothetical individual's sensory engagement with the world.[82]

Indeed, Neurath and Charles Morris, a U.S. defender of logical positivism and its compatibility with pragmatism and semiotics, had tried to fend off this misunderstanding at the pass. As George Reisch shows, the unity of science movement developed an activist and pluralist approach to science and political life in the 1930s and early 1940s. That gave way in the Cold War to icy logical analysis and apolitical professionalization.[83] The key transition point in this depoliticization was the charge by a former ally of Neurath's, Horace Kallen, on the eve of World War II in 1939, that the Unity of Science movement was a "totalitarian" movement that aimed to prescribe and enforce a single program for science that suppressed scientific diversity and intellec-

79. Quoted in Uebel, *Overcoming Logical Positivism*, p. 72. Compare Popper, "Three Worlds"; Hacohen, *Karl Popper*, p. 69.

80. Radder, *In and About the World*; Fuller, "Naturalized Epistemology Sublimated," pp. 292–93. Radder, "Which Scientific Knowledge Is a Common Good?," pp. 442–44, raises the same point as Carnap did about Neurath's material definition of knowledge, this time directed at Callon's localist, materialist definition of knowledge. Radder argues that scientific concepts have "non-local meaning," potentially extensible to new contexts, rather than confined to local contexts of meaning. The normative implication is that scientific knowledge can be understood to be a common good, in opposition to viewing it as a commodity.

81. Uebel, *Overcoming Logical Positivism*, p. 74.

82. Ibid., p. 96; Cartwright et al., *Otto Neurath*, pp. 163–64.

83. Reisch, *How the Cold War Transformed Philosophy of Science*, ch. 8.

tual freedom. The charge must have left Neurath scratching his head, given his efforts to oppose fascism via an ecumenical democratic socialism and a pluralist approach to science, as part of a movement of refugees from fascism. Kallen's accusation, first made at the Fifth International Congress for the Unity of Science at Harvard University immediately after Hitler's invasion of Poland, was revived after the war and debated in the pages of *Philosophy and Phenomenological Research* in 1946.[84]

What is interesting in light of our current discussion of science studies is the extent to which Kallen's critique resembles current misunderstandings of logical positivism by postmodernists, constructivists, and philosophical advocates of the "disunity of science" today.[85] Kallen, like most antipositivists today, rejected a demarcation criteria for science as inappropriately authoritarian, allowing only a minimal kind of unity for science by rejecting any interference with intellectual liberty, even for traditions rejected as pseudosciences like astrology or numerology.[86]

For Neurath, physicalism as a common language for science was not an attempt at reductionism, or a requirement to translate all scientific theories into the language of physics. Instead, physicalism referred to the common, everyday language of ordinary objects that all humans could agree upon that would allow them to coordinate their efforts to relate separate and distinct scientific programs to each other. He sought to do this not by advocating a theoretical synthesis, but by calling for a practical integration of distinct sciences to solve complex problems by relating them to each other. In this sense, it is not that far from Marx's "sensuous practice" applied to science and emerged as a goal for Neurath precisely in his attempt to convert Bavaria to a fully socialized economy as head of the Central Planning Office during the brief socialist revolution in 1919.[87]

The project for a unified science would be one to be accomplished in the future that would encourage pluralism in the present, while rejecting only "different magical expressions, theological sentences, or metaphysical formulations" that could not be expressed in a framework understandable to all. Only such parochial languages would require force to be imposed as a common language. The unity of science, then, was a "from the bottom to the top," interdisciplinary movement highlighting democracy in the intellectual and political spheres.

> I think that this gives a picture of the democratic attitude of the Unity of Science movement, which acknowledges from the start a multiplicity of possibilities. It

84. Ibid., pp. 174–86.
85. Galison and Stump, *Disunity of Science*.
86. Reisch, *How the Cold War Transformed Philosophy of Science*, pp. 170–71, 182.
87. Cartwright, *Otto Neurath*, pp. 171–79.

> is the problem of any democracy, which any actual scientific research organization has to solve: on the one hand, the nonconformists must have sufficient support; on the other hand, scientific research needs some cooperation.

By contrast, "[i]f priests and rulers have a language of their own, they become separated from the ruled masses, and it is just the unification of language that is a step forward to some democratic possibilities."[88]

In short, Neurath would agree with Lakatos and Feyerabend in rejecting the cognitive authoritarianism of Kuhn or Collins that stemmed from deference to experts and their specialized languages, and in requiring efforts to build relations between separate research programs while allowing a pluralism of perspectives and scientific liberty.[89] Neurath's anticipation of the naturalistic and sociological turns associated with contemporary schools of philosophers and sociologists shows how a concern with "materialism" already pulled in both directions at once—(1) producing a naturalistic "science of science" that eschewed foundationalism for an empirically grounded account of how science gains knowledge, drawing on findings in the scientific disciplines themselves, and (2) developing a historically and sociologically rooted account of the emergence of scientific knowledge that emphasized its engagement with the material world, rather than its reflection of an ideal-typical culture. It is the latter point that is connected to Neurath's Marxist-inspired desire to find grounds for universal justification of scientific claims, consonant with his commitment to "the social dimension of knowledge production and transmission within the framework of historical materialism."[90]

In Uebel's analysis, it is the constitutively social aspect of knowledge, based on shared language and Marxian sensuous material practice, which makes science possible:

> For Neurath, it was imperative, as it was for Carnap, that the knowledge claims of science be shown to be *in principle justified*; it also was imperative, as it was for Schlick that they be shown to be *in fact justified*. Above all, however, it was imperative that the rationality of science be shown *practically intelligible* in its actual pursuit. Neurath asked wherein lay the possibility and actuality of intersubjective *control* of scientific assertions. He demanded that the justification of scientific knowledge claims reflect the practical requirements of rational

88. Quoted in Reisch, *How the Cold War Transformed Philosophy of Science*, p. 178.

89. Reisch, "Planning Science," p. 172, notes that Popper's criticisms of Neurath's emphasis on planning misses the similarities in their perspectives: "Neurath expected unified science to grow gradually and cautiously, in much the same way that Popper's 'piece-meal engineers' would adjust society 'step by step . . . always on the lookout for the unavoidable unwanted consequences of any reform.' And, in its own way, Neurath's programme assumed just as much freedom as Popper's. It demanded an unrestricted production of ideas about how terms and statements of different sciences could be coordinated and made consistent."

90. Uebel, *Overcoming Logical Positivism*, p. 20.

> cognitive conduct and he stressed that the account of how epistemic agents came to be justified in their claims had to make those grounds *intersubjectively available*. That meant that any knowledge claim had to be justified on grounds that are intelligible to others beside the claimant. The objectivity of scientific knowledge had to be derived, in one way or another, from the intersubjective controllability of scientific statements, it had to build on the *practical condition of intersubjective understanding*.[91]

What this means for our discussion is that "materialism"—in this case, historical materialism—drove the demand to understand how expertise can in principle be verified by anyone so that knowledge claims are not the exclusive possession of the expert communities who produce them. Any hint of the self-warranting view of expertise would be anathema here as local—or provincial—special pleading.

More broadly, the Vienna Circle as a whole was allied with architectural modernism in opposing nationalism and other "locally" rooted ornamentations, as Galison has shown:

> Throughout their writings Carnap, Neurath, and others singled out modern architecture as the cultural movement with which they most identified; their interests were reciprocated as the logical positivists were more prominent as visitors to the Dessau Bauhaus than members of any other single group outside art and architecture. Further, the two movements faced the same enemies—the religious right, nationalist, anthroposophist, *volkisch*, and Nazi opponents—and this drove them even closer together, toward the conjoint life they had in mind. Both enterprises sought to instantiate a modernism emphasizing what I will call "transparent construction," a manifest building up from simple elements to all higher forms that would, by virtue of the systematic constructional program itself, guarantee the exclusion of the decorative, mystical, or metaphysical. There was a political dimension to this form of construction: by basing it on simple, accessible units, they hoped to banish incorporation of nationalist or historical features.[92]

Thus, in emphasizing physicalism, Neurath hopes to allow a genuinely scientific approach to society to be advanced.

In doing so, however, he accepts a highly implausible ontological assumption that Bhaskar has termed "actualism"—namely, the thesis that events in the world naturally occur in lawlike clusters.[93] Arguably, the emphasis

---

91. Ibid., 20.
92. Galison, "Aufbau/Bauhaus," pp. 710–11.
93. I wish to separate this criticism of empiricism from the criticism of the epistemic fallacy (that all ontological questions can be rephrased in epistemological form) with which it is typically conjoined, since non-realists typically find the work that is done with this term puzzling: the opposite charge of the ontological fallacy could be made. Whereas attacking the epistemic fallacy involves

on a *shared* physicalist observation language led to a neglect of the role of specific, historically and geographical delimited experiments in uncovering structures referred to in knowledge claims, structures not "verifiable" outside the community of those whose training and experience in specific scientific practices allow them to see what outsiders cannot. While Neurath rejects Carnap's attempt to build up shared and verifiable experience from methodologically solipsist assumptions, Neurath's own account presumes that it will be relatively straightforward to separate out each science's engagement with everyday physical reality from any "magical expressions, theological sentences, or metaphysical formulations" smuggled in by its practitioners. Neurath's actualism is evident from statements such as the following: "The problem is to create a closed and unified system of statements, by means of which we can make testable predictions about spatio-temporal processes, on the scientifically discoverable order. That is the task of unified science on a materialist basis, the task of physicalism."[94]

The focus on empirical regularities rather than highly unique, technologically mediated experimental interventions into the world (that are rarely replicated in detail, as Harry Collins points out[95]) leads Neurath and the rest of the logical positivists to make the question of the democratic justification of knowledge claims too easy by presuming that expertise is ultimately a dispensable category and that we all have potentially equal access to the mysteries of science once metaphysics is laid aside. The result in the social realm is that laws that "keep close to actual events and their peculiarities" of distinct social formations are needed.[96] The rejection of theoretical claims that are not, in principle, reducible to observation of everyday objects would disable the kind of deep structure explanations Marxists typically make when they appeal to entities such as capital and labor, which is one reason why contemporary Marxist or Marx-influenced philosophers of science like Putnam and Bhaskar pay more attention to the role of experiment in enabling our access to underlying causal structures than a fixation on empirical regularities can do. Putnam developed an alternative account of meaning from the logical positivists' verificationism that also moves meaning outside the head to the material world.[97]

---

pointing out that "ontological questions cannot always be transposed into epistemological terms" (Bhaskar, *Reclaiming Reality*, p. 18), the ontological fallacy would consist in failing to realize that epistemological terms cannot always be transposed into ontological terms. On this view, humanly obtainable knowledge tells us nothing about what the ultimate furniture of the world would be.

94. Neurath, *Empiricism and Sociology*, p. 417.

95. Collins, *Changing Order*, chs. 2–3.

96. Neurath, *Empiricism and Sociology*, p. 371.

97. Putnam, "The Meaning of 'Meaning.'" Neurath, *Empiricism and Sociology*, p. 417, does suggest that one can use social experimentation to test latent possibilities: "Hence we can compare large-scale practical political activity 'in the grand style' with the experiments of the physicist. That is to say, the more often one does something in the sense described by sociological theory, the more frequently one will test the sociological theory."

Karl Popper's objections at the time to Neurath's approach, and verificationism in general, reflect similar concerns with the link between epistemological transparency and political accountability, albeit coming from a very different philosophical and political standing point. Popper denies that verifiability serves the democratic objectives of intersubjective testing.[98] Indeed, coming from a quite different normative sensibility than Neurath, Popper argues that verificationism actually allows metaphysics to take root, especially that form of metaphysics hostile to the Open Society, Marxism.[99] Neurath's analogy between the neutral mechanical engineer and the neutral social engineer was particularly suspect for Popper.[100] This should not obscure the extent of agreement between positivists and Popperians that epistemology should enable non-scientists to evaluate the claims of scientists by a philosophy of science that was publicly accountable and that these issues had important connections with models of the good society.[101]

Popper objected to Neurath's Marxism as inappropriate in an open society, arguing that verificationism was both epistemologically and morally suspect.[102] Neurath did suggest that the proletariat might be in a privileged position to recognize ideology since they do not have the same stake in the status quo that led the bourgeoisie away from adequate historical knowledge, for instance.[103] However, he did not suggest that one had to be a member of the proletariat to apply verificationism. Verificationism was adequate for any observer whatever and it was simply an empirical claim that, in certain sciences, the proletariat would be more likely to apply it. Contrary to popular prejudice, neither Popper nor Neurath denied that science was a fully social enterprise. What they did deny is that binding procedures of justification could not be found to adjudicate between competing "local," historically contingent interests.[104]

---

98. Popper, *Logic of Scientific Discovery*.

99. But see Hacohen, *Karl Popper, the Formative Years, 1902–1945*, chs. 1–2, on the influence of Red Vienna on Karl Popper's early formulation of his political philosophy and philosophy of science. Hacohen shows that Popper's *Open Society* was primarily concerned with resisting fascism, and his criticisms of socialists in that work reflected his perception that they erred in not aligning with liberals to fight the erosion of constitutional democracy. During the Cold War, the work came to be seen as a cornerstone of the containment of communism, facilitated by Popper's own political shift rightward.

100. D'Amico, *Historicism and Knowledge*, pp. 152–53.

101. See Popper's comment (in Neurath, *Empiricism and Sociology*, p. 56): "Neurath and I had disagreed deeply on many and important matters which interested us both except one—the view that the theory of knowledge was important for an understanding of history and of political problems."

102. Karl Popper, *Unended Quest*, pp. 31–38, 78–87.

103. Neurath, *Empiricism and Sociology*, p. 356.

104. Galison, "Aufbau/Bauhaus," p. 714, has described Neurath's politics as "unpolitical." Even while heading the Bavarian revolutionary government central planning office, Neurath emphasized his desire to remain an "unpolitical administrator" (Neurath, *Empiricism and Sociology*, p. 21). Although his politics raise suspicion now in an academic environment that equates positivism and authoritarianism, it should be noted that Neurath's proposal called for voluntary cooperatives of

We can note how a concern with "placeless" justification appears to be connected with a surprising lack of concern with experiment, rather than mere observation, as a means of generating knowledge. Indeed, Carnap's *Aufbau* fails to consider experiment at all, focusing its efforts on showing how ordinary observation reports by individuals can be seen to be structurally related to each other, providing a shared language for describing the world. He does show, on that basis, that scientific referents from electrons to plants to psychological stages can be built up from structural relationships identifiable in experience of the world, but nothing is said about scientific experiments or practical methods of investigating these referents per se.

With the recognition of the resulting descriptive inadequacy of such decontextualist, empiricist approaches, alternatives emphasizing first, theory-ladenness (and hence "contextualism" of some sort) and, second, the importance of laboratory practice (and hence "situated" interventions and the need for expertise) for obtaining scientific laws become important. The recognition of the first point is conventionally attributed to Kuhn, although realists have pointed out that such a view differs little from the latter Carnap's holistic view, as he had argued explicitly in *The Logical Syntax of Language* (1934) for a principle of tolerance whereby the choice of language for representing the world is conventional and, therefore, distinct, referring systems are possible.[105] Within the Vienna Circle, in addition to Carnap's conventionalism and holism (endorsed by Neurath), Hans Hahn had had already argued in 1931 that all language statements were theory-laden and the isolation of basic statements was conventional. Neurath pushed this view further, arguing that basic statements understood as empirical givens did not exist.[106] In this sense, he would agree with the criticisms of positivist orthodoxy Feyerabend developed in the 1950s, discussed in chapter 4. Neurath was the first post-positivist.

---

peasants to be run separately from the state (drawing on Kropotkin) in order to avoid clashes between peasants and workers. Moreover, he was concerned to avoid bureaucracy by incorporating workers into the planning process on a rotating basis. Despite attempting to maintain his neutrality during political turnover, Neurath served a year and a half in jail for "assisting high treason" for continuing his attempt at full socialization of the economy during the brief Soviet Republic, April 14–27, 1919 (pp. 26–27). For Neurath's requirement of popular participation in the application of science to public policy, rather than a technocratic approach he called pseudorationalism, see Neurath, "Pseudorationalism of Falsification"; Cunha, "The Utopia of Unified Science," pp. 321–22.

105. Boyd, "The Current Status of Scientific Realism," pp. 41–82, 55–56. Friedman, "Carnap's *Aufbau* Reconsidered," argues that phenomenonalist reduction was not crucial for Carnap, but rather the construction of a system of interconnected concepts such that all concepts could be discriminated on purely formal grounds—hence Carnap *begins* with a holistic rather than atomistic approach, which he assimilates to a constructionist program (pp. 528–30).

106. Uebel, *Overcoming Logical Positivism*, p. 123.

LOCALITY IN CONTEMPORARY PHILOSOPHY OF SCIENCE 1: PUTNAM'S CAUSAL THEORY OF REFERENCE

Putnam has drawn upon Lenin and Engel's polemics against idealism to develop a causal theory of reference as an alternative to the theory of meaning of the logical positivists, which is identified as the hidden root of the concepts of incommensurable paradigms of Kuhn. For the logical positivists, meaningful sentences could only be those it was possible to verify empirically. Thus the meaning of a claim was taken to designate its extension, usually by specifying necessary and sufficient conditions for membership in a category. Putnam points out, following Engels, that the appropriate definition of fish includes "breathing through gills" even though not all fish breath through gills. The property "breathing through gills" is part of what Putnam calls the "stereotype" of a fish, which allows competent reference to take place among those who might disagree on the essential characteristics of fish and on the exact extension of the term.[107] Notice that this does the work of intersubjective intelligibility across theoretical differences, which was the stated concern of Neurath and Carnap, but that had been made impossible by Kuhn's concept of incommensurability.[108]

Although Kuhn grants that natural kinds are not associated with necessary and sufficient conditions, and his concept of exemplars resembles Putnam's "stereotypes," in other respects Kuhn sticks close to the logical positivist account of meaning. Different theoretical meanings of terms lead competing scientists to different "predictions" (understood as puzzle solving), which in Kuhn's case can lead them to different accounts of the empirical confirmation of these terms. Given the premise that meaning determines reference, Kuhn is led to speaking of different worlds. What Kuhn could have said is that competing approaches share the same world, but that a style of reasoning is needed before one can have a determinate truth value.[109] Alternatively, following Feyerabend, the (changing) theories providing an interpretation of specific observations may be in conflict, even while everyday, practical language is shared.[110] The world may admit of multiple approximately true vocabularies. This is one possible interpretation of the strong program, which argued that precisely because material reality is a shared shaping force for competing theories, it cannot discriminate between the true and false ones.

---

107. Putnam, "Explanation and Reference," pp. 171–85.

108. Note that this is equivalent to Kuhn's exemplar concept, which avoids much of the baggage found in his paradigm concept. For a defense of exemplars emphasizing the material embodiment of knowledge, see Rouse, *Knowledge and Power*, ch. 2.

109. See Hacking, "Language, Truth, and Reason," pp. 48–66. For a related argument see Margolis, *Pragmatism without Foundations*, pp. 86–89.

110. Oberheim, *Feyerabend's Philosophy*, pp. 61–63.

Kuhn's language of separate worlds produced by paradigms (with its idealist tinge) may reflect a misunderstanding about how concepts relate to their objects, which is the point of the alternative semantics of Putnam. This would make sense given Kuhn's later admission that he originally conflated the gestalt shift of the historian reconstructing past science with a putative paradigm shift in the actual history of science.[111] Whether or not some form of incommensurability remains a helpful concept in understanding science (as I have argued, borrowing Feyerabend's version), Putnam points to an important fact as to how common reference is maintained during theoretical controversy: the same everyday objects, by and large, are referred to by competing parties. Such a realist move shares much with constructivist work in emphasizing the importance of material culture, since it is the "sensuous material practice" at the heart of the scientific process that fixes reference on this view.

In order to make this picture work for the reference of scientific terms, unlike the widely available natural kind of "fish," Putnam puts forward the concept of an introducing event that fixes reference by interacting with the natural kind in an appropriate manner. Future uses of the term then depend on having an appropriate causal relationship to the introducing event. The appropriate type of causal relationship, Putnam argues, involves being the member of a collective that has had "contact" with the entity in question.[112]

One might provide a gloss that highlights the differences with logical positivism's "placeless" justification: While avoiding Kuhn's idealistic talk of different worlds, Putnam introduces the idea that reference might be a collective accomplishment dependent crucially upon "situated" expertise, in the sense that only experts in special sites (i.e., laboratories) may interact "directly" with an entity; these experts may disagree on how to account for this entity; they may coordinate their disagreement by reference to a "stereotype" that allows their common interaction with the entity to proceed despite this disagreement; and reference to such entities by the lay public makes sense by the fact of social dependence upon expert use of the term.

Here we have a picture where expertise and "situated" laboratory inquiry are ineliminable: epistemic access is unequally distributed, and no general

111. See Kuhn, "Possible Worlds in History of Science," pp. 9–32. It is not clear that one can always parcel these competing vocabularies out to different "domains" of an otherwise unitary representation of the world since disciplinary boundaries often are "crudely" conventional in terms of maintaining spheres of disciplinary authority, each of which finds a different way of describing the shared world. See Gieryn, "Boundary Work"; Woolgar and Pawluch, "Ontological Gerrymandering."

112. Putnam, "Explanation and Reference," p. 176. Kuhn, "Dubbing and Redubbing," pp. 309–14, shows that ensuring that even everyday objects share continuity of reference can be problematic, as prior to the eighteenth century, "water" referred only to liquid water, not steam or ice, included in the modern account of water as H_2O. What this means is that as the causal theory of reference has developed, theoretical descriptions have to be examined to demonstrate that reference is in fact shared, rather than assuming that to be the case, which can bring incommensurability back into play. See Wolf, "Reference and Incommensurability," pp. 215–21.

test, whether verificationist or falsificationist, can guarantee that one is using a vocabulary that genuinely refers. Theoretical controversy and change can be driven by different extensions of the stereotype among core-set scientists, just as the sociologists have argued. Expert disagreements about how to build upon the shared stereotype to best describe the object in question can go through the stages of controversy and closure, just as Collins argues, consistent with a continuity of reference based on the causal theory of reference. In this way, scientific realists hope to be able to take on board all the revisionist history and sociology of science conducted over the past half century and recast it to tell a story of progress toward a true representation of underlying reality.

To be sure, there remains much within the realist account that constructivists may continue to find objectionable, even if a general convergence between constructivists and realists on the role of situated expertise obtains. The realists use the causal theory to develop a picture of an accumulation of knowledge following the introduction of an approximately true vocabulary once one begins to actually refer to real entities (perhaps during the Scientific Revolution or with the establishment of a mature paradigm, in Kuhn's terms). This way of looking at things takes for granted that the relationship between a theory and the world can be discussed in isolation. Lakatos and Feyerabend both rejected that view.

As we saw in chapter 2, Lakatos developed an approach that rejected a static relationship between facts and theories, with progress instead being defined by the relationship between a series of theories and the facts generated conventionally within that series. As such, Lakatos rejects correspondence for a conception of progress internal to the dynamic of a research program. As Hacking put it, Lakatos sought "to provide a theory of objectivity without a representational theory of truth."[113] Feyerabend expands this to include the relationship between distinct research programs and putative facts about the world, as he argues that incommensurable theories can be used to provide an external basis for challenging an excessively close fit between dominant theories and the conventionally produced facts that they fit, what Oberheim calls his pluralistic-test model.[114] In other words, scientific realism would tend to block the operation of a critical approach that would be needed to see the limitations of current knowledge.[115]

---

113. Hacking, "Lakatos's Philosophy of Science," p. 129. For discussion, see Lynch, "Imre Lakatos and the Inexhaustible Atom," p. 28. A similar argument motivates Bachelard's introduction of a historical epistemology, as theory is no longer grounded in an unassailable experiment so theory and experiment may be out of step with each other at any one moment in time. See Lecourt, *Marxism and Epistemology*, p. 136.

114. Oberheim, *Feyerabend's Philosophy*, pp. 227–43; Lynch, "Imre Lakatos and the Inexhaustible Atom," p. 29.

115. See Kasavin, "Uniting the Cognitive and the Social," pp. 44–45, for the argument that realism leads to "the dogmatic reification of . . . scientific theories" by a process akin to commodity fetishism, blocking the admission of new, constructed objects to scientific ontology.

Moreover, arguments that realism provides the best explanation for the success of science depend upon the assumption that science has a unitary aim in the first place, something defenders of the locality thesis would deny. On Joseph Rouse's view, for instance, realism continues to suffer from "modernist" flaws by positing a global content/context distinction that allows an internal history of science to be abstracted from the multitude of diverse practices united under the banner of science.[116] As Fuller has pointed out, language does not function well as a vehicle to preserve content across contexts—indeed, it functions to bring about change in the meaning of concepts over time in ways that are not noticed by participants (and that the realist may obscure by normalizing their reference after the fact).[117]

The growing emphasis on the disunity of science reflects this postmodern trend as well.[118] However, Neurath's conception of the unity of science did not depend upon a reduction to the science of physics or any implication that inconsistent beliefs would be eliminated at any one time. His model was the encyclopedia, not a unified theory. Unity emerged from how distinct sciences were connected in explaining any specific process.[119] It is in Neurath's sense that I have endorsed a unity of science thesis as emerging over time despite the legitimate autonomy (and incommensurability) of perspectives at any given time in the history of science (see chapter 3). From a dialectical point of view, the totality of science may reveal an emerging, but never arriving, unity. Paraphrasing Feyerabend, potentially every scientific culture is all scientific cultures. The fixity of reference is not given by an engagement with the material world in a moment of time, however, and is confounded by Lenin's concept of an inexhaustible atom, as new scientific developments revise our understanding of the furniture of the world.

The notion of an initial "baptism" that *fixes* future reference once and for all is quite dubious historically. While we can identify Priestley's "deplogisticated air" with Lavoisier's (and our) oxygen, this only makes sense because we take for granted that Priestley mischaracterized a natural kind because we take for granted the modern account of oxygen. Should subsequent theoretical changes challenge that assumption, unlikely though that may be at this point, the question of whether and when past scientists were referring to the same objects as us would have to be determined anew. It makes as much sense to say that later theoretical changes in the extension of a term can "ret-

116. Rouse, "Philosophy of Science," pp. 151, 158.

117. Fuller, *Social Epistemology*, p. 84.

118. Galison and Stump, *Disunity of Science*.

119. Cartwright et al., *Otto Neurath*, pp. 179–88. See also Neurath, *Philosophical Papers, 1913–1946*, p. 59, where although different kinds of sciences have distinct laws, "it can *not be said of a prediction of a concrete individual process that it depend on one definite kind of law only*." Compare Klein, *Interdisciplinarity*.

rospectively" alter the original referential act, in turn defining a new trajectory of putatively common reference leading to a new endpoint.[120] Instead of baptism, we have rewriting based upon selective memory.

The causal history of reference still awaits the end of inquiry, which obviates realism's superiority as an explanation of contemporary science. Thus, with the discovery of subatomic particles possessing a fraction of a charge, it would be possible to reconsider the anomalous data Robert Millikan discarded in his oil drop experiments determining the value of the electron's charge, as well as the evidence for subelectron charges argued for by his opponent in controversy, Felix Ehrenhaft. What had been a failure to refer, even considered by some fraudulent efforts by Millikan to clean up the data, could be reinterpreted as referential in light of a later theory. (Although Paul Dirac has speculated about the possibility of the anomalous charges being measurements of quark charges, it is not considered likely based on current knowledge).[121]

The example also brings out the slipperiness of the concept of *approximate* truth insofar as Millikan's experiment was significant for having uncovered the basic unit of electrical charge, a conclusion that turns out to be wrong—or rather, as realists would have it, only approximately true—after the discovery of quarks with fractional charges. We can revise the contextual descriptions and say that Millikan discovered merely the charge of the electron, which is three times the basic unit of charge (quarks can have charges with magnitude of one-third and two-thirds of that of the electron). My point is that there is no end to the redescriptions that need to be carried out regarding what a scientist discovered at a point in time as theoretical change proceeds in science. Only a retrospective narrative trick makes everything seem like part of a continuous story of reference and winnowing of truth from error.

On this view, while the causal theory of reference can help historians figure out how scientists at different times and with different theoretical assumptions could be talking about the same thing because they share the natural world, its ultimate dependence upon the current state of knowledge begs the question whether science is converging on the one, best account of an underlying reality. Moreover, realism continues to represent a residual category that remains decidedly hypothetical, as realists reject any antirealist "reduction" of truth that equates truth with instrumental success or evolutionary fitness. In effect, realist arguments so rig the burden of proof by the way they describe their

120. Žižek, *The Sublime Object of Ideology*, ch. 3.

121. Niaz, "The Oil Drop Experiment," pp. 485–93. Holton, "Subelectrons, Presuppositions, and the Millikan–Ehrenhaft Dispute," pp. 164–65, especially n. 12; Silverman, "Raising Questions." Note that Feyerabend studied with Ehrenhaft, an experience that shaped Feyerabend's criticisms of empiricism. See Oberheim, *Feyerabend's Philosophy*, ch. 4; Feyerabend, *Killing Time*, p. 66.

opponents that ever-increasing accuracy in uncovering referential truth is shifted from the conclusion of an argument to an unquestionable premise.[122]

For realists, the "social construction of knowledge" is understood to mean either (a) the distortion of nature by social processes, or (b) the structuring of the actual real-world object to reflect one's theory. Thus, when Richard Boyd discusses social constructivism, he notes, "In 'socially constructing' racial differences, nineteenth-century biologists did not construct a world in which those of African descent are biologically suited to a subordinate role, however much they constructed theories to that effect, nor have their latter-day followers done so—any more than those same biologists (or we) have socially constructed a world in which the place of women is determined by biological necessity."[123] No social constructivist would ever suggest that that was the case. If some scientists incorporate racist and sexist assumptions in guiding their work in science, this reflects how they see the world and the limitations of such a view are free to be challenged (and have been) by other scientists. Only by assuming that the term "knowledge" as understood by the constructivist includes a correspondence to an underlying reality can the constructivist be said to believe that a process of social construction of racist and sexist science creates underlying gender and racial differences.[124] In other words, constructivist arguments must be interpreted to make realist assumptions for such a charge to hold. Constructivists, by and large, plead agnosticism about what underlying reality is since they believe that our knowledge constructions do not break down into social and natural components, something that is dismissed by constructivists as the sociology of error.

Of course, we have pointed out that strong programmers and actor-network theorists, each in their own way, recognize the contributions to science from nature but they deny that that contribution can be "located" in representations, as it were. Realists demand to understand the history of science as the continual replacement of knowledge claims with new ones that can be characterized as allowing a greater contribution from nature *rather than* from social construction. The point is that the relative contributions of nature and society for mature science cannot be decomposed into parts. "Nature" only speaks, to the extent that it does, in and through society's efforts to understand nature. If non-racist and non-sexist science is better than racist and sexist science, it is not unmediated nature itself that can show us that but rather the community

---

122. Fuller, *Social Epistemology*, ch. 3.

123. Boyd, "Constructivism, Realism, and Philosophical Method," p. 134.

124. Arguably, this conclusion could be seen to follow from Woolgar, *Science: The Very Idea*, where the idea that nature shapes our constructions of it is simply reversed. Kitcher's "modest realism" points out the continuing effects of past, repudiated classifications, such as belief in significant racial differences, can live on as historical constraints on scientific and societal practices that are difficult, but not impossible, to dismantle. See Kitcher, *Science, Truth, and Democracy*, pp. 52–53.

of scientists who end up replacing scientific claims held to be inadequate with new theories. Moreover, the niceties of approximate truth are ignored when examples like this are trotted out, since it is perfectly possible that racist and sexist science combine truly referential biology with non-referential and ideological constructs.[125]

Before the mythical end of inquiry, all approaches can be seen to be a mix of referential and imaginary concepts according to realism, so that realists are in no better position to show why the racist and sexist is wrong lacking the final vocabulary of nature. Like the antirealist, the realist must just appeal to other constructions of knowledge to explain where ideological science has gone wrong. The realist may demand a "principled solution,"[126] to the separation of ideological from true science, but in practice they can do no more than the antirealist would do: point to other reasons put forth by human beings why the ideologues are wrong. Realism in this strong sense may amount to little more than an exclamation point, which may be why there has been a growing trend to water down the realist account of reference among philosophers, once it has done its work of countering Kuhn's incommensurable worlds, as we will see below.

The whole notion of "approximate truth" (developed by Popper under the label verisimilitude) gives the realist position, what Fuller calls a "'pseudo-empirical' feel," in that different conceptions of the electron evident in the Bohr-Rutherford model and the later quantum mechanical model are held to be approximations to the same object by virtue of "playing many of the same conceptual roles."[127] Such a representation reflects a principle of charity that makes the history of science intelligible, which, however, these theoretical changes render moot without an assumption that the content of knowledge is preserved over time. The situation is only compounded when considering that truth can be "approximated" in different ways, even assuming a unitary truth that the history of science is converging on. So before the end times, we are stuck with competing and contradictory theories that at any one time more or less equally refer to the world. The views can be contradictory, as Neurath already showed before Feyerabend, since the non-referring elements of each theory have not yet been sifted out. Such a conclusion differs little from the strong program now that the real world shapes all theories and cannot by itself decide between them.[128]

---

125. Lynch, "Between Kin Selection and Cultural Relativism"; Putnam, *Realism with a Human Face*, introduction, makes a related point about decomposition into natural and social components.

126. Boyd, "Constructivism, Realism, and Philosophical Method," p. 134.

127. Fuller, *Social Epistemology*, p. 73. On the controversy in Popper's camp over whether verisimilitude reopened the door to inductivism and whether that was desirable, see Hacohen, *Karl Popper*, pp. 528–29.

128. Uebel, "Logical Empiricism and the Sociology of Knowledge."

Despite the debate over realism, contemporary realists and constructivists both put emphasis on the situated activities of scientists in the laboratory and the role of expertise in licensing scientific claims. Since the causal theory of reference is intended to be naturalistic, it is appropriate to examine empirically how local practices shape scientific claims. The divide that remains between work in naturalistic epistemology and sociology of science has much to do with which discipline of science is invoked to explain how science works empirically and whether universal or cultural explanations are sought.

Thus, experimental work in psychology on the use of natural kind terms connects closely to the causal theory of reference and considers universal developmental stages in conceptual development to shed light on concept use in science.[129] On the other hand, an emphasis on cognitive heuristics as pragmatic tools no longer oriented toward unitary truth (and connected to experimental demonstration of cognitive liabilities among laypersons and experts) can lead to convergence with sociological relativism.[130]

LOCALITY IN CONTEMPORARY PHILOSOPHY OF SCIENCE 2: ARTHUR FINE'S NATURAL ONTOLOGICAL ATTITUDE

The concept of locality gets taken a step further within philosophy of science by Arthur Fine's Natural Ontological Attitude (NOA). In effect, he introduces further "postmodernist" considerations in denying the need for a "global" realist legitimation of science (i.e., a "metanarrative" in Lyotard's sense), suggesting that such a move is an imposition upon the locally dependent adequacy of truth claims. Fine's position seeks to defeat realism and antirealism simultaneously by denying their central dispute over the "aim" of science:

> Does science aim at the truth, or does science merely aim at empirical adequacy? This is the springboard for the realism/instrumentalism controversy. NOA wants to pull back a bit from the question to ask, more fundamentally, whether science "aims" at all. Of course, there is a point to particular investigations, and certainly particular research groups have aims and goals. . . . But only a fallacy in quantifier logic would lead one from "They all have aims" to "There is an aim they all have."[131]

---

129. Markman, "Natural Kinds."

130. See Stich, "A Pragmatic Account of Cognitive Evaluation." For approaches developing a naturalistic approach to science that incorporate insights from both psychology and sociology in line with the original strong program, see Shadish and Fuller, *The Social Psychology of Science*; Fuller, *The Cognitive Turn*; Fuller, *Philosophy of Science and Its Discontents*; and Collin, *Science Studies as Naturalized Philosophy*, esp. ch. 1.

131. Fine, "Unnatural Attitudes," p. 173.

Both realism and antirealism are seen as "global" impositions upon the "local" character of reason giving. Rather than substituting some pragmatic criterion of truth, Fine argues one should avoid a general theory of truth altogether. Avoiding a general theory of truth requires treating scientific claims as adequate or not in the local contexts of their production and use. In this sense, he shares ethnomethodology's version of localism and rejection of skepticism. For Fine, science stands in need of no further justification than that it already provides in defense of particular claims in local contexts of justification.

Fine's account might seem to be a very useful way for social constructivists to avoid realism without committing oneself toward redefining truth in terms of "success," "consensus," or the like—that is, to avoid committing what philosophers call the epistemic fallacy.[132] But Fine has another concern, which he believes distinguishes his approach from social constructivism—namely, to oppose any sort of totalizing criticism of science as an institution. Thus, NOA "is a pro-attitude toward science, basically accepting although ready to challenge too-easy conclusions in specific cases." It "involves a trusting attitude. It trusts the overall good sense of science, and it trusts our overall good sense as well." It urges us to proceed "without intentions for attaching science to some ready-made philosophical engine."[133] It is not surprising, then, that when Fine turns to consider social constructivism he should object to its debunking character.[134]

It should be said at the outset that it is never clear what exactly social constructivism is alleged to "debunk," much less what effect such debunking is supposed to have on the institution itself, which Fine surely admits is *not* maintained by any particular legitimating account criticized by social constructivists. As we have seen, social constructivists are generally allied *with* Fine in debunking philosophical accounts of science rather than science itself. Fine's objection to constructivism is that he believes its sociological emphasis commits it to a consensus theory of truth—that is, yet another antirealist reduction of truth to empiricist terms. Instead of reducing truth to sensory observations as empiricist philosophers do, he argues that social constructivists reduce truth to the observable consensus of scientists. However, in general, social constructivists do not assert that social assent *is* truth, but that at most consensus is our best indicator of what is held to be knowledge and hence in need of explanation by the science analyst.[135]

---

132. Fine, *Shaky Game*, p. 140. For some good sociological reasons to avoid such a redefinition of truth by success, see Mulkay, "Knowledge and Utility."
133. Fine, "Unnatural Attitudes," p. 177.
134. Fine, "Science Made Up."
135. Barnes and Bloor, "Relativism, Rationalism and the Sociology of Knowledge."

Strictly speaking, a consensus theory of truth does not fit most constructivist sociologists since the practical difficulties of attaining consensus in the requisite manner typically ensure that the consensus reached is not a pure invention (where scientists just agree to favor certain accounts of the natural world) but a highly constrained outcome involving difficult and elaborate experimentation. Moreover, if the establishment of consensus is the focus of Collins's EMPOR, consensus is far from being the type of social explanation used by Latour, Callon, and Rip. Scientific knowledge is not a set of shared beliefs but statements that have been freed from modalities associated with contested science. In addition, many constructivists deny that there is a determinate fact of the matter as to what the state of consensus is, because statements of consensus are themselves dependent on different contexts and different audiences call for different assessments and because the boundaries of the community to which consensus obtains is one of the things which is at issue.[136]

The social explanations given for "what counts as knowledge" typically appeal to authority to speak for others, often backed up by a host of resources and institutional power, rather than the mental assent of all scientists concerned. Perhaps this could be a more subtle form of a consensus theory of truth, viewing social power rather than correspondence as the ultimate source of scientific knowledge. But most contemporary social constructivists allow "nature" to play a role, as we have seen, though they typically deny that this involves a knowledge product that can be decomposed into natural and social components. If nature can be seen to play a role on both sides of a scientific controversy, so too can power, as different groups within the core set appeal to their own preferred experts and institutional supports. Thus, the institutional "power" behind scientific claims would also be just as much of an outcome of the closure of a controversy rather than a cause of closure as Latour and Woolgar argued was the case for nature. Indeed, Latour is quite explicit that constructivism should be taken to debunk the traditional sociological account of the power of society as it does unmediated nature.[137] Other branches of SSK continue to appeal to the existing social science toolbox but likewise see the role of social and natural causes as inextricably mixed for all sides to a scientific controversy.[138]

It is still possible to overlay some account of scientific progress on this picture: it just won't take the form of a winnowing down of social contributions to the knowledge product in favor of natural components; it will, however, involve some different kind of assessment like I have suggested

136. Gilbert and Mulkay, *Opening Pandora's Box*; Gieryn, "Boundary Work."

137. This debunking view of traditional sociological explanation was evident already in Latour, "The Politics of Explanation" (1988) and received its canonical treatment in Latour, *We Have Never Been Modern* (1993).

138. Barnes, Bloor, and Henry, *Scientific Knowledge*.

in the case of a dialectical picture of scientific change drawing on Lakatos and Feyerabend. Also, it is possible to conceive of different conceptions of progress without entailing a rejection of the search for truth. This amounts to saying that the injunction "maximize truth" is not only mystifying as to what precisely is to be done but also is quite likely not possible. The attainment of some truths may come at the cost of disabling the pursuit of others, a key insight of Steve Fuller's program of social epistemology, where assessing the economic tradeoffs between different costs and benefits associated with alternative paths in the development of knowledge characterizes the task of the philosophy of science.[139]

What does seem clear is that there is a political function to Fine's critique of social constructivism in that any criticism of science is to proceed *piecemeal* rather than by appeal to a global theory. His defense of the integrity of scientific practices against totalizing theories, realist or antirealist, amounts to a kind of postmodern defense of the autonomy of science. But his use of the metaphor of "locality" is selective to the extent that he appears to commit himself to a commonsense version of the internal/external distinction. Philosophical realism and logical positivism are conceived as "external" impositions on science even though such philosophical impositions have often contributed significantly to scientific work itself (just as communist theory emerged from and shaped political practice for Marx). Perhaps Fine would consider such influence pernicious, but such a (global!) conclusion would still not solve the problem of how science's social authority is maintained, as Rouse points out,

> Fine cannot claim in the same way that scientific practice provides the resources for its own interpretation politically and culturally. While scientific work does require considerable attention to questions of evidence, reality and explanation, scientists are usually professionally unconcerned with the political or cultural understanding of their practices.[140]

LOCALITY IN CONTEMPORARY PHILOSOPHY OF SCIENCE 3: ROUSE'S POSTMODERN PHILOSOPHY OF SCIENCE

It is for this reason that Rouse extends Fine's NOA to allow for political critiques of science, such as those deriving from feminist epistemology, "so long as they situate themselves within particular historical contexts and the critical resources specific to those contexts."[141] But this also remains a curious

139. Fuller, *Social Epistemology*. The roots of this idea can be traced to Herbert Simon's economics. See Fuller, "Bounded Rationality in Law and Science."
140. Rouse, "The Politics of Postmodern Philosophy of Science," p. 618.
141. Ibid., p. 624.

interpretation of the injunction to promote "locality." Presumably, we are always already "situated" and require no special effort to achieve this. Rouse's point is that the rich body of contextually inspired feminist work ought to be the basis for intervention and not some global critique. But this seems historically and psychologically implausible: without the global critique in mind, why would one be motivated to do the contextually rich work to begin with? And are we really to believe that intervention waits for such research to be effective or is it more likely that such work is necessary largely to achieve credibility in a hostile environment? And to the extent that greater insight is achieved through such historical work, isn't it likely that such insight will have its effect by reconceptualizing global generalizations rather than by keeping in mind the infinite complexities of a huge body of research?[142]

In short, global "essentializing" is a crucial component of most "locally" situated bodies of knowledge. Our conception of the implications of such local contextualism must take into account the role of global essentializing within local practices and in mediating between different local practices. Conceiving of science as receiving its epistemic strength from its situatedness in place requires us to empirically examine how this occurs and what *relations* obtain between places. I suggest that this requires the kind of empirical examination that we find in social constructivist work coupled with an appreciation of the material links between different practices and locales. Laboratory practices themselves incorporate—if we may use that term—distinct research programs with globalizing theoretical ambitions developed "off-site" through scientific training, deriving from specific theoretical and cultural lineages, that shape what local practice will look like.[143]

CONSTRAINT IN PRACTICE

In the 1992 anthology, *Science as Practice and Culture*, the empirical examination of scientific practice is developed in a manner that gets beyond reductive sociological explanations in terms of "interests" or reductive post-positivist philosophical explanations in terms of theory. Instead, a renewed emphasis on practice leads to a recognition that despite the openendedness of scientific work that social constructivists have pointed to, an emphasis on the multiplicity of cultural elements that scientists deal with leads to a recognition of constraint as the result of "interactive stabilization."[144] Pickering summarizes this broad consensus:

142. Lynch, "Complexity, Natural Selection, and Cultural Evolution."
143. Lynch, "Between Kin Selection and Cultural Relativism," pp. 304–6.
144. Pickering, "From Science as Knowledge to Science as Practice," p. 10.

> [I]f we recognize the multiplicity of the technical culture of science, any number of somethings else beside interest become available to explain closure: any single cultural element may be open-endedly extended, but the task of fitting various extensions together, of bringing disparate elements into association, is not.[145]

In the first part of the anthology, labeled "Positions," a variety of approaches toward developing this insight are presented, while the second part, labeled "Arguments," opens up a set of disputes that no longer map neatly onto traditional disciplinary boundaries. Ian Hacking begins with a chapter titled "The Self-Vindication of the Laboratory Sciences," a phrase that suggests that his commitment to an "entity realism" combined with some affinity for constructivist themes results in some degree of convergence with Fine's denial of the necessity for either realist or antirealist legitimations. As Fine puts it,

> It may well be that "entity realism" is just a special version of NOA. That depends on how the truth of existence claims is to be understood (or not). I suspect that Hacking's (1983) version is like NOA, and that the realist label is just his way of trying to redirect philosophical discussion by giving the new direction an old and honoured name.[146]

For his part, Hacking agrees with Fine's denial of a unitary aim for science:

> There is only one way in which my thesis is contrary to a bundle of metaphysical doctrines labeled "realist." Realists commonly suppose that the ultimate aim or ideal of science is "the one true theory about the universe." I have never believed that even makes sense. The present picture suggests that there are many different ways in which a laboratory science could have stabilized.[147]

Hacking begins by noting that the Duhem-Quine thesis, widely understood as implying the underdetermination of theory by data, is better understood as showing how stable science can occur.[148] In a materialist spirit, Hacking extends the thesis to encompass not just theory and data but also different types of theory, apparatus, and analysis (in fact, fifteen different ideas, things, marks, and manipulation of marks, in his typology), which "are mutually adjusted to each other," the result being a manufactured set of increasingly stable elements:

145. Ibid., p. 9.

146. Fine, "Unnatural Attitudes," p.176, n. 22.

147. Hacking, "The Self-Vindication of the Laboratory Sciences," p. 31.

148. Hacking criticizes Duhem for focusing only on the relation between theory, data, and auxiliary hypotheses, but it is Duhem, and not Quine, who confines holism primarily to *physics* as a result of the use of experimental apparatus. See Duhem, *The Aim and Structure of Physical Theory*, pp. 180–90.

> They are self-vindicating in the sense that any test of theory is against apparatus that has evolved in conjunction with it—and in conjunction with modes of data analysis. Conversely, the criteria for the working of the apparatus and for the correctness of analyses is precisely the fit with theory. The theories of the laboratory sciences are not directly compared to "the world"; they persist because they are true to phenomena produced or even created by apparatus in the laboratory and are measured by instruments that we have engineered.[149]

While avoiding claims of semantic incommensurability that run afoul of Putnam's criticisms, and being sympathetic to a broadly realist picture, Hacking is quite clear that different stabilizations of a laboratory science are possible that would result in a quite real incommensurability: "Such imaginary stable sciences would not even be comparable, because they would be true to different and quite literally incommensurable classes of phenomena and instrumentation."[150]

What is interesting is that Hacking straightforwardly endorses the counterfactual possibility of stable, alternative sciences, something the sociology of scientific knowledge has danced around for some time without addressing explicitly. Moreover, he does so by *denying* that mature laboratory science is characterized by a great deal of openness. (Both these issues are developed in chapter 6.) Instead, an initial openness of options is increasingly constrained as developments proceed in one way rather than another, a theme David Gooding picks up in his contribution. Gooding, however, seems to leap from the recognition that "it is skilled agency that brings about the convergence of material and verbal practices" to a picture of scientific knowledge as "express[ing] the culture in which it occurs," which seems to suggest that any number of different expressions are possible and that the agency of the scientist is much more important than any fit with nature.[151]

This is not surprising: as a representative of the empirical program of relativism, he follows Harry Collins in substituting sociological for natural explanations for scientific theories, rather than promoting the picture of the impossibility of disentangling such contributions proposed by the strong program or the heterogeneous networks of humans and non-humans of Callon and Latour.[152] Thus, Gooding's inference is facilitated by conceiving the notion of "fit" as implying "a preordained fit," which can only lead to an emphasis on agency *rather than* natural constraint.[153]

---

149. Hacking, "Self-Vindication of the Laboratory Sciences," p. 30.
150. Ibid., p. 31.
151. Gooding, "Putting Agency Back into Experiment," pp. 104, 109.
152. Collins and Yearly, "Epistemological Chicken"; Latour, "Don't Throw the Baby Out with the Bath School!"
153. Gooding, "Putting Agency Back into Experiment," pp. 103, 141.

This view is evident as well in the contribution by Andrew Pickering and Adam Stephanides, which reconstructs individual mathematical practice in terms of "free" and "forced" moves. The contrast seems overstated in that the "free" moves might seem more likely than other conceivable moves if they had explored the conceptual resources available to the scientist. They seek to preserve both the objectivity and the contingency of mathematics, although they seem to hedge on the question of just how many different *stable* outcomes are likely: "These contingencies have no regular organizing principle lying behind them—they are truly historical—and they are, we believe, just as constitutive of the production of knowledge as the regular forced moves with which they intertwine. . . . Or at least, to offer a more guarded formulation of the same point, some special argument is required to show that in general such moves do not matter."[154]

Here Hacking's argument is valuable: he neither takes the position of the "unitary" realist who argues that only one right stabilization is possible nor does he so overemphasize the scientist's agency as to promote a picture in which it is difficult to conceive what would *not* be possible as an outcome. Hence Hacking can revive a useful notion of incommensurability without denying that science is in some sense adequate to the world, so long as what is meant by the "world" is carefully specified:

> Could two theories with no common measure, in the above literal sense, both be true? Is not at most one theory true, the old mature one or an aspiring new one that takes account of a new data domain? Only if we suppose that there is in the end only one true ultimate theory that corresponds to the world. Some philosophers who are halfway along this road find solace in saying that different theories are true of different aspects of reality, but then what work is "reality" doing here? We need say no more than this: the several systematic and topical theories that we retain, at different levels of application, are true to different phenomena and different data domains. Theories are not checked by comparison with a passive world with which we hope they correspond.[155]

This perspective allows Hacking to appreciate Latour's discussion of the need to "extend" the laboratory by recreating conditions in the world sufficiently similar to the laboratory to allow successful application of the science in the world at large.

For Latour, Pasteur does not succeed in displacing his scientific rivals merely by having possession of a scientific theory that better corresponds to an underlying, preexisting reality, but because that theory is also associated with Pasteur's efforts to modify the larger society in a way that facilitates

154. Pickering and Stephanides, "Constructing Quaternions," p. 164.
155. Hacking, "The Self-Vindication of the Laboratory Sciences," p. 57.

Pasteur's "extension" of his laboratory. Pasteur wins a series of trials and builds allies in his effort to extend phenomena that are "true to" the laboratory until they are "true to" the wider world. We need not suppose that such a stable extension—the practical use of pasteurization to control disease—is merely the additive result of a host of open-ended individual extensions of actor-networks, since not just any set of individually extendable elements can together lead to this stable result.

Other stable solutions were latently available at the time that could have duplicated the success of Pasteur's program up to a point. In one case, the hygienist movement that was part of Pasteur's success would have been able to achieve a large portion of success on their own: "Where would the hygienist movement have gone without Pasteur and his followers? In its own direction. Without the microbe, without the vaccine, even without the doctrine of contagion or the variation in virulence, everything that was done could have been done: cleaning up the towns; digging drains; demanding running water, light, air, and heat."[156] In another case, a different agenda might well have led to similar health improvements: "The fight against *degeneration* (which is not at all a fight against microbes) could have done everything that was accomplished with the hybrid Pasteurism-Hygiene."[157] Conversely, since mature laboratory sciences "are true to the laboratory phenomena," rather than nature *simpliciter*, there is no guarantee that successful application will obtain, and Latour's analysis of combined social/natural resistance to the extension of laboratory science is well motivated. Indeed, part of the point of Latour's enterprise is to assess what other resources have to be in place in order for the application of a science to be successful, allowing a better understanding of the actual role that the laboratory plays in any success story. As he puts it, "[i]f we do not restore the power ratio between the social movement at work throughout Europe and the few bacteriological laboratories, we cannot understand the real contribution of those laboratories, just as we cannot understand what Kutuzov did if we attribute to him the entire movement of his army."[158]

A RELATIONAL NOTION OF PLACE

In effect, Latour introduces a "relational" notion of place to our reflections upon the constraining and enabling function of the site of laboratory science. If science is in some sense "relative" to the conditions of its production, it certainly is not because any old place will do or that different truth claims

156. Latour, *The Pasteurization of France*, p. 23.
157. Ibid., p. 254, n. 5.
158. Ibid., p. 23.

are equal. In effect, epistemic access is not equally distributed, and as a result some places and claims carry more weight than others. But this is not *simply* the result of statements, technologies, or procedures carrying around their truthfulness within themselves, so to speak. The "force" of true claims obtains to the extent that an alignment of resources and allies are configured properly.[159] This insight is not necessarily incompatible with realism, but it does emphasize a number of additional factors in explaining the "success of science" than merely an increasing approximation to the truth. The laboratory has its worldly power not merely because of what the scientists in the laboratory discover but because of the resources drawn upon by the laboratory from other sites (funding, trained students and technicians, computing power, etc.) and of various "enabling" conditions allowing for extension of the laboratory (standardization of laboratory techniques, the capacity to rework other social practices to allow implementation, further allies contributing to the same goal, and, in general, resources to overcome resistance when it is encountered).[160]

To be sure, one must also consider the organization of the laboratory itself. Knorr-Cetina addressed this issue in her contribution in a manner that is compatible with Latour's relational notion of place, emphasizing how the laboratory reconfigures spatial and temporal elements of the natural world in order to change the "balance of power" existing in nature.[161] Knorr-Cetina argues that the laboratory is not just the physical space where experiments happen to be done but also is set up to overcome the limitations of objects in the natural world: one does not have to treat the object as it is, where it is, or study an event when it happens. The power of a laboratory thus depends upon "the *malleability* of natural objects":

---

159. Note here the parallel between Marx and Latour in thinking about practice in terms of a set of relations between humans and artifacts, with artifacts themselves only properly understood in terms of a web of relations, as internal relations constituting the objects as such (Descombes, *The Institutions of Meaning*, p. 187). Ollman, *Alienation*, p. 26, shows that "Marx conceives of *things* as Relations" and "social factors as Relations where physical objects are involved." Latour, "Where Are the Missing Masses?," p. 258, n. 23, for his part, notes that in addition to relations between humans and things, "there are also relations among things, and *social* relations at that." Law, "STS as Method," p. 41, speaks of actor-network theory as "radically relational, . . . mapping how every object or actor is shaped in its relations." Sterne, "'What Do We Want' 'Materiality!' 'When Do We Want It?' 'Now!,'" p. 121, supports a focus on materiality in communication technologies "so long as materiality refers to both physical things and the irreducibly relational character of reality." Haraway, "A Giant Bumptious Litter," argues for the need "to deauthorize the kind of possessive individualism that sees the world as units plus relations," substituting an understanding that "it's relationality all the way down."

160. On the role that a standardized package of theory and methods play in coordinating different "social worlds" leading to the success of the oncogene theory of cancer, see Fujimura's contribution titled "Crafting Science." For a "relational" theory of power, see Wartenberg, *Forms of Power*.

161. Knorr-Cetina, "The Couch, the Cathedral, and the Laboratory." For different models for connecting laboratories to the rest of society, see Knorr-Cetina, "Scientific Communities or Transepistemic Arenas of Research?" and Latour, *Science in Action*.

> What laboratory studies suggest is that the laboratory is a means of changing the world-related-to-agents in ways which allow scientists to capitalize on their human constraints and sociocultural restrictions. The laboratory is an enhanced environment which improves upon the natural order in relation to the social order.[162]

The organization of the laboratory and appropriation of social and natural resources enables epistemic power over objects that cannot be duplicated in non-laboratory settings.

Knorr-Cetina analyzes the social use of space in discussing the relative balance of power between experiments and laboratory sites. She traces three different laboratory "technologies" that correspond to different relationships between the space of the laboratory and experiments conducted in that space. The objects dealt with in the laboratory vary: "[I]n the first case, objects in the laboratory are *representations* of real-world phenomena; in the second, they are *processed partial versions* of these phenomena; in the third, they are *signatures* of the events of interest to science."[163] Such strategies are not unique to science; indeed, she is able to illustrate the strategies involved by analogy to the war game, medieval cathedral building, and the psychoanalyst's couch, respectively.

> In the first case, we deal with experiments such as found in social psychology, where subjects are given artificial tasks intended to stand in as representatives of real world group activities. Like the table top war game where generals can plan their strategy by attempting to accurately represent real world situations, such experiments depend upon a strategy of nonintervention. Blind and double-blind protocols are intended to avoid having experimenter-subject interaction interfere with the goal of representing an unmodified world. In this case, the "space" of the laboratory plays little role beyond the experiment itself: virtually any space will do and the space can revert to other uses when the experiment is over. The usefulness of the experiment is crucially dependent on the representativeness of the experiment's design and the command over space and time is comparatively minimal.[164]

In the second case, the paradigm science is molecular genetics. In this case, there are no injunctions against interference with the object under investigation. In fact, the object under study must undergo quite a bit of manipulation to render it useful for study. Here we have a technology of intervention, which Knorr-Cetina likens to the continual improvement of twelfth- and thirteenth-century cathedrals by builders traveling from site to site. A system

162. Knorr-Cetina, "The Couch, the Cathedral, and the Laboratory," p. 116.
163. Ibid., p. 123.
164. Ibid., pp. 123–25.

of surveillance of existing cathedral forms developed and a continual manipulation of the object took place, given that design drawings were not available. In some sense, then, successive cathedrals were processed versions of other cathedrals. In molecular genetics laboratories, natural materials are treated as objects appropriate for processing, and as a result the laboratory is a site where work proceeds continuously whether or not an experiment is currently taking place. Indeed, in some cases experiments become the virtual object and the laboratory the real space, reversing the previous case. The integrity and distinctness of experiments becomes less clear, and a two-tier system of management emerges: a laboratory level and an experiment-level social organization occurs in which different individuals may be primarily concerned with one to the exclusion of the other. In such a situation, a distinct advantage accrues to the researchers in terms of reversing the spatial and temporal balance of power found in nature.[165]

The paradigm science in the third case is high energy physics. In this case one does not directly manipulate processed natural entities in the laboratory at all, but rather processed *information* extracted from highly artificial events. Like the psychoanalyst, one begins, if you will, with pathology, and attempts to extract meaning. In high energy physics, further operations conducted on "raw data" take far more time than the experiment's beamtime.[166] Often the point is to extract a single "event" from such data processing. At this level, we see laboratories and experiments become separated yet again. In this case, laboratories become total institutions that are run separately from any group of experimenters. Indeed, experiments often have no "place" in that international collaboration is involved in any single experiment, although such experiments are dependent upon the existence of place-bound laboratories.[167]

Such a scenario bears a striking resemblance to the flexible capital accumulation that Harvey has argued increasingly characterizes capitalism since the 1970s. Harvey is a geographer who has introduced examination of the social management of space and time to a historical materialist examination of late capitalism, a shift he sees as reflected in the cultural shift labeled *postmodernism*. Unlike many commentators who see such a shift as involving a fundamental alteration in the character of capitalism, Harvey sees the shift as essentially involving an intensification of processes already present in capitalism.[168]

The shift involves a deemphasis on manufacturing economies of scale, with factories rooted in place, and an increased emphasis on economies of scope, in which international corporations are free to move from place to

165. Ibid., pp. 125–29.
166. See Galison, *How Experiments End.*
167. Knorr-Cetina, "The Couch, the Cathedral, and the Laboratory," pp. 129–34.
168. Harvey, *Condition of Postmodernity.*

place and take advantage of small batch production. Just as the international corporation is still dependent upon the existence of places where labor and infrastructure are readily available, HEP (high energy physics) experiments are dependent upon the infrastructure of particle accelerators rooted in place. However, just as the international corporation is not tied to any one place (though its profits tend to flow to the global North) and can mobilize resources as needed from a variety of locales, so can any given HEP experiment, since even if the most powerful existing accelerator is needed, the processing of data can draw on multiple resources from many locales. In a sense, high energy physics becomes the "postmodern" science *par excellence*.

This resemblance between the social uses of space and time in capitalism and science highlights the inadequacy of the logical positivists' "placeless" approach to scientific accountability. The epistemic power enjoyed by modern science is a function, in part, of its management of space and time, a power that the average citizen cannot duplicate by employing a special method to test claims. To some extent, science must proceed in contexts that are not directly scrutinizable, and recognition of this must enter our assessments of how to root out ideology in science, how to assess the relative benefits of various lines of research, and how to develop strategies to pursue the kind of knowledge of use and value to those not engaged in scientific research themselves. What is needed is both a better empirical understanding of the political economy and geography of science and a sense of what an appropriate political philosophy of science ought to look like. What rights do people within and without science have to dispute representations, technologies, and their uses? How can they do so effectively?

As the material infrastructure of science grows, how much capital is required for any new, incremental achievement? In *Capital*, Marx called this (somewhat confusingly) the growing organic composition of labor (as the "dead labor" of fixed capital grows relative to living labor). Marx understood the growing organic composition of labor to be the driving force behind the crisis of capitalism. Does the growing organic composition of labor in science lead perhaps to a falling rate of discovery?[169] Do the growing costs of infrastructure in big science lead to a lessened rate of innovation per unit cost?

Collison and Nielsen have tried to get a preliminary handle on this question by surveying scientists' own perceptions of significant scientific discoveries in their fields. Their conclusion is that costs have grown, while discovery remains flat:

> Over the past century, we've vastly increased the time and money invested in science, but in scientists' own judgement, we're producing the most important

169. Marx, *Capital*, Vol. 1, ch. 25.

breakthroughs at a near-constant rate. On a per-dollar or per-person basis, this suggests that science is becoming far less efficient.

The result is that "it's getting much harder to make important discoveries across the board. It's requiring larger teams and far more extensive scientific training, and the overall economic impact is getting smaller. Taken together, these results suggest strong diminishing returns to our scientific efforts."[170]

Moreover, since the investment locks in place one line of development, following Hacking, should we worry that Big Science's virtues come with a very real down side where alternative perspectives cannot begin to get a foothold and startup costs are too great? Call it the *monopoly stage of scientific discovery*. It was just this kind of worry that led Feyerabend to question too close an empirical fit with reigning scientific theories, calling for the development of incommensurable approaches via "counterinduction" as an alternative. The burden of this chapter has been to show that such a strategy requires access to a laboratory—or to be more precise, a network of material practices capable of developing the alternative. Just such a dilemma confronts health advocates who claim modern medicine does not serve their interests well. Will minority views fail to thrive, and will this be a long-run problem for the social ecology of science?

IS ADRENAL FATIGUE REAL? THE GEOGRAPHY OF REPRESENTATION

One way to begin to think about these issues is to see the obstacles that are placed before critics of scientific approaches who lack the material infrastructure of laboratories, training facilities, and professional journals. Even when avoiding a focus on anomaly mongering, and making a plea to attend to existing research programs, as Taubes did in calling for more support for researchers outside the saturated fat mainstream, critics can run into a wall when their preferred minority positions are left fallow from a material position.

To see this, consider the issue of whether "adrenal fatigue" exists and whether physicians are treating thyroid patients properly. Mainstream endocrinologists recognize disorders of endocrine function, including Addison's disease, where the adrenal cortex fails to produce sufficient cortisol and aldosterone, leaving one incapacitated and unable to mobilize the body to respond to daily stresses. In an extreme case, Addison's disease can lead to an acute adrenal crisis that can be life threatening. Diagnosis for Addison's disease

170. Collison and Nielsen, "Science Is Getting Less Bang for Its Buck." If we take Lenin's inexhaustible atom to heart (or Vannevar Bush's endless frontier), we would not necessarily see these diminishing returns as reflecting natural limits of science, as argued by Horgan, *The End of Science*.

depends upon blood tests, most notably an ACTH stimulation test, whereby synthetic ACTH is injected signaling the body to produce cortisol. In normal functioning, the pituitary releases ACTH to cause the adrenal glands to produce cortisol. If cortisol does not respond—if blood levels of cortisol remain below 18 μg/dL—Addison's disease is diagnosed and supplementation with cortisol for life is prescribed.

According to certain physicians and patient advocacy groups, a distinct disorder exists where cortisol levels remain inadequate but sufficient to "pass" the ACTH test. On this view, adrenal function slowly declines through four stages as the adrenal gland "wears out" from overuse as the result of persistent and chronic stress. In 1998, chiropractor and naturopathic doctor James L. Wilson coined the term "adrenal fatigue" to describe a pervasive illness brought on by the stresses of modern life, what he also dubbed a "21st-century stress syndrome." In arguing for diagnosis and treatment of a condition where adrenal function degraded over time due to stress, Wilson also explicitly revived an older, more qualitative, symptomatic approach to medicine less tied to laboratory value. If the patients had symptoms, then treatment was required based on those symptoms, not on arbitrary laboratory values. Wilson directly tied his discussion of adrenal fatigue to previously accepted medical practice and ideas that have been abandoned in the new clinical regime where everything turns on the identification of normal and pathological lab values. Likewise, the notion of "breakdown" of organ function is rejected for congenital or environmental conditions that directly specify why an organ ceases to work properly with little focus on a transition between normal and poor function.[171]

For most endocrinologists, the idea of adrenal fatigue is a dangerous myth that risks leading patients to treat potentially undiagnosed conditions with vitamin supplements and lifestyle adjustments. The Endocrine Society warns patients that

> [n]o scientific proof exists to support adrenal fatigue as a true medical condition. Doctors are concerned that if you are told you have this condition, the real cause of your symptoms may not be found and treated correctly. Also, treatment for adrenal fatigue may be expensive, since insurance companies are unlikely to cover the costs.

Where alternative health practitioners advise using saliva cortisol tests that are held to track fluctuating, daily cortisol levels more accurately than blood tests used by endocrinologists, they argue such methods have no scientific standing.

---

171. Wilson, *Adrenal Fatigue*, ch. 7; https://adrenalfatigue.org/about-james-l-wilson-dc-nd-phd/.

> There is no test that can detect adrenal fatigue. Many times, a person will be told he or she has adrenal fatigue based on symptoms alone. Sometimes, a blood or saliva test may be offered, but tests for adrenal fatigue are not based on scientific facts or supported by good scientific studies, so the results and analysis of these tests may not be correct.[172]

A systematic search of the scientific literature finds no reliable indication that such a condition exists despite patient orientation of their symptoms to the construct.[173]

The deadlock between patient groups demanding treatment for subclinical fatigue presumed to be caused by declining adrenal function and endocrinologists who treat adrenal function as an all-or-nothing question based upon standardized laboratory tests represents a clear case of incommensurability between the medical community and some of its patients. Doctors express exasperation with patients who come to appointments armed with website printouts who argue that "low normal" test results should trigger treatment based upon vague symptoms.[174] Some commentators note a middle position is possible, as scientists have shown that stress can have numerous and profound negative effects on the body, particularly when stress is chronic rather than episodic.[175]

As naturopathic doctor Saul Marcus puts it, "When it comes to adrenal fatigue, conventional medicine insists it doesn't exist. However, adrenal fatigue is essentially a stress reaction, and stress is very well understood as a cause of illness." Endocrinologist Theodore C. Friedman agrees that stress can cause ill health but naturopathic doctors fail to understand that stress induces the release of more cortisol, not less, and they neglect the role of the hormone aldosterone, probably as a result of the fact that they cannot legally prescribe blood tests for aldosterone. "It's the concept of the adrenals burning out that doesn't make sense."[176] Wilson, for his part, agrees that stress *initially* causes cortisol to rise, since this excess production of cortisol over time is what tends to deplete adrenal function. Therefore, he distinguishes four stages of adrenal

---

172. Hormone Health Network, "Adrenal Fatigue," https://www.hormone.org/diseases-and-conditions/adrenal-fatigue.

173. Cadegiani and Kater, "Adrenal Fatigue Does Not Exist." If human health is potentially affected by psychological states and the conscious framing of experience, it would be possible to consider adrenal fatigue as the result of what Hacking calls a "looping effect" between categorization by experts and patients, though differentiating underlying natural kinds not subject to looping from social kinds subject to it is problematic in this case. See Hacking, "Making up People."

174. Ross, Blockman, and Jones, "We Are Tired of 'Adrenal Fatigue.'"

175. Perhaps the most relevant body of scientific research potentially speaking to adrenal fatigue is the study of glucocorticoids and the effect on the hippocampus in primates. See Sapolsky, *Why Zebras Don't Get Ulcers*. For debate about whether high levels of stress can degrade brain function in PTSD patients, see Yehuda, "Stress and Glucocorticoid"; Sapolsky, "Response."

176. Brissette, "Is Adrenal Fatigue Real? Forget the Label and Treat the Stress."

fatigue with the last stage characterized by "exhaustion" of the adrenal gland as it fails to produce sufficient cortisol for basic needs.[177]

In a concession to Epstein's account of AIDS activists, Collins allows that patients can have a sort of knowledge that deserves consideration because they have tacit knowledge of their own experience of illness.[178] This is misleading in two respects. First, patient experience of illness is not itself knowledge. Indeed, patients do not spontaneously know how to describe their illness. Doctors often use standardized surveys of patient experience to attempt to discipline the unruly experience of patients and tutor them in distinguishing different types of symptoms. Experience of illness is not knowledge of illness, and Epstein's AIDS activists were not knowledgeable simply because they were ill. Second, when patients compare their experience of illness, they may begin to actively construct alternative accounts of their disease, but they do so through an active process of group construction, as fallible as any other scientific hypothesis, not through articulation of a tacit expertise flowing from their experience. Indeed, the collective process of sharing experience of illness and seeking better treatments is better understood as presenting an incommensurable alternative to professional diagnosis, which is why there is such consternation among doctors and professional societies about patient research on the internet.

Like the AIDS activists studied by Epstein, some endocrine disease patient groups develop alternative views with the help of minority scientists and internet coordination of patients. Can this succeed in challenging established views? Should it and what kind of case can be made and what kinds of demands can be lodged? Patient activists in the internet age build upon an alternative infrastructure for incommensurable claims: internet mobilization of patient experiences and activist rereading of the scientific literature (exactly what Collins and Evans want to ban as illegitimate).

Thyroid patients include activist and patient rights groups, most notably the very active "Stop the Thyroid Madness" website, based on a book by patient activist Janie Bowthorpe. For these patients, adrenal fatigue is held to often accompany and exacerbate the allegedly flawed treatment of thyroid disease by endocrinologists, which takes place along very similar lines to the adrenal fatigue controversy in denying that diagnosis should proceed by strict lab values and not by patient symptoms. Like proponents of adrenal fatigue, thyroid activists urge that treatment proceed based on symptoms, in line with how clinical practice proceeded before the invention of a tablet form of synthetic levothyroxine (T4) in 1955 replaced injected forms of synthetic T4 that were used in research since 1926. Physicians had been using desiccated

177. Wilson, *Adrenal Fatigue*, ch. 23.
178. Collins and Evans, *Rethinking Expertise*, pp. 113, 142.

sheep or pig thyroid, which included a combination of thyroxine (T4) and triiodathyronine (T3), as early as 1891. Treatment with desiccated thyroid was common in the first half of the twentieth century, with dosage managed clinically by symptoms. T4 is the storage form of the hormone, while T3 is the active form in the body. Thyroid activists contend that T4 levels can be normal, while T3 levels are inadequate. Consequently, they reject protocols that rely on testing serum T4 levels to dictate treatment.[179]

The result has been a rise in the use of "combination therapy," either via a combination of synthetic T4 and T3 or via desiccated thyroid. Endocrinologists worry that combination therapy, and treatment of subclinical hypothyroidism by symptoms, could lead to overtreatment. Such overtreatment can decrease thyroid stimulating hormone (TSH) and raise the risk of heart disease, osteoporosis, and dementia. While a meta-analysis showed no improvement from combination therapy for body pain, depression, anxiety, fatigue, quality of life, or body weight, patient surveys indicate a preference for combination therapy. Patients on T4-only monotherapy even reported symptoms worse than patients with untreated thyroid disease for fatigue, constipation, and pain. A Dutch survey found that media treatment of combination therapy spurred an increase in its use. Likewise, potential overtreatment was indicated by the fact that 28 percent of survey respondents adjusted their own medication levels, probably based on symptoms, and 14 percent had suppressed TSH levels.[180]

Stop the Thyroid Madness author Bowthorpe counters that when patients are treated by symptoms and not TSH levels, TSH can go below normal levels without triggering hyperthyroid symptoms, in effect denying that overtreatment is taking place. In support, she cites research that different organs in the body can convert T4 to T3 at different rates, with the liver lagging the pituitary, leading to variable hypothyroid symptoms in the body. In 2005, Alevizaki et al. had argued that this implied that TSH levels may not reliably measure thyroid activity throughout the body, titling their article provocatively "TSH May Not Be a Good Marker for Adequate Thyroid Hormone Replacement Therapy." In response, Staudenherz and Sinzinger declare themselves "astonished" at the title of their paper and object that if T3 is prescribed as well as T4, "the rate of (subjective and objective) side effects by far would outweigh relative tissue hypothyroidism." In return, Alevizaki, Alevizaki, and Koutras agree that the issue of prescribing T3 is unresolved, lacking "well designed prospective studies." They argue, however, that "the title of our publication reflects the increasing concern (as also shown by a

179. Bowthorpe, *Stop the Thyroid Madness*, pp. 33–34, 44–45.

180. Michaelsson et al., "Treating Hypothyroidism with Thyroxine/Triiodothyronine Combination Therapy in Denmark."

growing number of papers) over what actually constitutes euthyroid status and "normal" TSH range; the fact that no one has come up with a definitive answer to the problem (or a better marker than TSH) is, we feel, an additional reason for presenting and discussing evidence for the need of improving the means of monitoring thyroid status in our patients."[181]

Kalra and Khandelwal argue that this possibility of "tissue hypothyroidism" creates a dilemma for physicians: "Should we treat patients according to their TSH level, which represents the health of their hypothalamic-pituitary axis, or according to their symptoms, which represent the health of the whole body?"[182] Should these minority reports be seen as an example of a contribution that core-set scientists with tacit knowledge of whom to trust know to reject, in line with Collins's criticisms of internet activism? Or is it a harbinger of a coming reconsideration of the merits of monotherapy and the need for better biomarkers?

What comes across most clearly in the activist literature and on patient support boards is that monotherapy leaves many with continuing, debilitating symptoms. This has led to a reconsideration of established protocols by some practitioners and researchers. Brown et al. argue that there is evidence to suggest that, "for a proportion of sufferers, the standard medical treatment does not completely reverse the constitutional and neuropsychiatric symptoms brought about by this condition." Others have begun to recognize that as many 30 to 40 percent of patients have poor thyroid function while on T4 monotherapy, indicating that both undertreatment and overtreatment are concerns.[183] Endocrinologists often treat continuing patient symptoms as not real, either attributing their "rather diffuse symptoms" to an "inability of coping" with a chronic disease or to a tendency to "blame hypothyroidism for any symptoms." Studies also indicate that psychiatric conditions are more common in thyroid patients, both treated and untreated, reinforcing the tendency for many doctors to dismiss patient symptoms that do not show up on laboratory tests.[184]

Many patients, especially women who suffer disproportionately from hypothyroidism, feel that doctors dismiss their symptoms or complain that they

181. Bowthorpe, *Stop the Thyroid Madness*, pp. 66–67; Alevizaki et al., "TSH May Not Be a Good Marker for Adequate Thyroid Hormone Replacement Therapy"; Staudenherz and Sinzinger, "Letter to the Editor"; Alevizaki, Alevizaki, and Koutras, "Author's Reply," p. 366.

182. Kalra and Khandelwal, "Why Are Our Hypothyroid Patients Unhappy?"

183. Biondi and Cooper, "Thyroid Hormone Therapy for Hypothyroidism," pp. 20–21.

184. Michaelsson et al., "Treating Hypothyroidism with Thyroxine/Triiodothyronine Combination Therapy in Denmark," p. 12. On patient difficulties getting diagnosed when doctors want to blame psychiatric symptoms, see the study of Chernobyl victims by Petryna, *Life Exposed*, and the cases discussed in Brown et al., *Contested Illnesses*. The same issue of treating diffuse symptoms while rejecting standard laboratory protocols characterizes patients who advocate for a "chronic Lyme disease," which may be overtreated with antibiotics by "Lyme literate" practitioners. See Sanders, *Every Patient Tells a Story*, pp. 179–83.

were prescribed psychiatric medicines in lieu of treatment of an underlying physical illness. This tendency is reinforced by characteristics common to many chronic illnesses. Milliken and Northcott note that as a result of "a gradual and capricious onset, sufferers have difficulty recognizing their ambiguous symptoms as illness and convincing others, including their doctors, that they are sick."[185] In the case of hypothyroidism, this problem is particularly acute:

> A fundamental problem with hypothyroidism stems from its ambiguous symptoms, which tend to be vague and subjective, appearing gradually or in cycles. Initially, symptoms can have multiple meanings, such as being overtired, under stress, or sick. As a result, symptoms are easily discounted or blamed on other causes, for example, job stress, parenting demands, or marital difficulties.[186]

Often there is a long gap between the onset of illness and when lab tests give abnormal results. Patients in online support groups routinely report years of misdiagnosis and dismissal of symptoms by doctors, including being referred to a psychologist or psychiatrist or prescribed antidepressants.[187]

A LAB OF ONE'S OWN? THE CASE FOR PUBLIC DEBATE OVER RESEARCH DIRECTIONS

Patient activists leverage select publications in the scientific literature to make their case for symptomatic treatment and in the process attempt to advance minority positions in the literature. At the same time, sympathetic practitioners and researchers leverage patient dissatisfaction to advance their alternative hypotheses and methods. Lacking the support of dissenting scientists and physicians, activist groups are unable to develop new protocols or uncover new biomarkers.

In catering to them, many practitioners of "alternative" or "holistic" medicine may find a market for patients willing to pay for treatments not covered by health insurance without reliable tests of outcomes.[188] Such market diversification can act as a means of keeping alternative treatment plans going in the face of patient dissatisfaction, a kind of "breathing space" in Feyerabend's terms. Sometimes these alternative practices are facilitated by laboratory tests outside normal protocols, whether of vitamin and mineral levels or

185. Milliken and Northcott, "Seeking Validation," p. 204.
186. Ibid., p. 207.
187. Bowthorpe, *Stop the Thyroid Madness*, pp. 66–67, xxv–xxviii. For patient groups, see https://stopthethyroidmadness.com/talk-to-others/.
188. See the edited collection by Bowthorpe, *Stop the Thyroid Madness II*, with contributions by alternative practitioners and a foreword by holistic medicine guru David Brownstein.

saliva cortisol tests. At the same time, many alternative practitioners lack legal authority for other kinds of tests or are not knowledgeable about possible diagnoses that can explain their patients' symptoms. Clearly, keeping alternative perspectives out in the cold while insulating them from criticism while among true believers is inadequate, but so too is the unwillingness of orthodox medicine to explore continuing patient dissatisfaction and ill health. This practical form of incommensurability will not go away of its own accord or by educational propaganda by professional societies. In historical materialist terms, it is an "objective contradiction" that requires work to overcome.

In practical incommensurability of this kind, the side with the laboratory and institutional authority wins, unless and until patients can combine their collective experience with minority positions in science and establish viable research programs. Like the "autonomist" view of class conflict, patient activists can carve out small, incomplete victories ("breathing space") that protect their interest in exploring alternative treatments. But what they really need is to carry out a struggle for hegemony, to have their issues and concerns subject to examination and criticism. It won't do to be either ignored or condescended to. If orthodox endocrinology's treatment of adrenal insufficiency is the thesis, and popular movements for remedying "adrenal fatigue" the antithesis, then the synthesis may involve an integrated, interdisciplinary focus on the effects of stress on the body (with an understanding of mismatch based on evolutionarily novel stresses of modern life).[189]

Likewise, the possibility of tissue hypothyroidism is a thesis that requires significant laboratory investigations to explore further.[190] Treatment by symptoms may be a stopgap until better biomarkers and further understanding of individual variability is addressed. To accomplish this, patient activists and endocrinology researchers need to work together to establish research programs that would be satisfactory to both. Conferences that bring together researchers and patients suffering from specific illnesses, and dedicated organizations that raise funding for targeted research, have become much more common in recent decades and have even led to significant discoveries. For example, families with children who have the full genetic mutation for Fragile X Syndrome described movement and cognitive symptoms experienced

---

189. Sapolsky, *Why Zebras Don't Get Ulcers*. The role of synthetic chemicals like DDT and PCBs in disrupting thyroid function and producing neurological deficits in humans, especially in utero, remains a viable hypothesis requiring further research. See Brucker-Davis, "Effects of Environmental Synthetic Chemicals on Thyroid Function" and discussion in chapter 6 below. See also the suggestion by de Punder and Pruimboom, "Stress Induces Endotoxemia and Low-Grade Inflammation by Increasing Barrier Permeability," that chronic stress can disregulate the HPA-axis (hypothalamus, pituitary gland, and adrenal glands), exacerbated by high-fat and high-carb diets. For the role of patient activists in pushing mainstream science to address underlying causes, including systemic, societal causes going beyond individual behavior, see Brown et al., "'A Lab of Our Own.'"

190. Zulewski et al., "Estimation of Tissue Hypothyroidism by a New Clinical Score."

by older male carriers of the premutation for the gene in conferences and medical consultations. This led clinical researcher Randi Hagerman to discover Fragile X–associated tremor/ataxia syndrome (FXTAS).[191]

As we have seen, Hacking has argued that "any test of theory is against apparatus that has evolved in conjunction with it" and that it is possible to imagine alternative, counterfactual, stable laboratory sciences given different, possible historical paths science could have taken. If we could compare such alternative paths in the multiverse of science, they would not be directly comparable: they would be incommensurable in Feyerabend's sense. This, in effect, is what patient activists are trying to imagine: what would it be like if medical science had developed in a different pattern that did not foreclose their concerns.

Laboratory science closes off options at the same time that it generates insights. If Feyerabend is right, the generation of alternative, incommensurable approaches might provide the basis for dislodging entrenched perspectives that lock in dominant approaches and provide the possibility for new and hybrid approaches. In Neurath's terms, the practical condition of intersubjective understanding can break down between patients and medical experts, an undesirable form of disunity, but actively facilitating communication across incommensurable perspectives without reducing one to the other is possible and desirable. Such disunity cannot be overcome by an intellectual critique but by building new infrastructure that materializes—or not—such possibilities. The next chapter looks further at this issue of alternative paths and counterfactuals in science and technology.

---

191. Hagerman et al., "Intention Tremor, Parkinsonism, and Generalized Brain Atrophy in Male Carriers of Fragile X"; Hagerman and Hagerman, "The Fragile-X Premutation"; Partridge, "Raise the Bar on FXTAS."

Chapter Six

Counterfactuals and Second-Guessing Scientists and Engineers

Green Chemistry, the Precautionary Principle, and Social Movements in the History of Science

A key difference between a correspondence view of science and a dialectical one is that the first only admits degrees of truth, while the second realizes that truth and obscurity are locked together in essential ways that require specific changes over time to sort out. Popper's notion of approximate truth, or verisimilitude, weakens the punch of his early articulation of a falsificationist denial of proof by telling us that we can, after all, be confident that our current theories have some degree of truthlikeness.[1] The focus nonetheless is on the current theory's relationship to the world. For Lakatos, a research program obscures while it reveals. The perceived balance might change over time, which can affect our assessment of which programs deserve our support and pursuit. But research programs come and go and their virtues wane and their vices wax. That is why Lakatos believed that science should encourage a proliferation of approaches and an assessment of their track records over time.

Moreover, as Feyerabend pointed out, the adoption of new, promising programs may be blocked if the dominant program locks in a particular trajectory in a way that blocks spontaneous consideration of its weaknesses. Without forcing a comparison with an external rival via an incommensurable picture of the evidence, dominant perspectives may survive past their sell date by simple institutional inertia. Feyerabend's objection to Kuhn's philosophy of science was precisely that it would facilitate this kind of institutional-cum-cognitive inertia. Finally, as research programs reflect their historical and social origins, such inertia might be reinforced by powerful interests outside the institution of science, such as powerful corporations or governments. Feyerabend's concern for the "separation of science and the state" reflects

1. Popper, *Conjectures and Refutations*, pp. 219–33; Oddie, "Truthlikeness," § 1.2.

this concern about a potential mutual reinforcement of dogmatic experts and powerful interests.[2]

In this sense, Lakatos and Feyerabend anticipated the idea that the closure of controversies lead certain facts to be "black boxed," treated from there on out as unquestioned facts. Unlike symmetrical STS, however, they coupled this recognition with a critical, normative approach that saw grounds for dislodging established orthodoxies that had stopped being fruitful or that were actively blocking competing possibilities. Black boxes may need to be opened back up. Tacit in their approach is the idea that the actual history of science has not been the best of all possible worlds but one possible path, and potentially a suboptimal one compared to where science might end up. To truly explore these ideas, we need to consider the relevance of counterfactual reasoning to the history of science and technology. Why were greener approaches to the synthetic chemistry of the twentieth century foreclosed? Could a different approach have won out?

THE EMERGENCE OF SYNTHETIC CHEMISTRY

After 1900, the identification of naturally occurring vitamins and hormones in the body, followed by the development of novel synthetic chemicals, set the stage for the development of the modern chemical industry in the first few decades of the twentieth century. With expanded war production during World War I, the U.S. chemical industry began to catch up with the Europeans. In the interwar period, research quickened as novel synthetic compounds were developed for a variety of uses, including plastics, pesticides, and pharmaceuticals. Following World War II, production expanded dramatically using growing supplies of petroleum as a feedstock. This expanded industry finally aroused broad public and regulatory concerns that went beyond workplace hazards to include consumer exposures through pesticide residues, packaging materials, and food additives.[3]

While the U.S. government had sought greater control of air and water pollution from factory emissions in the first three decades of the twentieth century, targeting the production of specific, harmful chemicals lagged, as the owners of chemical companies aggressively sought to limit regulation of their activities.[4]

At this time, the science of toxicology was funded by chemical manufacturers, and its practitioners took for granted that safe levels of workplace exposure

2. Feyerabend, *Science in a Free Society*, pp. 106–7.
3. Ross and Amter, *The Polluters*, pp. 3–5, 17–21, 25.
4. Ibid., pp. 10–16, 21–27.

could be defined for economically important chemicals.[5] While the 1906 Pure Food and Drug Act regulated adulterated and mislabeled drugs, it wasn't until the 1938 Federal Food, Drug, and Cosmetic Act, implemented in response to 105 poisoning deaths associated with the medicine Elixir Sulfanilamide, that the Food and Drug Administration (FDA) was given the ability to regulate the safety of drugs. In addition to regulating drugs, the 1938 act banned poisonous substances from food. However, the new law incorporated industrial toxicology's belief that "the dose makes the poison" by establishing a regulatory approach mandating acceptable tolerance levels for "unavoidable" poisonous substances in foods. The 1958 amendment to the law required premarket testing of medicines and new food additives, restricted unsafe levels of harmful chemicals, and banned carcinogens in food outright.[6] Congressional hearings on the safety of new plastics, fertilizers, and other chemicals led to the 1958 bill. In 1962, the publication of Rachel Carson's *Silent Spring* raised further concern with the health and environmental effects of chemicals. Still, U.S. chemical manufacturers were big business by this time, with sufficient power to resist new regulatory efforts. Moreover, provisions and applications of the 1958 bill were shaped by pressure from manufacturing associations, most notably in allowing use of chemicals known to be toxic under the assurance that safe levels of exposure could be established and regulated.

Could twentieth-century chemical synthesis and production have proceeded in a way that caused less damage to human health and the environment? The option of pursuing safer chemicals was precluded, in part, as the result of technical and regulatory assumptions that regulation could and should proceed by identifying safe and unsafe levels of chemicals, rather than distinguishing between safe and unsafe chemicals tout court. These industrial chemicals could have been developed with less harmful properties for physiological and ecological health, in particular as the result of adoption of a research program to design chemicals that were "benign by design" (in the words of contemporary "green chemists" Paul Anastas and John Warner) rather than the path that firms and regulators chose: designing toxic chemicals with an eye to safe levels of use.

COUNTERFACTUALS IN THE HISTORY OF TECHNOLOGY

Contemporary "green chemists" Anastas and Warner claim that twentieth-century chemistry took a wrong turn in targeting safe levels of toxic

5. Vogel, *Is It Safe?*, pp. 34–35; Ross and Amter, *The Polluters*, p. 34.
6. Wax, "Elixirs, Diluents, and the Passage of the 1938 Federal Food, Drug and Cosmetic Act"; Vogel, *Is it Safe?*, pp. 20–22, 34–35; Cavers, "The Food, Drug, and Cosmetic Act of 1938," p. 27.

chemicals—a fateful step that could have been avoided. They hold that the Paracelsian view that "the dose makes the poison" served as an excuse to justify inaction: "At some point when one is dealing with substances of high toxicity, unknown toxicity, carcinogenicity, or chronic toxicity, it becomes problematic, if not impossible, to set appropriate levels that are tolerable to human health and the environment."[7] Anastas and Warner envision an alternative path that chemistry could have pursued, a greener approach to chemistry that would have avoided many of the harms associated with twentieth-century "brown chemistry."[8]

With careful use of counterfactual reasoning, historians of science and technology can take on board the criticisms some scientists and engineers make of their own traditions and evaluate the likelihood that alternative approaches could have prevailed, while also understanding better why they ultimately did not. Recent examination of counterfactuals in the history of technology has focused on the role of contingency, which fits well with the anti-Whiggish tradition within history of science and technology.[9] However, merely emphasizing contingency leaves unclear the long-term consequences that could have resulted from proposed counterfactual scenarios. Counterfactuals also help us understand the multiple causes of closure defining a historical trajectory. Above all, counterfactuals are a means to get clear about causal reasoning, so a successful counterfactual exercise should help us understand better the actual causes of the real historical development.

Addressing counterfactual possibilities requires addressing two questions: (1) whether a change is *insertable* in real history—that is, whether it could have happened given plausible changes in history as it occurred, and (2) whether a postulated change would have led to a branching path in history or would have been compensated for by substitute causes. The latter requires examination of *second-order counterfactuals* to see whether consequences of the proposed counterfactual would have reinforced a new historical trajectory or have been compensated for by other causes reverting historical development to the actual path it took (or one close to it). Such *amplifying* or *reversionary* second-order counterfactuals can lead to either *underdetermination* (a branching path hinges on the postulated change) or *overdetermination* (the postulated change is insufficient to overcome other causes of the trajectory as it occurred).[10]

---

7. Anastas and Warner, *Green Chemistry*, pp. 17–18. Other bad choices Anastas and Warner discuss include the neglect of synergistic effects and the failure to provide a full-cost accounting of the life cycle of a chemical.

8. Woodhouse, "Change of State?"

9. Brown, "Not the Eads Bridge"; Brown, "A Different Counterfactual Perspective on the Eads Bridge"; Schatzberg, "Counterfactual History and the History of Technology"; Vinsel, "The Value of Counterfactual Analysis."

10. Tetlock and Parker, "Counterfactual Thought Experiments"; Elster, *Logic and Society*.

UNDERDETERMINATION VERSUS OVERDETERMINATION

The tacit emphasis on underdetermination by constructivist historians of science and technology was first made evident in Shapin and Schaffer's argument that the political climate in Restoration England favored Boyle's experimental program over Hobbes's deductive science modeled on geometry. Given a more favorable political climate for Hobbes, not impossible given his role as a tutor of Charles II, science could have been institutionalized in a way that discounted experimentalism. However, given the multiple ways that even those pursuing a mathematical, deductive approach incorporated experimental approaches, it seems unlikely that Charles II institutionalizing Hobbes's approach would have significantly derailed the momentum that experiment had already attained. Even if insertable in history, the counterfactual of political legitimation for Hobbes's approach would have been swamped by reversionary second-order counterfactuals rather than amplifying ones.[11]

Historians prone to emphasizing underdetermination focus on contingency (Hobbes vs. Boyle), without explaining where this alternative branching point would have led. In the extreme case, believing that contingency runs rife in history makes it difficult to see how an alternative, Hobbesian science could be *sustained*, as new contingencies make historical change open and changeable at every moment.[12] Alternatively, many historians believe that even if a counterfactual event could have occurred, the historical context includes a multitude of reasons why the same historical outcome would result no matter what, as multiple causes shape any historical trajectory, not single events. In the extreme case, this view reflects the "essay question" approach to history, where one gets credit for providing the most comprehensive list of causes for a big historical event, like the French Revolution or World War I. Instead, I will argue for a more balanced perspective on historical contingency, one that highlights the role of historical contingency while also delineating how likely it was that alternative outcomes could have been sustained in practice.

Could non-toxic, or less toxic, chemicals have been produced in the twentieth century? I aim to show that there existed developed, alternative approaches to understanding the safety of hazardous substances in a number of fields. These alternatives emphasized eliminating inherent toxic substances on the assumption that low levels of toxic substances likely remained harmful, a view called the per se standard. By contrast, the dominant approach

---

11. Lynch, "Arguments for a Non-Whiggish Hindsight"; Lynch, "King's Evidence"; Dear, *Discipline and Experience*.

12. But see Brown, "A Different Counterfactual Perspective on the Eads Bridge."

sought to identify safe levels of any substance on the assumption that "the solution to pollution is dilution," a view called the *de minimus* standard. In essence, one could seek to distinguish toxic from benign substances, or one could identify safe and toxic levels of any substance.

The latter view won out, with partial exceptions to be discussed, but the arguments for the former were (and remain) compelling, so that understanding how a different outcome was possible is well supported by historical documentation. At the same time, science does not operate in isolation from larger societal forces. In the context of powerful institutional support for the toxicologists' *de minimus* standard, the postulation that the eclipsing of the per se standard was a near run thing within science must be supplemented by the recognition that an alternative scientific trajectory would not be sustainable without countervailing public pressure and alternative sources of financial support.[13]

HINDSIGHT BIAS AND THE SOCIAL CONSTRUCTION OF TECHNOLOGY

Methodological developments on the use of counterfactuals in recent political history have tried to counter the "hindsight bias" identified by experimental psychologists. After the fact, political historians have tended to assume that outcomes that would have been seen as highly improbable before they occurred (rise of the West, negotiated settlement of the Cuban missile crisis, fall of the Soviet Union) are seen as inevitable after they happened. To develop critical alternatives to established historiographical questions on such topics, this typical, three-stage pattern of development is helpful: (a) an early open stage showing initial unpredictability of outcomes, followed by (b) a growing inflexibility as developments lock in one trajectory, and (c) closure locking in suboptimal arrangements.[14]

These stages parallel the stages of Harry Collins's empirical program of relativism (EMPOR), which identified three stages for analysis of a scientific controversy, with the approach extended to technology by the social construc-

13. Comparative history suggests that the strength of political opposition to chemical pollution plays a role. Jones and Lubinski, "Historical Origins of Environmental Sustainability in the German Chemical Industry, 1950s–1980s," pp. 19–20, 25–28, 36–37, argue that until the 1970s, German chemical companies Bayer and Henkel resisted regulation just as much as U.S. chemical companies. At that time, the emergence of a politically successful Green Party and environmental protests directed at each company's headquarters, where the firms were susceptible to pressure as the result of stronger regional ties than U.S. corporations possessed, led to a more proactive and precautionary approach.

14. Tetlock and Parker, "Counterfactual Thought Experiments"; Tetlock, Lebow, and Parker, *Unmaking the West*; Lebow, *Forbidden Fruit*; Lebow and Stein, *We All Lost the Cold War*.

tion of technology (SCOT). First, the analyst focuses on the interpretive flexibility associated with the open character of interpreting data or designing a technology. Second, the program focuses on the causes of closure, where the range of interpretations or design options narrow. Third, the analyst focuses on the role of the wider social and political context in shaping outcomes.[15] This approach helps delineate how certain suboptimal outcomes in modern science and technology came about—especially when the analysis incorporates Collins's argument that expert core sets that are artificially constrained can lead to *premature* closure by ignoring qualified experts from different fields.[16] This provides a way to develop a model of suboptimal outcomes in technological developments as well, along the lines of Diane Vaughan's influential treatment of the "normalization of deviance" leading to the *Challenger* launch explosion.[17]

Indeed, the minority view supporting the per se standard did achieve scientific support in the study of radiation and carcinogenic substances. However, the dominant paradigm of understanding risk by identifying safe and unsafe levels won out in the American chemical industry and can be seen as a suboptimal outcome in the sense that closure proceeded by regulatory adoption of one side of a scientific debate as the result of corporate pressure to maintain economically important chemicals in production. Essentially, a premature closure of scientific debate (and resulting chemical design desiderata) was effected as the result of stage three economic and regulatory influence. Thus, even as we can see the moment of contingency that made an alternative approach to chemical synthesis and production possible, we can also identify those further causes that helped ensure that the outcome occurred as it did. The result is to see more clearly just how strongly the economic interests of chemical manufacturers shaped modern chemistry and chemical engineering.

DE MINIMUS AND PER SE STANDARDS

Toxicologists distinguish two approaches to regulating chemicals: a *de minimus* standard that identifies a threshold below which chemicals are presumed safe and a per se standard where no safe level is defined for chemicals found to be toxic. With some important qualifications to be discussed, toxicology adopted a *de minimus* standard, an approach that was solidified from a regulatory point of view by the 1958 Federal Food, Drug, and Cosmetics Act. The approach reflects both industry interest in defending safe levels of chemicals

15. Collins, "An Empirical Relativist Programme in the Sociology of Scientific Knowledge"; Pinch and Bijker, "Social Construction."

16. Collins, "Public Experiments and Displays of Virtuosity," pp. 739–42.

17. Vaughan, *The* Challenger *Launch Decision*.

deemed economically essential, as well as disciplinary closure of scientific controversy in toxicology despite continued objections to the *de minimus* standard by outside experts in two fields: cancer researchers studying the effects of chemical substances, especially plastics and pesticides, and geneticists researching low-dose radiation in nuclear fallout.

The *de minimus* standard applied to most chemicals as the result of the 1958 act, except those judged carcinogenic, which were deemed unsafe at any level as the result of the Delaney amendment. Representative James Delaney had chaired the Select Committee to Investigate Chemicals in Food Production beginning in 1950. His committee examined the effect that the growing use of synthetic fertilizers and plastic packaging had on food safety, hearings that were the driving force behind the 1958 act. In the end, however, the language of the act reflected the demands of industry for a tame regulatory approach.[18]

CARCINOGENESIS AND THE PER SE STANDARD

The scientific background for the Delaney amendment was shaped by the testimony of cancer researchers, especially Wilhelm Hueper, a significant influence on Rachel Carson and the director of the environmental cancer section of the National Cancer Institute. His testimony reflected research practices and questions distinct from those in toxicology. Whereas animal experiments in toxicology proceeded by large doses and examination of acute effects, with identification of safe levels on that basis, Hueper described research on the effects of synthetic estrogens at levels judged non-toxic that nevertheless led to cancer. He also argued for the significance of exposure during fetal development, anticipating the approach of Theo Colburn and proponents of the endocrine disruption hypothesis.[19]

High-dose experiments facilitated speedy identification of toxic substances, but extrapolating to lower doses was fraught with difficulty. Extrapolating from high-dose experiments while assuming a linear dose-response curve would imply no threshold of absolute safety. Toxicologists sought to define safe levels of economically important chemicals using a model where linear effects begin after a threshold, in effect blocking consideration of low-dose effects. In a situation where cost considerations prohibited reliance upon low-dose, longer-term studies, how to extrapolate from high-dose animal studies reflected political differences over regulatory philosophy.[20] In

18. Vogel, *Is It Safe?*, ch. 1.
19. Ibid., pp. 24–30.
20. Proctor, *Cancer Wars*, ch. 7.

this context, toxicology's role as a discipline shaped by the financial support of chemical companies determined the kinds of risks that were considered worthy of research, neglecting environmental causes of cancer and focusing on acute effects of significant doses in order to define safe levels of workplace exposure.[21]

Indeed, early research on the hormonal impact of plastics provided significant evidence of carcinogenic effect that was neglected for a generation, partly as the result of scientific suppression.[22] Proctor tells the remarkable story of the suppression of Hueper's research by his superiors, first at DuPont Corporation and then at the National Cancer Institute (NCI), part of the National Institutes of Health. His superiors at DuPont often prohibited publication of his research results on the grounds of industrial secrecy, and he was not allowed to carry out longer-term studies to test his hypothesis that chronic exposure was more significant than acute effects. Hueper was fired in 1937. Thereafter, threatened lawsuits discouraged him from presenting information based on his research at DuPont.[23]

After leaving DuPont, Hueper continued his research at a pharmaceutical firm and published his *Occupational Tumors and Allied Diseases* in 1942 before being hired by the NCI to direct its new environmental cancer section. While ostensibly free to conduct research, he encountered frequent restrictions on publication and travel, with carbon copies of his papers submitted surreptitiously to DuPont for commentary before review. He was forced to withdraw a presentation on his research on Colorado uranium miners before the Colorado State Medical Society (CSMS) in 1952 when Atomic Energy Commission (AEC) oversight led to its prohibition.

Shields Warren, the AEC's director of biology and medicine asked the NCI to fire Hueper after he circulated the paper to the CSMS president. While not fired, he was subsequently banned from cancer research on humans.[24] Even though Hueper was influential in securing legislation putting forth a per se standard for carcinogens in the Delaney amendment, in practice, regulatory capture of the FDA process for implementing the Delaney agreement took place. Jerome Heckman, a lawyer for the Society of the Plastics Industry, argued that when quantitative evidence of significant contamination of food by polymers in packaging was lacking, such chemicals did not constitute a food additive under the law.

In effect, Heckman was able to turn the Delaney *per se* standard into a *de minimus* standard, while grandfathering chemicals already in use and

21. Vogel, *Is It Safe?*, pp. 7–9, 94, 201–2.
22. Ibid., pp. 26–30.
23. Proctor, *Cancer Wars*, pp. 39–41.
24. Ibid., pp. 44–45.

streamlining the regulatory approval process.[25] The Manufacturing Chemists' Association (MCA) and other manufacturing associations produced doubt about claims of harm under the guise of "trade association science," while corporate literature reviews in toxicology routinely ignored evidence of carcinogenicity.[26] Funded by industry and using experimental techniques that blocked consideration of low-level effects, toxicology could itself be seen as a scientific field captured by industry (call it *disciplinary capture*).

THE FALLOUT DEBATE AND THE LINEAR NO-THRESHOLD MODEL

During the height of atmospheric nuclear weapons testing in the 1950s, public debate on the effects of radioactive fallout informed debates about chemical toxicity as well. Rachel Carson drew on the invisible and insidious character of strontium-90, its bioaccumulation in the food chain, and public doubts about governmental forthrightness to frame her discussion of the dangers of pesticides.[27] Debate about whether nuclear fallout presented a real danger to the U.S. public split along disciplinary lines. Among geneticists, the idea that genetic damage from radiation increased linearly without a threshold (the linear no-threshold model or LNT) dominated, while most physicians involved in the nuclear fallout debate held that low-level fallout was safe.[28] Physicians were concerned that public dissemination of the no-threshold model would cause public panic and make medical use of radiation impossible. Hermann Muller's demonstration in 1927 of radiation-induced genetic mutations led him to express caution on the use of X-rays for medical therapies and provided the foundation for the linear model. Since that time, geneticists have complained that the implications of the LNT model were ignored by medical practitioners and regulators. Just as with the study of carcinogens, studies of long-term chronic exposure to radiation were neglected in favor of focusing on acute effects.[29]

Evidence for the neglect and suppression of the geneticists' position is strong, which called for a precautionary approach based upon emerging understanding of the role of radiation in inducing genetic damage (and possibly cancer). Geneticists also recognized that scientific demonstration of the exact

25. Vogel, *Is it Safe?*, pp. 138–42, 1–3.

26. Proctor, *Cancer Wars*, pp. 125–32; Oreskes and Conway, *Merchants of Doubt*; Vogel, *Is It Safe?*, p. 38.

27. Carson, *Silent Spring*, pp. 5–7; Lutts, "Chemical Fallout."

28. Jolly, *Thresholds of Uncertainty*; Kathren, "Historical Development of the Linear Nonthreshold Dose-Response Model as Applied to Radiation."

29. Jolly, *Thresholds of Uncertainty*, pp. 7, 50, 217–41, 76–82.

level of harm for human populations was not provable by current methods. Geneticists were largely excluded from the Manhattan Project's assessment of radiation risk, where physicians, physicists, and radiologists took a toxicological approach focused on immediate damage produced above a threshold.[30] The Manhattan Project did fund studies by geneticists at the University of Rochester. Both Curt Stern's research on fruit flies and Donald R. Charles's research on mice ultimately supported the linear interpretation, though an official Manhattan Project retrospective implied that Stern's research had demonstrated the existence of a threshold. The Los Alamos Scientific Laboratory's 1950 publication, *The Effects of Atomic Weapons*, ignored these studies altogether.[31]

The consensus among geneticists following World War II, and particularly during the "fallout debate" about atmospheric nuclear testing in the 1950s, was that each additional amount of exposure to radiation brought additional mutations, so that the introduction of any new sources of radiation exposure constituted a significant hazard. By contrast, physicians and AEC regulators compared new exposures to naturally occurring background radiation in order to judge new exposures as negligible.[32] These contrasting views on linearity and thresholds were reflected in differing assessments of radiation by the genetics panel and the pathology panel of the 1956 National Academy of Sciences (NAS) Committee of the Biological Effects of Atomic Radiation (BEAR).[33] In 1955, Muller's paper on the genetic risks of radiation, developed for the United Nations International Conference on the Peaceful Uses of Atomic Energy, which promoted Eisenhower's "Atoms for Peace" program, was rejected by the AEC because of doubts about Muller's political loyalty to the United States and his public airing of fallout dangers—the paper discussed radiation damage received by victims of the Hiroshima and Nagasaki bombs and supported linearity.[34]

Later, in a parallel case to Hueper's of significant scientific work that was suppressed by governmental agencies, the AEC cut funding and suppressed John Gofman's efforts at the Lawrence Livermore National Library (LLNL) to extend the geneticists' LNT model for low-dose radiation from genetic damage to somatic effects, particularly long-term damage leading to cancer.[35] The concept of a tolerance dose had become accepted by the 1930s and identified thresholds of radiation exposure above which somatic damage,

30. Ibid., p. 229.
31. Ibid., pp. 78–83, 106–7.
32. Walker, *Permissible Dose*, pp. 10–11; Jolly, *Thresholds of Uncertainty*, pp. 162–266.
33. Jolly, *Thresholds of Uncertainty*, pp. 314–454.
34. Carlson, *Genes, Radiation, and Society*, pp. 356–67; Jolly, *Thresholds of Uncertainty*, pp. 253–66.
35. Semendeferi, "Legitimating a Nuclear Critic," pp. 259–65.

such as reddening of skin, would take place.[36] Thresholds for somatic effects of radiation were endorsed by the NAS BEAR committee in 1959 and were widely accepted through the 1960s.

Gofman reported chromosomal alterations in cancer cells in 1967, supporting the hypothesis that cancer was caused by genetic damage to somatic cells. By 1969, he had presented evidence that low-dose radiation could induce chromosomal damage.[37] The AEC pressured LLNL to rein in Gofman after he argued that Federal Radiation Council (FRC) standards were ten times higher than a precautionary approach would require. Gofman's funding was cut, he was threatened with dismissal, and prepublication approval was required.[38] Gofman's no-threshold approach to cancer did finally make it into the 1972 NAS Biological Effects of Ionizing Radiation (BEIR) report, but the ensuing public controversy ended his research.[39]

The LNT model finally gained traction as a way to estimate radiation risk quantitatively, but regulation of low-dose exposure remains elusive. In regulating radiation exposure, the AEC (and its successor the Nuclear Regulatory Commission [NRC]) finally adopted the LNT model along with vague language that exposure should be "as low as practicable" to encourage nuclear plant designers to lessen exposures below formal threshold levels.[40] The ascendency of the LNT model reflected a compromise between realist and instrumentalist interpretations of the model, as the 1972 NAS BEIR report argued that the model represented "the only workable approach to numerical estimation of the risk in a population" despite the fact that the model might overstate low-level risks and neglect the existence of cell-repair mechanisms.[41]

Nonetheless, in contrast to threshold models, all levels of radiation exposure needed to be balanced against specific benefits. At the same time, the document incorporated the view of critics who felt that risks of nuclear emissions were being exaggerated compared to energy sources such as coal.[42] A developed, if contested, alternative to the threshold model had emerged for handling radiation exposure, albeit within a technological milieu where increased risk was seen as unavoidable and rejection of the technology as

36. Walker, *Permissible Dose*, pp. 2–10; Hacker, *The Dragon's Tail*, pp. 14–19; Kathren, "Historical Development of the Linear Nonthreshold Dose-Response Model as Applied to Radiation," pp. 11–12.

37. Semendeferi, "Legitimating a Nuclear Critic," pp. 263–65, 267–69.

38. Ibid., pp. 271–73, 277–78.

39. Ibid., pp. 297, 278; Walker, *Permissible Dose*, pp. 47–51.

40. Walker, *Permissible Dose*, pp. 31–35, 44–56.

41. Quoted in ibid., p. 49. See also Kathren, "Historical Development of the Linear Nonthreshold Dose-Response Model as Applied to Radiation," p. 14, who notes that linear extrapolation "was really a statement of the upper level of risk in the low dose region of the dose-response curve—the very region of interest from a protection standpoint and the very region in which dose-response data were not available."

42. Walker, *Permissible Dose*, p. 50.

unacceptable. Gofman, however, would go on to question publicly whether nuclear power was worth the risks.

It is certainly possible that no-threshold approaches might be suitable for some areas of study, like carcinogenic chemicals and low-dose radiation, while being inappropriate for toxicology more generally. However, the way these arguments played out, different extrapolations from high-dose animal experiments reflected different political assumptions about risk.[43] Toxicology's emphasis on threshold harms was used as a shield to prevent further investigation of low-dose harms, as well as to focus attention on acute harms rather than long-term effects. The no-threshold approach was implemented in the Delaney amendment but was quickly rendered moot by regulatory capture, as pressure from the chemical industry shaped the FDA's interpretation of the amendment.

The most significant outcome of the threshold model was that industrial chemicals, synthetic pesticides, and plastics used in consumer goods continued to be developed and manufactured even where evidence of toxicity existed. A further investigation of a second-order counterfactual would be necessary to judge the extent to which substitute chemicals or alternative manufacturing processes could have been found had regulatory intervention followed a more preemptive approach. Moreover, even given greater acceptance of the no-threshold approach within the scientific community, corporate pressure would likely have continued to prevent a significant shift in the type of chemicals manufactured, at least absent a larger environmental movement to challenge business as usual. What counterfactual analysis does reveal is that serious challenges to business as usual did occur. But those challenges were overwhelmed by the singular importance granted to industrial chemicals and pharmaceutical products under the reigning American political economy of the day. Lack of knowledge of harms alone did not result in the proliferation of toxic chemicals. That dangerous history resulted from the power of corporate interests in shaping the kinds of chemicals that would be developed, as well as the methods to evaluate and regulate their safety.

ENGINEERING ETHICS AND TECHNICAL PRACTICE

Recent discussion in engineering ethics has focused on the role of engineering practice and culture in shaping outcomes of ethical relevance. The shift to considering "macroethics," oriented to the professional responsibility of engineering to society, in addition to the traditional focus on "microethics," regulating individual engineers' relationships with clients, employers, and

43. Proctor, *Cancer Wars*, ch. 7.

colleagues, has broadened consideration of the social impact of engineering work.[44] "Social responsibility" has become a focus, and the teaching of "Science, Technology, and Society" courses has addressed this need through a deeper engagement with the social consequences of engineering practice and the causes of engineering disasters.[45] The emphasis on engineering practice and culture has provided an opening for contributions from Science and Technology Studies (STS), examining how the social dimensions of science and engineering practice shape the knowledge and technologies produced.[46]

In 2000, Ron Kline and I sketched out a way to apply insights from Science and Technology Studies to engineering ethics education. We sought to challenge two of the predominant approaches to engineering ethics education: (1) hypothetical ethical dilemmas and (2) engineering disaster case studies. The latter emphasized that engineers should be prepared to act as whistleblowers, substituting concern with public health and welfare for management concern with profit. Both approaches took for granted that the ethical implications of engineering practice were relatively straightforward: the goal was to exhort engineers to consider ethical obligations that followed from abstract moral theory, professional obligations to protect public health and safety, or simply the need to stand up for what's right against impending disaster.[47]

While we supported the effort to get engineers to think about moral issues, we were concerned with a prior question: how do engineers identify potential moral issues in the first place? The engineering ethics literature seemed to take this for granted, what with its prepackaged, sterile ethical dilemmas ("Should you lie about test reports if your boss tells you to?") and tales of heroic engineers like "near whistle-blower" Roger Boisjoly who were ignored by management for reasons of money or schedule.[48]

In drawing upon Diane Vaughan's analysis of the *Challenger* launch accident, we sought to use it as a model for how case studies could focus on the latent ethical implications of ordinary engineering practice well before disaster strikes. This made sense since Vaughan had demonstrated that, on the eve of the launch, engineers' hands were tied not by the "amoral calculation" of managers vetoing engineering judgment for organizational imperatives but by the long process of incremental judgments that had constructed risk as-

44. Herkert, "Future Directions in Engineering Ethics Research"; Herkert, "Professional Societies, Microethics, and Macroethics"; Herkert, "Ways of Thinking about and Teaching Ethical Problem Solving"; Davis, "Engineering Ethics, Individuals, and Organizations."

45. Zandvoort et al., "Editors' Overview"; Mitcham, "A Historico-Ethical Perspective on Engineering Education"; Ozaktas, "Teaching Science, Technology, and Society to Engineering Students"; Basart and Serra, "Engineering Ethics beyond Engineers' Ethics."

46. Fisher, "Editorial Overview"; Schuurbiers, "What Happens in the Lab Does Not Stay in the Lab"; Wynne, "Lab Work Goes Social, and Vice Versa."

47. Lynch and Kline, "Engineering Practice and Engineering Ethics."

48. Ibid., p. 200.

sessment in ways that were consistent with safety rules and defined what was known about the effects of cold on O-rings and the safety of the solid rocket booster (SRB) joint.[49]

Given the evidence before them and the organizational culture emphasizing that "emotional" arguments based on judgment can't trump the accumulated body of evidence seemingly demonstrating no correlation between cold and O-ring damage on flights with field joint problems, Boisjoly was left with no leg to stand on in his efforts to stop the launch. On the eve of the launch, as engineers debated the significance of the expected cold weather for the launch employing ambiguous data, the teleconference between NASA contractor Thiokol, Marshall Space Flight Center, and Kennedy Space Center was put on hold so Thiokol engineers and managers could caucus. When Thiokol senior vice president Jerry Mason asked senior Thiokol manager Robert Lund to "take off his engineering hat and put on his management hat," joining the other three senior managers in a decision to launch, he relented and the launch proceeded. The "managers" were actually experienced engineers who understood their role: to make a decision when data was lacking. The decision was made based on the belief that the data supported the conclusion that damage to the primary O-ring would be within acceptable parameters and the secondary O-ring would ensure redundancy.[50] While commentators have misinterpreted comments about replacing engineering with management "hats" as a sign that management callously substituted schedule pressures for safety, Vaughan shows us that the engineers' own efforts at defining risk over many years had become objectified and alienated from their activity and tied their hands during the teleconference.

Vaughan calls this process of *incremental change* away from a "goodness of fit" that is only recognized at the last moment or in hindsight the "normalization of deviance." Unfortunately, this term has led to confusion. On the one hand, we were criticized for letting the engineer or manager off the hook by emphasizing the complexity of engineering practice—in short, for not being sufficiently moralistic.[51] On the other hand, we were criticized for focusing on individual responsibility to avert the normalization of deviance in an institutional setting where individual agency is lacking—in short, for being

49. Vaughan, *The* Challenger *Launch Decision*, pp. 47–49, 55–56, 106–14, 177–78.

50. Ibid., pp. 290, 315–26. But see Pinkus et al., *Engineering Ethics*, pp. 318–19, on the possibility that the engineers could have found a correlation of O-ring damage with temperature across all flights, not just those with field joint problems, if the engineers had thought to do so.

51. Doing, "Applying Ethnographic Insight to Engineering Ethics"; Davis, "Engineering Ethics, Individuals, and Organizations." On the tendency for students reading Vaughan to excuse engineers from responsibility, as well as pedagogical techniques for overcoming this tendency, see Kline, "Using History and Sociology to Teach Engineering Ethics."

too moralistic.[52] Coupled with this concern for the role of individuals either evading responsibility for their actions or being burdened with responsibility for institutional factors beyond their control, there is the related issue of the extent to which engineers can ameliorate potential problems before they occur.

From the point of view of engineering ethics, it was easy to substitute the call for engineers to avoid normalizing their deviance for earlier calls to heroically resist amoral calculation.[53] From the point of view of Science and Technology Studies (STS), the term implies an inappropriate use of hindsight and an asymmetrical delineation of deviance based upon outcomes that could not have been predicted at the time.[54] In short, the skeptical side of science studies emphasizes that focusing upon embedded engineering practice, while interesting, will not produce lessons applicable to *new* contexts. Instead, each new case can be expected to have its own locally produced constructions of knowledge and risk. Reaction to this piece has either driven the field back to abstract moral handwringing unlikely to impact practice or has denied that we can produce the kind of knowledge that can be applied to engineering practice.

To rebuild this unstable alliance between engineering ethics and STS, we need to confront directly the obstacles to incorporating "reflection" into engineering practice and articulate a place for engineers in building institutions that challenge corporate and governmental imperatives currently shaping engineering practice. The situation parallels Feyerabend's concern that dominant perspectives can make it difficult to challenge an established observation language without outside intervention. In thinking about how to promote critical reflection on established practices, I argue that (1) engineers already reflect upon their own practice and seek to alter it when they have reason; (2) STS can provide a resource for engineers in altering their practice by focusing attention on how constructions of risk become black-boxed and alternative approaches are cut off; and (3) scientists, engineers, and STS researchers can be crucial participants in larger social and political movements to challenge accepted directions and choices in technology.

By focusing on interventions by scientists and engineers themselves, I am not trying to "stand in" for the actors themselves, representing them from afar, but to intervene *at the same level* as reflective engineers themselves do when they try to assemble social and technical resources to better public health and welfare. In doing so, many engineers reflect upon the past history

---

52. Swierstra and Jelsma, "Responsibility without Moralism in Technoscientific Design Practice"; Conlon and Zandvoort, "Broadening Ethics Teaching in Engineering."

53. Li and Fu, "A Systematic Approach to Engineering Ethics Education"; Doorn, "Responsibility Ascriptions in Technology Development and Engineering"; Herkert, "Future Directions in Engineering Ethics Research"; Herkert, "Ways of Thinking about and Teaching Ethical Problem Solving"; Jablokow, "Engineers as Problem-Solving Leaders."

54. Doing, "Applying Ethnographic Insight to Engineering Ethics," p. 246; Swierstra and Jelsma, "Responsibility without Moralism in Technoscientific Design Practice."

of technological choices leading to harmful outcomes and draw lessons about how things should be done differently now. I believe that analyzing how past choices with bad outcomes were made can offer strategies for shaping current technological decisions, even if no participant in the past could have changed the outcomes. Rather, just as engineers build new knowledge of safe plane design from the contingent circumstances leading to a particular crash, so too can we help engineers themselves to "second-guess" what they have done in the past to help shape what they might do in the future. The expertise that STS scholars have to offer to engineers and scientists to reconsider historical trajectories include analyzing (a) how technical practice shapes constructions of risk and (b) how technical practice is influenced by larger organizational and cultural contexts.

After discussing how ethical reflection can build upon STS research on the sociotechnical practice of engineers and scientists, I explore a case where scientists and engineers themselves incorporated ethical reflection on the social consequences of their work in developing new approaches to their field. Paul Anastas and John Warner proposed that chemical synthesis proceed in a less polluting and more sustainable fashion in their 1998 book, *Green Chemistry*, and Anastas and Julie Zimmerman later extended this approach to engineering.[55] The U.S. Environmental Protection Agency (EPA) and the American Chemical Society (ACS) promote the "rules" of green chemistry and engineering originated by Anastas, Warner, and Zimmerman (see figure 6.1). Both research programs have expanded their role in graduate education and introduced sustainable and less polluting techniques to the marketplace.[56] Crucial to the development of these approaches by Anastas, Warner, and Zimmerman is ethical reflection upon the past unsustainable and polluting practice of chemistry and engineering. In Vaughan's terms, they criticized the "normalization of deviance" in their field.

WHO DEFINES DEVIANCE?

In considering Vaughan's concept of "normalization of deviance," a crucial question is who defines deviance? On this point, Vaughan is unfortunately on the side of skeptical STS scholars: there is no way to define deviance from desired safety standards except retrospectively, once the ongoing constructions of known risk have solidified into a series of decisions that have become precedents. There was no readily accepted way of challenging the decision at

55. Anastas and Warner, *Green Chemistry*; Anastas and Zimmerman, "Design through the 12 Principles of Green Engineering."

56. Marteel-Parrish and Abraham, *Green Chemistry and Engineering*; Allen and Shonnard, *Green Engineering*.

that time. Since prior work had black-boxed the cold safety of O-rings, appeal to engineering judgment based on an unproven but suspected temperature connection was not deemed sufficient to counter established documentation of safety. And while it might be argued that the impossibility of proof and the inevitability of judgment in technical practice, as a finding of STS, might have given pause about this quick dismissal of judgment, any solution emphasizing the need to always valorize engineering judgment would have its own problems. For example, Doing has argued that engineering judgment was used as a justification for avoiding examination of corporate responsibility for the BP Gulf oil spill.[57] Key here is the problem of the generalizability of the lessons gleaned from a post hoc analysis of a disaster.

On this point, it is necessary to argue against the skepticism inherent in Vaughan's own position, which reinforces the skepticism inherent in most STS work. Any process leading to an undesirable outcome will require a close study of the engineering practices that led to it. Any aspects of that practice that led to the ill-fated outcomes can be expected to be sufficiently idiosyncratic to that practice as to offer no guidance in some different context. In a section titled "The Inevitability of Mistake," drawing on Perrow's concept of "normal accidents" in complex technological systems, Vaughan makes clear that the sociologist cannot offer a "fix" that can be applied in new situations that would prevent potential signals of danger from being ignored.[58] Indeed, the normalization of deviance is "normal practice" for all engineers: one can't build and refine constructions of risk unless one is also closing down or ignoring alternative scenarios. Some kind of deviance must be normalized in order to do any work at all—so that the next time, the processes leading to disaster will be different, perhaps even facilitated by the ill-considered advice of the sociologist reflecting on the last case.

In opposition to management consultants who claim the ability to reform organizations to prevent such occurrences, Vaughan's microfocus on practice allows her no such easy out, precisely because it was the results of just such a reform that failed in this case: the implementation of Flight Readiness Reviews (FRRs) that sought to circumvent organizational secrecy and reinforce the independence of engineers had been the very mechanism by which Biosjoly's hands were tied. The engineers had signed off on the cold safety of past flights as part of a process explicitly designed to give engineers grounds for vetoing management imperatives. In Vaughan's words,

---

57. Doing, "Applying Ethnographic Insight to Engineering Ethics"; Jasanoff, "Acceptable Evidence in a Pluralistic Society," observes that the British regulatory regime incorporates engineering judgment, while the U.S. approach excludes it for quantitative proof.

58. Vaughan, *The* Challenger *Launch Decision*, pp. 415–22. More generally, citation of Perrow, *Normal Accidents*, has functioned within STS to reinforce skepticism that anything can be done to prevent disaster in complex systems.

> NASA had a well-conceived system for regulating decisions about risk that worked in every case but one. Moreover, the eve-of-launch teleconference was exactly the sort of decentralized decision process in response to crisis that high-reliability theorists call for. We saw how unseen history, political bargains, culture, and structure undermined decision making that night.[59]

Generalizing from what went wrong that night, it is possible to impose new rules that might require clearer "go" or "no go" signals or greater reliance upon skeptical outsiders, but these rules can be expected to come with their own set of problems. Like engineers who balance tinkering with a known-but-flawed design with starting from scratch, the social engineer must worry that bureaucratic changes will introduce new unknown flaws:

> An aspect of culture may simultaneously enhance safety in some situations and undermine it in others. If we try to change it in response to an accident, are we undermining some part of the system that formerly was functioning in an acceptable manner?[60]

In fact, Vaughan points out that in the aftermath of the *Challenger* explosion, the number of C 1 designations that require safety attention actually increased, bringing more formal attention to system components that need additional safety attention, but at the cost of routinizing what was supposed to be a signal of alarm. In addition, new burdensome forms requiring engineers to pass bad news up the hierarchy further distracted engineers from their core work.

LEARNING FROM THE PAST

So what hope is there to learn from the past? And how can post hoc analysis of engineering disasters help us do better in the future? Here is it worth pointing out that Vaughan's skepticism comes from awareness of how poorly new bureaucratic requirements actually "latch on" to real-time engineering decision making: rules must be invoked in practice, and participants must decide which rules to invoke, when, and how. This issue has been addressed in STS work, going back to the early strong program emphasis on the role of interests in shaping science.[61] Where the strong program explained science sociologically by invoking interests that shaped scientific work, Woolgar argued that imputations of interest were them-

59. Vaughan, *The* Challenger *Launch Decision*, p. 417.
60. Ibid.
61. Barnes, *Interests and the Growth of Knowledge*.

selves a disputed component of technical practice itself and so could not stand as an independent analyst's account of that practice.[62] The problem with that move is the same as the problem with Vaughan's skepticism about reform: it retreats to a higher level of analysis rather than back into the fray at the same level as the actors under study. The analyst finds that putative explanations of practice—bureaucratic rules or social interests—themselves are invoked and manipulated by the participants, so they cannot stand as independent causes of that action. Aside from the logical error of this move—it is possible for actors to impute interests to others even as they are shaped by interests—this misses the opportunity to intervene precisely at the level of participant representations.

Since the engineers are already (more or less) reflective practitioners, introducing more reflection, especially to captive student engineers-in-training or civically active engineers, can inform and alter practice in ways that are mediated by the engineers themselves, and not by another layer of bureaucratic reform. As we argued, "[t]he problem with [Vaughan's] way of looking at the issue is that the entire burden is placed on sociologists or other outsiders to reform organizational practices."[63] The sociology of scientific knowledge began with a rejection of external philosophical judgment of technical experts, and STS has followed suit by retreating to its own disciplinary focus on redescribing (rather than changing) technical practice. So far, STS has only interpreted technical practice in various ways; the point is to change it.

While it is possible that an engineer educated about the normalization of deviance might appeal to such a concept in an attempt at whistleblowing, this is not a likely scenario. However, engineers do much more than passively inhabit organizational structures. Many build technological systems that link them to affected communities, regulators, and other engineers. Some use professional organizations to promote changes in accepted practices. A smaller number link themselves to social movements concerned about the environment, the impact of risky technologies on public health, or efforts to shape technological design democratically. To engage with such civically active scientists and engineers, STS needs to finally carry out Collins's third stage in the empirical program of relativism by linking engineering design to wider social and political structures.[64]

Collins's first stage focused on the interpretive flexibility of research findings and emphasized the openness of scientific work, leading to competing interpretations of the data. The second stage addressed the mechanisms

62. Woolgar, "Interests and Explanation in the Social Study of Science."

63. Schön, *The Reflective Practitioner*; Lynch and Kline, "Engineering Practice and Engineering Ethics," p. 202.

64. Collins, "Stages in the Empirical Programme of Relativism."

bringing about consensus and sought to identify how the inherent openness of scientific practice was narrowed to arrive at closure of controversy. The third stage aimed to connect scientific work to the larger social and political context. The first two stages have been examined in detail, but most STS work does not connect scientific or engineering practice directly with social and political contexts informing it.[65]

HISTORICAL COUNTERFACTUALS AND NORMATIVE CRITIQUE

Attention to the three-stage analysis of controversial technologies can illuminate how history could have been different given different conditions. Often times, consensus formation by experts is insulated from direct public participation but shapes the options available to policy makers. As Latour argues, engineers play a role in helping realize—literally making real—imagined technologies in their negotiations with each other and other social actors. In Mackenzie's analysis of the growth in accuracy of ICBMs (intercontinental ballistic missiles), engineers and mid-level policy advisers shaped Cold War policies toward the first-strike, preemptive doctrine of counterforce rather than the official doctrine of countercity associated with mutually assured destruction. Each improvement in missile accuracy (rather than a focus on reliability) would make possible the targeting of Soviet missiles, which would make a first-strike attack more conceivable by the United States. As a result, the Soviet Union and the United States were more prone to launching missiles in a crisis to "use them or lose them." Mackenzie, just before the fall of the Soviet Union, offered a scenario whereby nuclear weapons become

65. Some work has already been done, particularly on the way national cultures have shaped the organization of engineering work, with Americans oriented to industry and the French to government, for instance, or the ways in which educational requirements served to bolster the low status of engineers in Britain. See Downey, "Low Cost, Mass Use"; Downey and Lucena, "Knowledge and Professional Identity in Engineering"; Downey, Lucena, and Mitcham, "Engineering Ethics and Identity"; Han and Downey, *Engineers for Korea*. For the classic discussion of the conflict between progressive and conservative engineers in the early twentieth-century United States, see Layton, *The Revolt of the Engineers*; with critical discussion by Kline, "From Progressivism to Engineering Studies," and Meiksins, "The 'Revolt of the Engineers' Reconsidered." For a start at "fine-grained historical work delineating the ways in which engineers have served specific social agendas as they have reworked the material world," see Brown, Downey, and Diogo, "The Normativities of Engineers," p. 738; Wisnioski, "'Liberal Education Has Failed'"; Bassett, "Aligning India in the Cold War Era"; Valderrama et al., "Engineering Education and the Identities of Engineers in Colombia, 1887–1972." The impact of warfare on engineering generally and the role of Cold War politics on directing military innovation in particular have received much attention: Walton, *Instrumental in War*; Mitcham, "A Historico-Ethical Perspective on Engineering Education." Finally, it is a truism within engineering ethics that engineering is a "captive profession" in that engineers working in corporations will be much more susceptible to corporate concern with maximizing profit than with professional obligations to protect public welfare. See Davis, *Thinking Like an Engineer*, p. 22; Noble, *America by Design*.

"uninvented" by the dismantling of the social and material infrastructure that keeps these technologies alive, just as nuclear test-ban treaties sought to limit further technological innovation in nuclear weapons. Such analysis provides us with a start toward understanding how technological trajectories can be made the subject of explicit democratic debate.[66]

Mackenzie offers us a scenario whereby politically and sociologically reflexive engineers could help advise peace activists on the best way to alter technological trajectories effectively. Much depends, however, on what changes our historical or sociological analysis suggests might have led to different outcomes. In Mackenzie's analysis of the engineers promoting accuracy over reliability, the implication is that the series of decisions Mackenzie traced among the "guidance mafia" associated with Charles Draper's MIT Instrumentation Laboratory were crucial: had the guidance mafia not promoted accuracy or received less support among the network of decision makers at crucial moments, countercity strategy and reliable missiles with modest accuracies could have prevailed, which may have lessened political tensions during the Cold War.

Notice, however, that Mackenzie's focus on interviews and documents related to these decisions naturally leads him to see this level as more significant; while political historians tend to favor broad delineation of policy by new administrations. Cultural studies analysis, like that in Paul Edwards's *The Closed World*, emphasizes another level, the reinforcing discourses that shape otherwise unconnected decisions in different contexts in similar ways. On this analysis, the kinds of counterfactual scenarios Mackenzie envisions seem less plausible, as individual decisions would fail to displace distributed "closed world" discourse that favored aggressive posturing.[67] In effect, the contingent decisions Mackenzie looked at might have gone differently, but would the overall trajectory have been disrupted or would reinforcing support for a first-strike capability have taken place?

This should make clear that STS scholars are not agreed upon what processes have in fact shaped past technical decisions and consequently can be expected to differ on what sites of intervention would be most likely to alter current trajectories. Moreover, it looks like the selection of a research focus brings with it tacit assumptions about causes. Mackenzie reads the paper trail and talks to the engineers and concludes that their ideas about what technologies they wanted to build shaped Cold War politics, while Edwards traces out-of-control metaphors spanning computer design, information theory, social science, and popular movies, suggesting a more cultural and therapeutic

66. Latour, *Aramis*; Mackenzie, *Inventing Accuracy*.
67. Edwards, *The Closed World*.

mode of intervention: work through destructive narratives and start telling new stories.[68]

The skeptic would conclude that STS can hardly advise political movements when it cannot establish where the lever is to pull to change the world. The more constructive—and constructivist—reading would be to point out that all of these works actively *construct* new linkages and possible sites of interventions by the way that analyze complex social formations. We can take these alternative scenarios as potentially generative of new linkages with social movements, engineers, and policy makers. Connecting an engineering disaster to normalization of deviance involves identifying additional constraining factors besides the missed whistleblower moment that has drawn our attention. As a result, new possibilities for corrective action are raised.

Recall that Thiokol engineer Brian Russell felt that on the eve of the *Challenger* explosion, he and the other engineers were being asked to prove that the flight was unsafe rather than prove that it was safe, a switch that Vaughan attributes to the longer history of constructions of risk that the engineers had carried out. The outcome of the teleconference was not radically contingent but relatively constrained by the accumulated, past contingent decisions defining safety margins. The emphasis on the openness of stage-1 history within STS treats counterfactual scenarios as cases of "underdetermination" (at branching points, history could have been otherwise) rather than "overdetermination" (history as it occurred was inevitable, as other causes compensate for any postulated change).[69] In emphasizing interpretive flexibility in science and technology, STS scholars explore possible alternative trajectories rather than a predetermined trajectory. Such a focus highlights the significance that contingent events have on historical development.

Yet in most cases, overdetermination is likely to be the more plausible scenario as we consider the multitude of historical factors shaping any given outcome. Cases of underdetermination—labeled "contingency" in STS—imply that a small historical event has huge consequences for historical outcomes, that a decision at a moment in time that could have gone a different way led to a significantly different trajectory. More typically, however, any "closure" of controversy one way rather than another will need to be reinforced by additional historical factors to maintain the selected trajectory. In effect, closer examination of the multiplicity of historical factors contributing to a recognizable trend in history reveal that the thought experiment of removing one factor will make little difference in the overall historical trajectory in most cases.[70]

68. See also Haraway, *Primate Visions*.
69. Elster, *Logic and Society*.
70. For arguments identifying branching points in political history, see Lebow, *Forbidden Fruit*; Harvey, *Explaining the Iraq War*.

Most historical developments in science and technology are more plausibly considered overdetermined, meaning that select individual decisions or events are not likely to be decisive in determining future outcomes (stage-2 causes of closure trump stage-1 openness). In order to change the technological trajectory in significant ways, we need more than quick-thinking, ethically and epistemologically aware engineers. Cases of underdetermination may exist where one individual event or decision has a lasting consequence because circumstances were arranged such that the event or decision tips the balance of historical forces in one way rather than another that leads to a radically different trajectory as a result. Battlefield decisions can sway a battle (and possibly a war) based on contingent events on the ground that could easily have turned out differently. Disruptive events, like political assassinations or crises, may bring about world wars or political changes that may not have been inevitable.[71] While serendipity in discovery can play a similar role in science and technology, the social organization of inquiry channels inquiry in certain directions that may prevent similar disruptions in the historical trajectory of science by singular events or individuals.[72]

To see the application of this distinction between overdetermination and underdetermination, consider how Vaughan's view differs from the view that managers could have prevented the disaster if they had only listened to Boisjoly. Defenders of the idea that engineers should be primed to engage in whistleblowing tacitly accept that there are often events like the teleconference the night before launch where a little more persuasion or a little more ethical high-mindedness could prevent disaster. By contrast, Vaughan suggests that a closer examination of the processes leading up to the disaster show that the engineers' hands were tied by past decisions over a long period of time that incrementally closed off an alternative analysis of the O-ring problem. The disaster was probably overdetermined, though it may be possible to envision addressing elements of these larger causes shaping the disastrous outcome in a way that points to different sites of possible intervention.

The result is that closer examination of historical causation is likely to lead to a rejection of any simple, underdeterminist event as the sufficient cause of

---

71. Lebow, *Forbidden Fruit*.

72. Isaac Asimov's Foundation trilogy introduces the role of a "Mule," an unexpected individual or event that disrupts an overdetermined historical trajectory. See Phillips and Zyglidopoulos, "Learning from Foundation," p. 601. Polanyi, "The Republic of Science," argued that because science and technology introduce novel discoveries, their negative effects cannot be anticipated in principle, using the "unexpected" result of atomic weapons issuing from Einstein's physics as an example. Guston, "The Pumpkin or the Tiger?," shows that Polanyi had in fact been exposed to this possibility on numerous occasions, including directly from Leo Szilárd, whose efforts at controlling nuclear energy provide a clear example of the possibility of a precautionary approach and anticipatory governance of technology. A special section of *New Scientist* (2005), 187 (2513): 34–46, was dedicated to counterfactuals in the history of science.

a historical trajectory. Such research may also suggest different sites where multiple decisions were made that reinforced the trajectory that occurred. While it may be harder to envision altering these reinforcing causes, it is possible that a larger systematic change to established practices over a longer period of time may have a greater chance of being effective than pinning one's hope on heroic intervention at the last moment. What this suggests is that the general thrust of close historical or ethnographic examination of a process leading to an undesirable outcome may simultaneously lead to a rejection of simplistic counterfactual scenarios that were presumed to have been able to prevent the disaster and the suggestion of more complex changes that would have been more likely to prevent the disaster in the first place.

As we will see, the development of green chemistry and green engineering as self-conscious new fields is unlikely to be sufficient for radically altering our relationship to chemicals and pollution, but it can operate in concert with other countervailing approaches to current trends, such as environmental activism, policy initiatives, and cultural change. This suggests a more complicated and difficult "alternative trajectory" than one where a simple change in knowledge results in a fix for long-standing practices leading to environmental damage. This option of a "complex counterfactual" tends to be ignored because the way we disagree about branching paths in history tends to highlight select actors, who then are taken to have either brought about, or failed to bring about, radical changes in direction (see table 6.1).

Table 6.1. Characterizations of Green Chemistry's Impact

	internalist *knowledge development* *stage 1 and 2 STS*	*contextualist* *sociopolitical, institutional* *stage 3 STS*
rhetorical (outcome overdetermined)	Networking among like-minded chemists; no new approach to chemical knowledge. Chemistry overdetermined by internal dynamics of knowledge. Linthorst (2010)	Public relations for business as usual among chemical companies. Pollution overdetermined by corporate context of chemical production. Roberts (2005)
substantive (outcome underdetermined)	Revolutionary change in how chemistry is done. Alternative trajectory of knowledge launched by new ideas about sustainability and chemical synthesis. Anastas and Warner (1998)	Representation of a social movement against pollution within chemistry. Alternative approach to chemical synthesis brought about by public pressure. Woodhouse and Breyman (2005)

While Mackenzie's countercity approach—let alone disarmament—may have been doomed regardless of which engineers with alternative design ideas were there on the ground, it is possible that persistent tendencies operating across different local contexts might have turned out different in another way. With alternative social and political institutions, patterns of interactions, and dominant discourses, one could expect not just variant branches but a different tree. Such a complex counterfactual is, however, less "insertable" into history than the small, contingent decisions that are the focus of a lot of our case studies. The postulated changes are further from history as it actually happened and more difficult to envision having occurred than highly contingent battlefield decisions. Consequently, when considering the likelihood of alternative paths, we need to consider both the consequences of the postulated counterfactual and how plausible such a counterfactual event was in that context.[73] Less insertable events can still tell us about the significance that larger cultural or institutional factors may have had in shaping history as it occurred. The overdetermined set of causes making a trajectory difficult to change can also suggest the multiple sites of intervention that conscious change would require. Interest in counterfactual thought experiments illuminates both kinds of possibilities, small changes that might make a big difference and larger, persisting historical causes that are harder to change but that shape multiple outcomes.

SUSTAINABILITY, GREEN CHEMISTRY, AND GREEN ENGINEERING

The key starting point for evaluating the prospects for green chemistry and green engineering is to notice how authors make assumptions about the ease with which changes can be made to how we synthesize chemicals and develop technology. Disagreement about whether initiatives like green chemistry and engineering represent a real departure from past practices depend upon both assumptions about the plausibility of cheery counterfactual alternatives and beliefs about the relative autonomy or pliability of science and technology within the larger polity. Table 6.1 maps out these possibilities for green chemistry based upon four representative approaches in the literature. While Anastas and Warner identify their new approach to chemistry as fundamental to the solution of environmental problems, Linthorst, in an interdisciplinary journal addressed to "conceptual and fundamental issues in chemistry," denies that a significant new approach to chemistry has been developed. Anastas and Warner believe that articulating explicit rules of green

73. Elster, *Logic and Society*.

chemistry (figure 6.1) can lead chemists to synthesize chemicals with fewer toxic by-products or solvents, use reagents that are incorporated into the end product or reused, and rely upon renewable feedstocks and energy efficient methods. "Green Chemistry," for Linthorst, is a label that makes coordination among like-minded chemists possible, without being associated with the alteration of the autonomous development of the field of chemistry. In his view, Anastas and Warner have essentially repackaged autonomous developments in knowledge, like the concept of atom economy developed by Trost or an emphasis on catalysis, where reagents are sought that are not used up by the chemical reaction.[74]

A similar difference of opinion divides evaluations of green chemistry among STS scholars considering the larger social context shaping chemistry. Woodhouse and Breyman argue that green chemistry is one effect of the social movement of environmentalism and can lead to significant change in established practices. Although Roberts expresses some sympathy with this

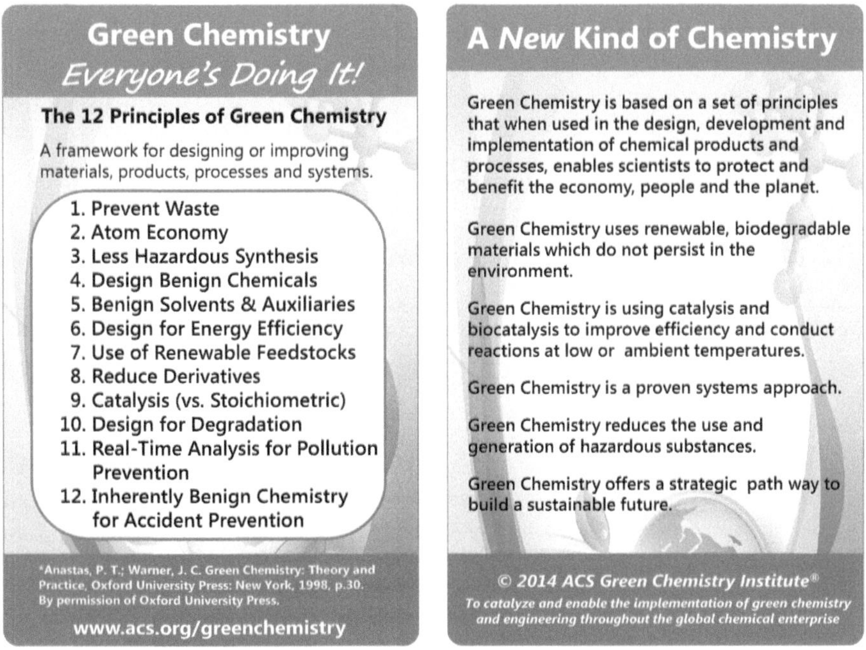

Figure 6.1. The 12 Principles of Green Chemistry. *Source*: 12 Principles of Green Chemistry. Reprinted with permission from ACS Green Chemistry Institute®. Copyright © 2019 American Chemical Society. https://www.acs.org/content/acs/en/greenchemistry/principles/12-principles-of-green-chemistry.html (accessed November 27, 2019).

74. Linthorst, "An Overview"; Trost, "The Atom Economy—a Search for Synthetic Efficiency."

idea (in particular, seeing green chemistry as a *scientific* movement emerging from a regulatory initiative), he also expresses skepticism that green chemistry will serve as more than a convenient tool of public relations. He notes that green chemists themselves discount their green solutions if the economic benefit is not favorable and favor technical fixes over multifaceted understanding of environmental problems. Whereas Woodhouse and Breyman hold out hopes that social movements can shift the direction of chemistry and engineering toward more sustainable solutions, Roberts expresses skepticism that green chemists can overcome the influence of the corporate context of their work.[75]

STS scholars typically argue that developments in science and technology are complex and that both developments internal to problem-solving traditions of knowledge and larger sociopolitical and institutional contexts are involved in any historical process. So to some extent, STS scholars will want to incorporate internal and contextual causes and think outside the box of table 6.1 (similar arguments can be made for the causes of the development of "brown chemistry," modern synthetic chemistry, which neglected green chemistry concerns—see table 6.2).[76] While occasionally proposing provocative branching points in history because of a commitment to historical contingency (e.g., the electric car could have been adopted rather than the

Table 6.2. Characterizations of Brown Chemistry's Development

	internalist *knowledge development* *stage 1 and 2 STS*	*contextualist* *sociopolitical, institutional* *Stage 3 STS*
overdetermined	Chemistry developed as appropriate to the level of knowledge that existed at the time. Lack of knowledge made pollution inevitable.	Anticipated harms were ignored to support profit-oriented development. Lack of power or opportunity made inevitable the neglect of knowledgeable critics.
underdetermined	Modern chemical methods relied on harmful chemicals. Pollution the result of contingent developments in knowledge that could have been otherwise.	Industry pushed brown chemistry. Toxicology limited by regulatory demand to specify threshold harms. Political economy led scientists away from considering green chemistry.

75. Woodhouse and Breyman, "Green Chemistry as Social Movement?"; Roberts, *Creating Green Chemistry*, ch. 4. For a similar critique of green engineering and green business initiatives, see Jamison, *The Making of Green Engineers*.

76. Woodhouse, "Change of State?"

internal combustion engine[77]), the sheer weight of all the contingent factors shaping change over time also drives many scholars to consider history as overdetermined by a multitude of causes, so skepticism about the value of counterfactual reasoning is strong. Accompanying this skepticism, is a skepticism about whether prevailing trends can be consciously challenged, with an accompanying retreat from the prescriptive approach taken in science and engineering ethics.

While complex counterfactuals don't fit the category of "almost happened," considered instead as a diagnosis of what went wrong, they can shape future ideas for technological design. Such complex counterfactuals that accept that single branching points of significant change are unlikely can nonetheless focus on mutually reinforcing tendencies across contexts that push in a particular direction. Such overdetermined causes may highlight common themes that need to be challenged in a variety of contexts for change to occur. The function of epistemological, moral, and political critiques of science and engineering practices may be to provide intellectual ammunition to those acting in a variety of settings who are already looking to shift priorities and trends. The concept of sustainability and the rules laid down for green chemistry and green engineering are examples of concepts that can play this role. To be sure, they are subject to interpretive flexibility in different contexts, so they cannot guarantee specific outcomes across all contexts as they are adopted. They can, nonetheless, link together attempts at reform into a wider intellectual or social movement.

Consider the current interest in sustainability within engineering design, which itself shows the impact of ecological science and politics on engineering. Harris, Pritchard and Rabins trace the history of efforts to modify the professional code of the American Society of Civil Engineers (ASCE), beginning in 1977, to take into account the impact of engineering work on the environment.[78] Engineering attention to sustainable design goes much further in seeking to change the rules of the game regarding how design gets done. Graedel traces a concern with sustainability to the 1987 report, *Our Common Future*, put out by the World Commission on Environment and Development and the 1999 report, *Our Common Journey*, of the Board on Sustainable Development of the U.S. National Research Council.[79] Not only can we trace the impact of environmental scientists here, but this development could be seen as a (delayed) impact of the social movement of environmentalism since the 1970s. The impact of such broader social and political influences is mediated through the decisions about how sustainability will be implemented in pro-

77. Kirsch, *The Electric Vehicle and the Burden of History.*
78. Harris, Pritchard, and Rabins, *Engineering Ethics.*
79. Graedel, "Green Chemistry and Sustainable Development."

fessional design contexts; these decisions involve creative extensions of the prior disciplinary framework and conceptions about what it means to be an engineer. Constructivist, open-ended practice that nonetheless incorporates wider societal impact describes this kind of development.

So, for instance, in the practice of "green chemistry" and "green engineering," sustainability requires rates of natural resource use that don't deplete supplies and levels of chemical products and waste that can be sustained by the environment.[80] These ideas directly reflect changing regulatory ideas within the U.S. Environmental Protection Agency (EPA). The 1990 U.S. Pollution Prevention Act sought to prevent pollution at the source rather than regulate end-of-pipe emissions. Anastas coined the term "green chemistry" while employed by the EPA, seeking to minimize pollution throughout the life cycle of the chemicals used. His 1998 collaboration with John Warner, *Green Chemistry: Theory and Practice*, laid down an approach to green chemistry based upon twelve principles (figure 1), an approach Anastas extended to green engineering in 2003.[81] Anastas founded the Green Chemistry Institute at the American Chemical Society (ACS) in 1997, which has developed grants, awards, and educational initiatives, along with an annual conference. In 1995, the Clinton administration set up the annual Presidential Green Chemistry Challenge Awards, and other governments and some U.S. states followed suit. The Royal Society of Chemistry established a journal, *Green Chemistry*, in 1999. The chemical industry has responded by implementing green chemistry approaches, sponsoring conferences in the field, and receiving presidential awards.[82]

Anastas and Warner argued that a driving force for the incorporation of benign or green chemistry techniques within industrial practice was the growing cost of regulation and pollution control. Research delineating such costs, drawing on approaches such as life-cycle analysis, become crucial epistemic inputs for fields like green chemistry.[83] Much of this analysis of sustainability depends upon a critique of *unsustainable past practices*, a topic where STS research can add depth to the potted history of environmental harms in Anastas and Warner's book.

---

80. Ibid.

81. Anastas and Warner, *Green Chemistry*; Anastas and Zimmerman, "Design through the 12 Principles of Green Engineering."

82. https://www.acs.org/content/acs/en/greenchemistry.html. Numerous links can also be found at the Michigan Green Chemistry Clearinghouse at https://www.migreenchemistry.org, including reference to state initiatives in California, Michigan, Minnesota, New York, Oregon, and Washington. Michigan offers a Green Chemistry Governor's Award, announced at an annual conference. See Kumar, "Historical Development of Green Chemistry."

83. Anastas and Warner, *Green Chemistry*, pp. 10, 15; Martin, "Green Catalysts for Industry"; Azapagic, "Life-Cycle Assessment."

On the other hand, many critics find the claims on behalf of green chemistry overblown, suggesting that industry is more interested in greenwashing its image rather than fundamentally changing polluting practices, and that the existence of green chemistry helps them to do so. Moreover, proponents of green chemistry typically treat the solution to environmental pollution as a narrowly technical problem and ignore larger sociopolitical and institutional factors contributing to the problem, including their own employment by companies that put profit first and pursue deregulation politically rather than implementing green chemistry across the board.[84] On this view, green chemistry can seem ineffective or even positively misleading.

Proponents of green chemistry typically emphasize that new developments in knowledge make possible new solutions. Quoting from Anastas and Warner's textbook, green chemistry is premised on the idea that it is now possible to "not only design a substance to have certain characteristics during its useful life, but . . . also design what that substance will become (or break down into) after its useful life is over." "[G]reen chemistry is an approach that provides a fundamental methodology for changing the intrinsic nature of a chemical product or process so that it is inherently of less risk to human health and the environment."[85]

While this expansion of criteria for chemical synthesis is made possible by greater toxicological knowledge of harms and the chemical mechanisms that produce it, their history of the field also emphasizes the impact that significant disasters and harms had on motivating the new focus. Anastas and Warner trace the rise of new waves of environmental regulation based upon the impact of Rachel Carson's exposé on the harmful effects of pesticides, the birth defects of thalidomide babies, dioxin at Times Beach, Love Canal pollution, the Cuyahoga River in Ohio catching fire, the identification of CFCs as depleting atmospheric ozone, and the Bhopal explosion. They argue that the dialectic of disaster, public outrage, and new laws and regulations will tend to force corporations to incorporate green chemistry techniques that substitute safer design over the life cycle of the chemicals involved to avoid the increased cost of end-of-pipe pollution control. They also argue that current practices are already limiting innovation by diverting funds from research to pollution control.[86]

While they laud improvements in knowledge that have enabled chemists to better prevent pollution, they also identify misguided assumptions where criticism could have prevented harms. These mistakes include the ideas that "dilution is the solution to pollution," that harms can be managed by identifying

84. Roberts, *Creating Green Chemistry*, ch. 4.
85. Anastas and Warner, *Green Chemistry*, pp. v, 8.
86. Ibid., pp. 2–6, 31.

maximum concentration guideline levels, that synergistic effects were not a concern, and that full-cost accounting was not necessary. Anastas and Warner treat these assumptions as something that could have been challenged at the time they were introduced, as "insertable" in history. The implication is that if challenges to these ideas had been more widely aired at the time, a greener approach to chemistry could have emerged that preempted at least some of the harms that emerged from nineteenth- and twentieth-century "brown chemistry."[87]

Just as with the *Challenger* case, it would be possible to blame the external pressures of management and disavow the responsibility of scientists and engineers, an approach Anastas and Warner reject. Their list of faulty assumptions highlight the "normalization of deviance" produced by the chemists themselves. None of these problems were impossible to anticipate in theory; and indeed, most have been questioned by minority views in the field and by relevant outside experts. We know that initial, back-of-the envelope calculations can become taken for granted very quickly as the normal basis for technical work, even though it would be easy for researchers to question them if sufficiently motivated. Moreover, other disciplines may possess relevant knowledge that is never accessed when these assumptions were black-boxed. Anastas and Warner note that chemists were just not trained in toxicology, so an "insertable" counterfactual would have been to implement such training or to require other experts to sign off on the standards.[88]

They also point out that most of the toxicological knowledge that exists now is the result of legislation requiring that acceptable levels of emissions be set. Anastas and Warner believe that identification of "acceptable levels" of toxic substances and the associated paradigm of minimizing exposure to hazards is flawed and should be replaced by the design of chemicals that are designed from the start to be "environmentally benign."[89] They reject the Parascelsian view that "the dose makes the poison." A focus on thresholds is misleading "when one is dealing with substances of high toxicity, unknown toxicity, carcinogenicity, or chronic toxicity." Setting thresholds on this basis cannot protect health and the environment.[90]

The legal context reinforcing past identification of acceptable levels is significant in two ways: it shows the impact of other social processes besides the corporate environment on the knowledge produced, and it demonstrates one way in which initial assumptions of clear-cut safe levels made for temporary reasons of workability become solidified by the impact of legislative language. Such "temporary" decisions become solidified and difficult to change

87. Woodhouse, "Change of State?"
88. Anastas and Warner, *Green Chemistry*, p. 9.
89. Ibid., pp. 18, 10. See also Voutchkova et al., "Towards Rational Molecular Design."
90. Anastas and Warner, *Green Chemistry*, pp. 17–18.

or reconsider by the ongoing work of scientists and engineers identifying safety standards and the legal requirement to do so.

STS research tells us more about how this interaction of scientific debate and legislative action took place. Toxicological emphasis on specifying safe levels of chemicals, while tracing to Paracelsus's view that the dose makes the poison or more proximately to animal experiments in the early twentieth century, was codified in the 1958 Federal Food, Drug, and Cosmetics Act and reflects both industry interest in defending safe levels of chemicals deemed economically essential as well as disciplinary closure of a scientific controversy over whether safe levels of radiation existed.[91] The controversy concerned low-dose radiation in nuclear fallout during the Cold War, as geneticists tended to argue that damage from radiation was cumulative and that there was no safe dose. Most in the medical sciences rejected this idea and argued that public dissemination of this idea was irresponsible as it inflamed public worries without cause.[92] While the concept of a threshold below which exposure was safe came to dominate toxicological work, geneticists studying radiation continued to presume there was no safe level, extrapolating risk via a linear no-threshold (LNT) model.[93] As a result, consensus within toxicology to identify safe levels of toxic substances can be seen as a premature consensus dependent upon a narrow, disciplinary definition of the relevant core set of experts, reinforced by a regulatory requirement to do so.[94]

Critics could still be found in genetics and elsewhere who questioned this assumption. Indeed, in Rachel Carson's *Silent Spring*, she drew an analogy with the dangers of radioactive fallout to help the reader understand that human-made pesticides could be dangerous as well.[95] Later, Theo Colburn and other researchers reconsidered the rationale behind the identification of safe levels for endocrine disrupting chemicals, which they argued had significant developmental effects at low doses.[96]

In terms of Collins's three-stage model, the stages are connected. The stage-1 interpretive flexibility in methods and standards for assessing risk are solidified into a stage-2 consensus as the result of stage-3 political and regulatory decisions. The need for scientists to settle on a workable set of standards for assessing toxicity ensures that an early consensus becomes more difficult to change as scientists are asked to contribute their expertise to carry out the regulatory requirements. The initial open and unsettled debate between the

91. Vogel, "From 'the Dose Makes the Poison' to 'the Timing Makes the Poison.'"
92. Jolly, *Thresholds of Uncertainty*.
93. Kathren, "Pathway to a Paradigm."
94. See Collins, "Public Experiments and Displays of Virtuosity," on premature consensus achieved as the result of a restriction of the core set.
95. Carson, *Silent Spring*.
96. Colborn, Dumanoski, and Myers, *Our Stolen Future*; Vogel, "From 'the Dose Makes the Poison' to 'the Timing Makes the Poison.'"

view that threshold levels of harm be identified and the view that threshold harms cannot be specified becomes settled as the result of the need for regulation and the early decision to regulate by identifying safe and unsafe levels. This initial decision was facilitated by the greater tractability of regulating by clear thresholds and the resulting organizational requirements react back on the work of the scientists themselves by providing an organizational incentive to identify threshold harms even if one "knows better." Before long, stage-3 requirements have forced premature closure of stage-1 scientific debate. Stage-2 closure has emerged not from the autonomy of expert debate but as the result of stage-3 demands for scientific input on regulatory decisions. The result is the "path dependence" of developments in "core-set" science set off by political decisions.

These wider political considerations in turn are constrained artificially by the limited political input into the regulatory process by the wider public. Given the opportunity to question the framework of regulation, later environmental critics sought to reopen this scientific debate, though working against the inertia to change built up by routine technical and regulatory practices that had required the identification of safe and unsafe levels. The "normalization of deviance" in this case made it just as difficult as it did for Boisjoly to challenge harmful decisions "at the last moment." In cases extending from the dangers of low-level nuclear waste, release of industrial pollutants into the environment, the side effects of pharmaceuticals, and the impact of novel chemicals with endocrine-disrupting effects in consumer products, environmental critics were faced with claims that safe levels of these substances had been identified that blocked further examination.[97]

This brief discussion of STS research relating to Anastas and Warner's claim that setting acceptable levels of toxic substances worked against chemistry that was "benign by design" illuminates the complex causes preventing an earlier development of green chemistry techniques (table 6.2). This suggests that examination of the relationship between regulatory regimes and ongoing scientific assessments of risk is needed. STS can also examine contemporary examples of pragmatic closure of unresolved safety questions, in effect diagnosing current processes of the normalization of deviance. The identification of significant decisions that become precedents and shape incremental change on that basis can be used to raise debate about other areas where the precautionary principle may be warranted. An example would involve examination of current safety standards and regulatory frameworks

97. Castleman and Ziem, "Toxic Pollutants, Science, and Corporate Influence"; Semendeferi, "Legitimating a Nuclear Critic"; Shrader-Frechette, "Technological Risk and Small Probabilities"; Daemmrich and Greene, "From Visible Harm to Relative Risk"; Colborn, Dumanoski, and Myers, *Our Stolen Future*; Vogel, "From 'the Dose Makes the Poison' to 'the Timing Makes the Poison.'"

for nanotechnology, looking precisely for the kind of reasonable, but misleading, assumptions that Anastas and Warner identify in chemistry's past. STS examination of potentially premature black-boxing in current scientific and regulatory regimes could help scientists and engineers in shaping their future practice. In short, STS work could become part of the conversation with scientists and engineers about the future of their field.

LIMITATIONS OF GREEN CHEMISTRY AS A SOCIAL MOVEMENT

The very effort of Anastas, Warner, and others to define a new field, to establish journals and research funding, and to promote knowledge of green chemistry in university and corporate environments demonstrate how analysis of past normalization of deviance can be used to challenge accepted practices and *to shape future standards of the field*, even though many of the constraints inhibiting conversion to green chemistry in corporate environments continue to exist and limit adoption of greener approaches.[98] While waiting for environmental costs to force the hand of chemical companies (while deregulation tends to push in the other direction), green chemists have already put in place new standards for judging success in chemical design. This takes place by filling in the larger context obscured by a chemical equation that ignores the role of auxiliary reagents and solvents and energy inputs. Instead of judging efficiency by how many moles of product are produced per mole of raw material, ignoring waste materials, green chemists focus on "atom efficiency," which looks at all the reactants used in a process and assesses the extent to which they are incorporated in the final product. Since some of the most toxic components in chemistry are solvents like volatile organic compounds (VOCs) or heavy metals needed to oxidize the petroleum products that make up 98 percent of feedstocks, developing solvent-less chemistry, incorporating supercritical fluids as solvents, using non-petroleum feedstocks, or incorporating visible light for oxidation are identified as ways to produce usable products without leftover toxic substances. Moreover, energy costs and energy sustainability are considered, leading to planning for the use of non-petroleum feedstocks, such as sustainable agricultural or biological feedstocks that also avoid the need for polluting oxidation processes in the first place.

Is green chemistry just a convenient label for packaging and funding work already developed as part of the normal trajectory of science, or does it

98. On green chemistry's limitations for fundamentally altering how chemistry is done, see Roberts, *Creating Green Chemistry*, ch. 4; and Woodhouse and Breyman, "Green Chemistry as Social Movement?"

represent a new departure? Linthorst has argued that "green chemistry" has become a significant label for chemical work since it consolidated already existing approaches like catalysis and atom economy into a label that served to link different researchers together into a recognizable community. In particular, the shift in regulation from end-of-pipe pollution cleanup to pollution prevention developed within the EPA in the 1980s, given legal support and funding by the Pollution Prevention Act of 1990, solidified green chemistry as a distinct philosophy. While Anastas had argued that the field "had been driven by new knowledge," Linthorst suggests instead that the term did not involve the development of a new field but instead facilitated cooperation between industry and government-funded researchers looking to apply existing approaches to the problem of pollution prevention.[99]

Anastas and Warner developed a set of criteria to influence methods of chemical synthesis. Key components of this philosophy built in particular on prior work on catalysis (which increases the efficiency of reactions with catalysts that are not depleted in the process) and on the concept of atom economy (arriving at chemical reactions that don't produce by-products or waste) developed by Barry Trost in an article in *Science* in 1991.[100] Linthorst's implication is that the significance of green chemistry is primarily organizational rather than epistemic, as journals and institutes were established in the emerging field. Many early contributors to the journal *Green Chemistry*, founded by the Green Chemistry Network within the Royal Society of Chemistry used the term "green chemistry" but did not discuss the approach developed within the EPA.[101] Two separate networks became linked by a common term for the field but did not share novel or deep commitments that would be indicative of a revolutionary shift in the field.

Viewed from a different perspective, however, significant changes often involve this kind of subtle coordination of distinctly different kinds of work associated by a marriage of convenience.[102] According to Roberts,

> The general idea of green chemistry discussed within these contexts can be summarized broadly as: an attempt to reconstruct the chemical sciences and their practices (industrial and academic) so that they internalize the need to ameliorate growing environmental and human health problems, taken to be the

99. Linthorst, "An Overview."
100. Trost, "The Atom Economy—a Search for Synthetic Efficiency."
101. Linthorst, "An Overview," pp. 63, 66, argues that green chemistry "is not a scientific specialty with a coherent conducted paradigm." In contrast to Linthorst, Roberts, *Creating Green Chemistry*, p. 79, argues that *Green Chemistry*'s first editor, James Clark, "drew heavily from the mission established in [Anastas and Warner's *Green Chemistry: Theory and Practice*]" and credited U.S. colleagues for why the journal was titled *Green Chemistry* despite intense debate on the question.
102. Fuller and Collier, *Philosophy, Rhetoric, and the End of Knowledge*.

root of public concerns and scrutiny, and to design both chemical processes and products so that they serve this purpose.[103]

STS scholars examine the extent to which rhetorical and institutional coordination of different research subcommunities facilitates transformed research practices, rather than simply distinguishing institutional and epistemic components of science as Linthorst does. Reform movements within science—in this case, reflecting the mediated impact of larger social movements concerned with the environmental consequences of modern technology—may alter research practices in a variety of ways that go deeper than surface organizational changes. While green chemistry can be criticized for its narrow focus on technical solutions for complex environmental problems, as Roberts notes,[104] the organizational work of Anastas, Warner, and others to refocus chemical synthesis on sustainable alternatives represents a real effort to alter existing practices in chemistry and chemical engineering.

My point is that green chemistry needs to be studied as an effort to reform science and engineering practice in line with newly emerging values. As such, it is an example of the role of social movements in reforming science and engineering practice, albeit in a significantly mediated fashion.[105] The impact of the larger environmental movement on green chemistry is clearly mediated by the professional context of green chemistry and the corporate context of chemical work. As such, it is not surprising that green chemistry has a tendency to address complex environmental problems with narrowly technical solutions that fail to challenge the economic and political structures that may be responsible for the existence of the problems in the first place. In short, green chemistry can indeed function as a form of "greenwashing" that corporations can rely on to avoid regulatory costs and bad publicity, and companies that are often criticized for pollution like Dow and DuPont have been heavily involved in Green Chemistry work.[106] However, the efforts to arrive at different approaches to chemical synthesis nevertheless offer possibilities for pollution prevention and control that otherwise would not be possible so that the reductionist tendency of green chemistry is still relevant to addressing the source of the problem in a way that more holistic environmental critics may be unable to accomplish.

103. Roberts, *Creating Green Chemistry*, p. 4.

104. Ibid., pp. 114–15.

105. Frickel et al., "Undone Science"; Frickel and Moore. *The New Political Sociology of Science*; Fuller and Collier, *Philosophy, Rhetoric, and the End of Knowledge*; Fuller, *Thomas Kuhn*, pp. 400–408; McCormick, "Democratizing Science Movements"; McCormick, "From 'Politico-Scientists' to Democratizing Science Movements"; McCormick, *Mobilizing Science*.

106. Roberts, *Creating Green Chemistry*, ch. 4; Ross and Amter, *The Polluters*. Howard-Grenville at al., "'If Chemists Don't Do It, Who Is Going To?,'" identify three different frameworks chemists appeal to in order to justify green chemistry: a normalizing framework emphasizing its consistency with mainstream chemistry, a moralizing framework emphasizing it as an ethical obligation, and a pragmatic framework emphasizing its usefulness as a tool for practical work.

One way to look at this is that scientists and engineers are our *representatives* for otherwise esoteric activities, representatives who are potentially in a position to address public concern with environmental sustainability but who also may fail us just as our political representatives do. An attempt to lay out new guidelines for chemical synthesis represents an attempt to translate broader political concern with environmental damage caused by chemical industry into new methods and practices of chemistry. In this sense, the work of scientists and engineers in translating public concern with the environment into efforts to direct science and engineering in a more environmentally sustainable fashion is simultaneously a political and epistemic move. Woodhouse and Breyman, quoting Anastas, draw the connection explicitly:

> "The moment a chemist puts pencil to paper to design the way a chemical product will be made, he is also intrinsically deciding whether toxic chemicals will be used or generated, [and] what the waste disposal cost will be," Paul Anastas told the first green chemistry symposium, held at the 1993 meeting of the American Chemical Society. One might rephrase that to say that chemists in certain respects are humanity's representatives in making decisions fraught with public consequences for human health and environment. If some of those with technical training in chemistry are trying to convince others to make changes in humanity's "policy" for chemicals, it is difficult to miss the parallel with other aspects of the environmental movement in which some persons and organizations try to convince others to change environmental policies and outcomes.[107]

In Collins's language, third-stage efforts to demand solutions to chemical pollution led to epistemic and organizational changes implemented by chemists as green chemistry.

ANTICIPATING NEW KINDS OF HARM

Green chemistry presents a constructive model for how ethical implications of engineering practice can be incorporated into an epistemological project of revisioning methods and standards in the field to remedy the normalization of deviance. STS can aid in endeavors like this through its close focus on the microconstruction of knowledge about risk and its contextualization within larger social settings. Engineering education should promote examination of undesirable engineering outcomes to understand the social processes that reinforced them and to consider counterfactual possibilities.

The goal is not merely to second-guess engineers but to help engineers "second-guess," or reconsider, the foundations of their fields with an eye to

107. Woodhouse and Breyman, "Green Chemistry as Social Movement?," p. 217.

a better future. The key contribution we can make is to focus on organizational and regulatory constraints on the process of core-set negotiations leading to a narrow or premature consensus. By examining cases like this, STS scholars may facilitate a hearing for minority views that could otherwise be marginalized.

Consider, for example, that while green chemists are second-guessing past approaches in synthetic chemistry, questions have been raised about the potential toxicity of nanotechnology's manipulations at the level of single molecules. The irony is that nanotechnology is considered one option for carrying out green chemistry, even while it may present unknown dangers of its own. It is known that nanoscale construction of molecules may be more reactive, potentially causing free radical damage to the body and having a greater bioavailability than aggregate chemicals of traditional synthetic chemistry. In 2004, the Royal Society concluded that nanoscale constructions require separate regulation and testing for toxicity than bulk forms of the same molecule; that is, "regulation will need to be adapted to reflect the fact that the safety of substances in the form of nanoparticles cannot be inferred from knowledge of their hazard in larger form."[108]

THE PRECAUTIONARY PRINCIPLE AND ANTICIPATORY GOVERNANCE OF SCIENCE

David Guston has called for a proactive approach to potential threats raised by nanotechnology, synthetic biology, and geoengineering under an approach known as "anticipatory governance." Anticipatory governance seeks to anticipate the outcomes of new technologies by developing a number of possible scenarios for discussion rather than requiring decisive predictions of harm to act. As such, it represents a form of the precautionary principle, which also brings discussion of the issue to public attention and integrates relevant natural and social scientific approaches.[109] Like Neurath's approach to unity of science, concrete problems require that distinct scientific practices be brought together for a full scientific, ethical, and political solution.

Polanyi's elitist authoritarianism, the target of epistemological criticism by both Lakatos and Feyerabend, is also the obstacle to distinctly ethical and political reflection on the processes and products of science and technology.

108. The Royal Society and the Royal Academy of Engineering, *Nanoscience and Nanotechnologies*, p. 82.

109. Guston, "The Pumpkin or the Tiger?," pp. 374–75. See also the discussion of the multidisciplinary nature of most environmental problems and the conflict between market approaches and democratic deliberation in Anstey, "A Feminist Perspective on the Precautionary Principle and the Problem of Endocrine Disruption under Neoliberal Globalization Policies," pp. 62–70.

Polanyi developed his approach to emphasizing the role of tacit knowledge and autonomous scientific communities in the philosophy of science by way of opposition to the emphasis on scientific planning by Bernal and the British Marxist scientists influenced by Boris Hessen and the Soviet delegation to the Second International Congress of the History of Science in 1931. Polanyi had argued that scientific discovery could not be planned, so that scientists were not responsible for anticipating negative consequences of their work, an approach that converged with the opposition to economic planning by Friedrich Hayek and the Austrian economists.[110]

Guston shows that despite Polanyi's exposure to the potential issues raised by atomic energy, he publicly denied that any negative consequences could have been anticipated prior to the development of the atomic bomb. Frederick Soddy and Leo Szilard, by contrast, had done just that by endorsing a vision of the scientist's social responsibility diametrically opposed to Polanyi's emphasis on autonomy and lack of responsibility for social consequences. Soddy wrote popular accounts detailing his work on what he called "atomic energy," potentially millions of times more energetic than chemical reactions. His account of the release of energy by radioactive substances like radium inspired H. G. Wells's account of nuclear war in his 1914 novel *The World Set Free*. He also took the lesson from World War I that governments would use scientific findings for destructive purposes, as they had diverted the Haber process from fertilizers to explosives and sought to address the "frustration of science" by capitalism.[111]

Likewise, Szilard anticipated that an atomic bomb could be made in 1933 and, worried that the Nazis would develop one first, contacted Einstein to write the letter to President Roosevelt that ultimately brought about the Manhattan Project. After the war, he campaigned for international control of atomic weapons. Polanyi had been exposed to Wells's book, Soddy's arguments, and had even discussed Szilard's proposal to research the possibility of a chain reaction directly with him. When asked about the potential practical consequences of Einstein's relativity in a 1945 radio program, he denied there were any. In his 1962 defense of the autonomy of science in "The Republic of Science," he likewise claimed that atomic energy and nuclear weapons could not be anticipated and denied that scientists should attempt to anticipate the long-term consequences of their work.[112]

The precautionary principle can best be defended by focusing on both the potential dangers raised by new technologies and the relative plausibility of

110. Polanyi, "The Republic of Science"; Beddeleem, "Fighting for the Mantle of Science," chs. 1–2; Mirowski and Plehwe, *The Road from Mont Pèlerin*.

111. Guston, "The Pumpkin or the Tiger?," pp. 367–68; Soddy, "Foreword."

112. Guston, "The Pumpkin or the Tiger?," pp. 365, 369–72; Sclove, "From Alchemy to Atomic War," pp. 169–71, 176–79; Polanyi, "The Republic of Science," p. 62.

our knowledge of harm. In most formulations of the precautionary principle, the goal is to act with as much foresight as possible given imperfect knowledge, rather than requiring proof of harm before acting. Understood in this way, not just any conceivable harm can be grounds for halting technological development, but given some evidence of harm, it is better to "be safe than sorry."[113] It was precisely this kind of approach that informed Rachel Carson's *Silent Spring* (1962), as she mobilized the scientific literature in genetics with the principle that one should "use foresight rather than hindsight." Her approach was anticipated by Charlotte Auerbach's *Genetics in the Atomic Age* (1956), which argued that our knowledge of how radiation-induced mutations was sufficient to suggest that we needed to limit exposure to nuclear fallout.[114]

In contrast, most reports of the social and ethical implications of emerging technologies call for after-the-fact regulation but reject consideration of stopping or altering developments.[115] Very much like Polanyi's take on atomic energy, the view is that we can begin to regulate a technology once its dangers are clear, but we cannot anticipate them in advance and should not see this as part of the professional responsibility of scientists and engineers. Others completely reject precaution, arguing that the opportunity costs of foregoing technological development will likely exceed the harm done, so that an alternative "proactionary" principle is warranted. The thought is that taking risks will have better payoffs than caution.[116]

The development of brown chemistry and nuclear technologies suggest that such an approach has high costs, costs that could have been assessed more explicitly and consciously at the time. The fact that the research programs of Wilhelm Hueper and Hermann Muller look prescient in hindsight suggests that a full consideration of the variety of competing research programs is needed to anticipate future harms. Even if their research had not been suppressed, we would not have expected that their work could prove decisive harm or that competing approaches emphasizing thresholds would

113. Lewens, *The Meaning of Science*, pp. 135–36. Critics of the precautionary principle argue that it is always possible to construct scenarios based on any action whatsoever such that the precautionary principle cannot manage uncertainty. See Hanekamp, "Neither Acceptable Nor Certain," pp. 249–51. The argument only works because entirely fanciful scenarios are juxtaposed to those dependent upon productive research programs (e.g., climate change treaties lead to global depression followed by nuclear war). How different negative scenarios relate to each other is something that requires interdisciplinary cooperation along the lines of Neurath's approach discussed in chapter 5. On this view, one must incorporate scenarios of the economic consequences of precautionary action. This requirement is already part of the Rio Declaration's precautionary principle and is also part of the modeling of the effects of climate change by the IPCC.

114. Richmond, "Women as Public Scientists in the Atomic Age," pp. 350–51, 359–63, 376–79. See also Walker and Walsh, "'No One yet Knows What the Ultimate Consequences May Be.'"

115. See, for example, Roco and Bainbridge, *Nanotechnology*, pp. 4–7, 16–21.

116. Fuller, "Precautionary and Proactionary as the New Right and the New Left of the Twenty-First Century Ideological Spectrum"; Fuller and Lipińska, *The Proactionary Imperative*. For criticisms, see Lynch, "After the Gold Rush."

be displaced.[117] However, an emphasis on the existence of multiple, progressive research programs at any one time would help provide clarity about when caution is in order.

The most influential formulation of the precautionary principle is Principle 15 of the Rio Declaration resulting from the 1999 Earth Summit:

> In order to protect the environment, the precautionary approach shall be widely applied by States according to their capabilities. Where there are threats of serious or irreversible damage, lack of full scientific certainty shall not be used as a reason for postponing cost-effective measures to prevent environmental degradation.[118]

Here the goal is to act in ways that anticipate significant harms when decisive proof is lacking, an approach that can be as supportive of new technologies that promise to be—but are not proven to be—safer.[119]

Developing inherently benign chemicals based upon green chemistry principles is not an anti-technology approach, but one where we make preliminary judgments about the best ways to synthesize useful chemicals in ways that are protective of human health and the environment. Prior to testing the chemicals for safety after they are developed, we cannot know for sure that there will not be unanticipated problems with the new technology as well. The precautionary approach should explore all the possible consequences of new technologies that viable research programs suggest may offer potential harms, which would certainly include consideration of how effective packaging would impact food safety or how pesticides could enhance our ability to feed a growing population.[120] What should not happen is that possible scenarios are never considered and debated publicly either because the research programs supporting them have been actively suppressed or because decisive proof of harm is demanded before we can act.

Recognizing the existence of multiple, conflicting research programs in science and technology is necessary if we are going to be able to anticipate future threats posed by our own ingenuity as a species. The exclusion of minority research programs from consideration represents a significant threat to our ability to engage in foresight rather than just hindsight, whether that exclusion results from the autonomous judgments of a self-selecting core set

---

117. Davies, *The Rise of the U.S. Environmental Health Movement*, ch. 7, shows that there are good scientific reasons why we should not expect to decisively identify environmental causes of impaired human health.

118. *Report of the United Nations Conference on Environment and Development*, p. 3; Zander, *The Application of the Precautionary Principle in Practice*, p. 28.

119. Lewens, *The Meaning of Science*, pp. 137–39.

120. Hanekamp and Bast, "Hormesis in Precautionary Regulatory Culture," argues that the concept of hormesis—that low levels of otherwise harmful chemicals or radiation are actually beneficial—challenges the EPA's implementation of a precautionary principle.

or from active suppression by government or industry. With hindsight we can see that foresight was possible.

We can also highlight the importance of reversibility, the other key component of the Rio Declaration's formulation of the precautionary principle. Precisely because any one research program is likely to obscure some risks while constructing knowledge of others, a pluralism of approaches is inherently valuable as a means of resisting the normalization of deviance. Moreover, understanding that research programs ebb and flow over time based upon changes in the totality of science and varying patterns of motivation supports pluralism. It tells us that keeping minority programs going that show at least intermittent progress can be a good way of hedging our bets. The dialectical approach to science suggests that science needs to do a better job of anticipating threats to human health and the environment, something that an emphasis on decisive closure of scientific controversies can obscure.

Chapter Seven

Commodification and the Erosion of the Scientific Ethos

> And, as in the capitalist development of production, a parallel process has started, of which the essence is to deprive and to alienate workers from their instruments of production, so the development of science has led, through the abolition of private libraries and of private laboratories, to scientific factories owned by the bourgeois state or a certain tycoon.
>
> —Imre Lakatos, 1947[1]

In chapter 1, we looked at one of the founding myths of Science and Technology Studies (STS): that sociologists in the 1970s finally worked up the nerve to do what no one had done before—a few forerunners like Kuhn and Fleck aside—namely, to examine the content of science sociologically. We have seen that key to this account was the mistaken view that philosophers and sociologists had agreed to a clear division of labor and a hands-off attitude to the content of science.

On this view, all philosophers of science denied that science depended upon a sociological foundation, seeing reason instead as autonomous and directed through its own internal logic so long as it was not interfered with from the outside. We have seen that logical positivist Otto Neurath and falsificationist Karl Popper, who could agree on little else, both saw the foundation of science as conventional, social at its core, with the basic, empirical building blocks of science dependent on specific practices that determined what could count as observations (to use the current de rigueur turn of phrase).[2]

1. Kutrovátz, "Imre Lakatos' Hungarian Dissertation," p. 368.
2. Reisch, "Planning Science," p. 172, argues that Popper and Neurath's views were much closer than their controversy over social engineering would suggest.

Lakatos had already developed a thorough-going, dialectical Marxist sociology of scientific knowledge in his Hungarian dissertation in 1947. After emigrating to England, his earlier work continued to influence surreptitiously his work with Popper, first through a dialectical and non-foundationalist approach to mathematics, then through a philosophy of science designed to synthesize Popper's critical rationalism and Kuhn's historicism.[3] In fact, his earlier work put him in a good position to do just that, as scientific method was understood as based upon conventional hard cores insulated from falsification and positive heuristics that shaped constructive work. Multiple research programs existed at any one moment, in opposition to Kuhn's view of mature science, while their fruitfulness waxed and waned over time, giving rise to an ever-changing view of science.

Feyerabend similarly developed a view conducive to both critical thinking and a fundamentally social-historical account of science. Beginning already in his 1951 dissertation, Feyerabend set out to systematically dismantle the logical empiricist view of science by showing that general theories saturated any alleged observational foundation for science. By 1962, Feyerabend had developed his account of incommensurable theories, which differed from Kuhn's by opposing the dominance of single paradigms.

Feyerabend's concept of incommensurability was key to facilitating criticism of established points of view by challenging the cozy relationship between a dominant theory and the facts it was held to explain by employing an external standpoint. By the mid-1970s, Feyerabend had turned his earlier pluralistic test model into a radical critique of any attempt to shoehorn the history of science into any kind of convergence on a single picture of the world. On the other hand, he later developed a critique of the closed-off nature of traditions associated with relativism consistent with his beliefs that pluralism should not preclude criticism and change.

On the standard story of the emergence of a constructivist sociology of scientific knowledge, earlier structural-functional sociologists had provided their own defense of an autonomous scientific reason by showing how the social institution of science provided the right kind of normative structure within the right kind of society to allow science to thrive. On this view, Robert Merton's account of the normative structure of science served to both legitimate the institutions of science and democratic societies and to underscore the autonomy of science after its emergence in early modern Europe.

---

3. Motterlini, "Reconstructing Lakatos"; Kadvany, *Imre Lakatos and the Guises of Reason*, pp. 286–88; Kutrovátz, "Imre Lakatos' Hungarian Dissertation"; Kutrovátz, "Lakatos' Philosophical Work in Hungary"; Dusek, "Lakatos between Marxism and the Hungarian Heuristic Tradition."

MERTON'S NORMS: AN APOLOGY FOR SCIENCE?

A cursory reading of Merton's "The Normative Structure of Science," reprinted in anthologies with its name changed from its original 1942 "Science and Technology in a Democratic Order," became part of most sociology of science courses. Usually this was assigned followed by Mulkay's argument that the normative structure of science was an ideology, setting the stage for the emergence of sociological explanations of scientific controversies. Merton's view was taken to be that science naturally aligned itself with democracies and that norms operated unproblematically on scientists to ensure behavior most conducive to allowing the unbiased search for knowledge to proceed.

Rare violation of norms proceeded from outside interference, especially in totalitarian dictatorships. Valid scientific knowledge was held to require no sociological explanation at all once science had been established; only interference with scientific autonomy required sociological explanation in an asymmetrical fashion. Merton's dissertation on the capitalist and Puritan roots of modern science looked uncomfortably similar to Boris Hessen's infamous Marxist, externalist explanation for the emergence of Newtonian science. However, Merton's foray into history of science could be dismissed by assimilating it to his account of the normative structure of science. Merton was taken to have shown the social origins of the institution of science that, once established, would generate true knowledge in perpetuity.

Collins called this period the first wave of science studies, characterized by its apologetic character:

> The sociology of science in those days was looking at the way the scientific community had to be organized and what its values had to be for it to do its work successfully and to allow the testing to be carried on without bias or outside influence. It was said that democracy nurtured these values and that is why we, rather than the fascist dictatorships, had the science to win the war.

Sociology of science developed sophisticated theories but "without ever challenging either the pre-eminence of science or the basic idea that, somehow, our scientific ideas were good because they were testable against reality."[4]

Second-wave sociologists opened up this black box to show that society could be found not just outside science but inside as well. If the first wave treated scientists as a priestly caste with special access to truth, the second wave profaned the priests, showing them to be ordinary, fallible, and interested humans like everyone else. Having said that, Collins argues that a third-wave sociology of expertise is now needed to show how expertise is a limited

---

4. Collins, *Are We All Scientific Experts?*, p. 22.

skill, not something available to all. He also appeals back to first-wave sociology to explain a growing sense that parts of science may be corrupted.

> One way to reconcile the special nature of science with an accurate portrayal of the scientists' world is to describe and analyse the skills, experiences and expertises of its practitioners. . . . Another and perhaps ultimately more important way to minimize the impact of the profanity of the priests is to put the evidence together differently. In their moment-to-moment actions scientists may act in a mundane way but the *spirit of science* is still "divine." It is what science is aiming for that matters and so long as this ideal drives the actions of most scientists most of the time, then Wave 1 of science studies was not so far out in its respect for the norms and values of science. That, indeed, is why we are so offended when we find that scientists have been bribed by the tobacco and oil companies to produce bespoke findings tailored to fit the company's financial interests. We know, immediately, that something wrong is taking place and this shows that we know how science ought to be and mostly is. Contrast this with the arena of politics where there is nothing unusual at all about lobby groups being funded by commercial companies—that is simply how the political world works. If it ever comes to be the case that it is accepted as a matter of course that is how the scientific world works, then science will cease to exist.[5]

Latour, too, has used the example of powerful corporate lobbies eroding trust in climate science as a reason to qualify a purely symmetrical approach that flattens expertise.[6]

I have sought to show that a serious answer to the question of expert credibility can be given that does not eliminate the possibility of criticism and the development of alternative points of view. This can be done by paying attention to the track record of multiple research programs, without simply taking the dominant program's presentation of the situation as accurate. A case can be made that promising alternatives should receive some institutional support so that they develop new lines of research. The key issue is the extent to which we can support alternatives without eroding the aspects of the social organization of science that work well. To answer this question, we need some account of the institutional structures inside and outside science that function to support the pursuit of knowledge.

My contention is that we should not assume that the institutional structure and support of science is already optimal, and the cases we have examined support this contention. If a dialectician always begins with the science we

5. Ibid., pp. 81–82.

6. Latour, *An Inquiry into Modes of Existence*, pp. 2–8; Latour, *Down to Earth*; Kofman, "Bruno Latour, the Post-Truth Philosopher, Mounts a Defense of Science." The topic has also motivated philosophers to interrogate the place of values in science and science policy. See Kitcher and Keller, *The Seasons Alter*.

have, not the one we wish we had, it is also true that this is not the best of all possible worlds. Reliance upon a laissez-faire approach to science is likely to leave us short in developing science to its maximum potential and meeting urgent needs for knowledge that we face in modern society.

As Feyerabend pointed out to Kuhn in comments of the draft of *The Structure of Scientific Revolutions*, the fact that science was set up to pursue truth does not mean it will continue to do so as more prosaic motives crowd out the initial goal. Ultimately, the danger he warned about was a growing dogmatism within science and a tyranny of experts over society as a whole.[7] On the other hand, it is hard to see how science can continue to exist if we simply let any alternative point of view set itself up as a science without regard for its ability to develop a fruitful research program or its potential integration with science as a totality. Science cannot open itself to every conceivable point of view even if that can cause strain between different institutions within society.

But this means we must have some way of understanding and evaluating how well science is functioning to support the values we attach to it and how well it is integrated within the larger societies of which it is a part. Here Merton's account provides a preliminary basis for discussion, precisely because the stereotype that Merton's account of the scientific ethos was an apology for science and democracy is flawed. First, the connection he drew between science and democratic forms of governance was unstable, in part because of attitudes within science, including a counterproductive "pure science" ideal that did not recognize the connection between the technical and the moral. Second, given changing circumstances, different forms of regulation are necessary to ensure the place of science within democratic societies. At the time he wrote, Merton recognized that the norms of science were not self-regulating, such that his approach to science could be seen as a kind of sociological Keynesianism requiring intervention and reform short of the planning called for by Marxists like Bernal.

To get a sense of how the normative ethos of science has been showing signs of strain and conflict with encroaching commercial imperatives, we can look at the growth of fraud and misconduct in science, the rise of predatory journals, the increase in retractions, publication and funding pressures, the growing lack of accessibility and sharing of research materials attendant to the rise of university patent offices, and the emergence of ghostwriting in pharmaceutical research.

The frequency of fraud and misconduct has generally been downplayed in the history of science. While it may be that it was a problem that was suppressed for reasons of public relations, as we have found to be the case

---

7. Hoyningen-Huene, "Two Letters of Paul Feyerabend to Thomas S. Kühn on a Draft of the Structure of Scientific Revolutions," pp. 356–57; Feyerabend, *Science in a Free Society*, pp. 73–91.

with other social problems like sexual harassment and child abuse, there are reasons to believe that the scale and significance of the problem has increased in recent decades. Consider that in the second edition of his textbook on the practice of science in 1992, Frederick Grinnell added a chapter on "Scientific Misconduct: Science at Risk" that covered high-profile accusations of scientific fraud such as that directed at Thereza Imanishi-Kari in David Baltimore's laboratory. Grinnell's general conclusion was that too much emphasis on fraud by outsiders like the politicians behind three new subcommittees investigating scientific fraud in the U.S. House of Representatives could damage the scientific community by equating it with error, a normal and ineliminable part of science. Science functioned well at rooting out and eliminating fraud due to its reliance upon the eventual correction of error by independent researchers and the neglect of research findings not believed to be true, a process that proceeds the same whether the error is intentional or not.[8] Indeed, as it turned out, a later investigation vindicated Imanishi-Kari, attributing the problems to sloppy laboratory records rather than any intent to deceive.[9]

Leading voices in STS underscored the point, challenging scientists' polemics on the cold fusion debate by rejecting the idea that the failure of the scientific community to replicate the cold fusion findings of chemists Stanley Pons and Martin Fleischman indicated that they had done something improper.[10] If science was a golem, vigilance should be coupled with indulgent understanding. In essence, an understanding of core-set interpretive flexibility meant that separating scientific researchers into distinct categories of the honest and dishonest was a similar violation of symmetry as distinguishing truth from falsity or rationality from irrationality.

In truth, the cold fusion case did throw up for view the impact that university technology transfer offices had begun to play in pressuring researchers to make priority claims in light of the encouragement of university patents by the Bayh-Dole Act in 1980. The law was only one significant contribution to a decades-long shift to greater commercialization of academic science, spurred by the biotech industry. The scientific persona shifted in kind, as ascetic and altruistic pictures of scientists gave way to hedonistic and self-interested ones.[11] Already by the late 1970s, concern over fraud in the scientific and mainstream press blamed the competition for grant money that

---

8. Grinnell, *The Scientific Attitude*, p. xi, ch. 6. Compare the journalistic investigation of fraud as an elite scandal in Broad and Wade, *Betrayers of the Truth*. On the conflict of views about fraud between politicians and scientists, see Booth, "A Clash of Cultures at Meeting on Misconduct."

9. Beardsley, "Profile: Thereza Imanishi-Kari—Starting with a Clean Slate."

10. Collins and Pinch, *The Golem at Large*, pp. 76–77; Close, *Too Hot to Handle*; Huizenga, *Cold Fusion*.

11. Shapin, *The Scientific Life*, pp. 80–89, 215–22; Mirowski, *Science-Mart*, pp. 148–49.

led to overstating the significance of new findings, a trend also facilitated by the rise of for-profit commercial journals who looked to create "venue[s] for scientific blockbusters."[12]

However, STS belief in the relative paucity of fraud in the history of science did suggest that constructivists had a tacit belief in the workings of a normative ethos for science despite their disavowals of Merton's structural functionalist approach to science. However, any cooperative social system can be undermined when individual incentives to cheat grow more powerful than group monitoring of social behavior, when disinterestedness lessens and organized skepticism breaks down. Foremost among these perverse incentives are increased competition for jobs, publications, and funding. A greater population of scientists with flat funding levels puts the emphasis on publishing in highly ranked journals. Moreover, productivity is no longer enough. Faculty salaries are correlated with visibility and specialization as well. The sociology of science contributed to these trends, as Eugene Garfield's development of an "impact factor" shifted from a tool for sociologists to study social networks in science to a measure of success tracked by administrators and scientists.[13]

The rising tide of cases of scientific fraud in the 1990s led to an explosion of formal retractions of scientific journal articles after 2000, something unheard of before this time.[14] Arguably, the targeting of mere error as borderline fraud (often under the equivocal label "misconduct") haunts this development as well, so it is a little hard to distinguish a real social problem from a moral panic here. Nonethelesss, the higher the impact factor of the journal, the greater the number of retractions, indicating a perception that retractions are "essential for correcting the literature and maintaining trust in the scientific process."[15]

Edwards and Roy worry that "[i]f a critical mass of scientists become untrustworthy, a tipping point is possible in which the scientific enterprise itself becomes inherently corrupt and public trust is lost, risking a new dark age with devastating consequences to humanity." While the rhetoric may be overblown, their approach is sober and builds on Merton's concept of unintended consequences of intentional action by showing that reward systems in science are undermining their stated goal through perverse incentives resulting from competition, quantitative measures of productivity, and the growth of

12. Weinstein, "Fraud in Science," pp. 639–41; Buranyi, "Is the Staggeringly Profitable Business of Scientific Publishing Bad for Science?"

13. See Leahey, "Not by Productivity Alone." The need to maintain an uninterrupted stream of publication to justify grant renewals essential to keeping laboratories running is often mentioned as a cause for scientific misconduct. See Carafoli, "Scientific Misconduct."

14. Mirowski, "The Future(s) of Open Science," pp. 183–84.

15. Fang and Casadevall, "Retracted Science and the Retraction Index," pp. 3855–56; Cokol, "How Many Scientific Papers Should Be Retracted?"

a business model for academic administration. Researchers have responded to productivity demands by short, incremental, and shoddy papers that often manipulate statistical techniques to generate statistical significance (known as p-hacking), while peer reviewers ensure their own visibility by demanding that their own articles be cited.[16] Even reputable scientific journals exploit the system to enhance their impact factors by promoting trendy contributions and blocking new ideas outside of the mainstream.[17]

As with retractions, similar worries are growing about the existence of a "reproducibility crisis," where the inability to replicate key laboratory findings in psychology and medical science spurred a concern that the empirical method was bankrupt or non-existent in certain sciences. By and large, this literature fails to appreciate the difficulties of replication that Harry Collins underscored as a normal part of science, especially since science's reward system does not prioritize replications but novel findings (a Mertonian concession by the relativist sociologist!).[18]

Feyerabend's endorsement of counterinduction and the development of incommensurable observation languages as a spur to conceptual novelty suggest that an obsessive focus on the failure to replicate may misconstrue how science works. Citing Feyerabend on this point, Carafoli notes that significant advances by scientists like Galileo or Einstein flow from "the willful violation of generally accepted principles and/or methodological constraints." Such contributions would be judged fraudulent were it not possible to see retrospectively that "at their origin there invariably were formidable intuitions which perhaps could not be possibly verified experimentally within the technical limitations of the times."[19]

Then there is the rise of predatory journals looking to make a buck (or a yuan) from desperate academics trying to circumvent a highly competitive peer review system on the way to promotion. Some companies will even sell you a coauthorship on articles already accepted in good journals.[20] "Ghost writing" of medical journal articles by pharmaceutical companies has eroded the independence of scientific expertise in the drug approval process, with as many as 40 percent of medical articles on recently patented drugs written

16. Edwards and Roy. "Academic Research in the 21st Century," pp. 51–52; Merton, "The Unanticipated Consequences of Purposive Social Action."

17. Schekman, "How Journals Like *Nature, Cell* and *Science* Are Damaging Science."

18. Begley and Ellis, "Raise Standards for Preclinical Cancer Research"; Prinz, Schlange, and Asadullah, "Believe It or Not"; Ioannidis, "Why Most Published Research Findings Are False"; Collins, *Changing Order*, pp. 18–19.

19. Carafoli, "Scientific Misconduct," p. 370.

20. Hvistendahl, "China's Publication Bazaar." Peer review has by no means been practiced uniformly in science, especially before World War II, so a case can be made that it is a key contributing cause to current problems rather than a system that ensures quality control in science. See Baldwin, "Scientific Autonomy, Public Accountability, and the Rise of 'Peer Review' in the Cold War United States."

by hired publication planners. Elsevier even created made-to-order medical journals tailored for marketing pharmaceutical drugs in collaboration with its subsidiary publication planning firm, Excerpta Medica.[21] Once pharmaceutical companies get new drugs approved, billions of dollars are spent on marketing, crucially including drug reps who use sophisticated marketing models and prescription tracking for influencing prescribing doctors. Physicians known as "Key Opinion Leaders" (KOL) help drum up business through continuing education seminars with presentations authored by drug companies.[22]

Brown argues that there has been particularly rapid deterioration in the reliability of pharmaceutical research, as comparative drug studies consistently support the sponsor's drug over competitors. The pursuit of patents by universities has exploded since the 1980 Bayh-Dole Act allowed private companies and universities to take out patents on federally funded research. Contract researchers often do not have access to the full data, and cherry picking has become rampant. Leading biomedical journals have responded with a common policy requiring that clinical trials be registered in advance, researchers have access to full data, and conflicts of interest be exposed. At the same time, many journals have loosened conflict-of-interest rules because it is becoming harder for journal editors to find scientists unencumbered with pharmaceutical funding. Government-funded clinical trials are finding it harder to recruit subjects since they do not pay finder's fees to general practitioners who sign up their patients, a practice often skewing criteria for inclusion.[23]

Industry collaboration with university scientists has also put a damper on sharing of research materials, showing a direct distortion of Merton's norm of communism under the influence of corporate emphasis on secrecy.[24] Commercial patents for academic discoveries may have inhibited the development of knowledge. Research on genetically modified seeds has been suppressed as the result of licensing agreements Monsanto requires for researchers to use their seeds.[25] University Technology Transfer offices have played a similar role in restricting access to research materials traditionally shared among labs. "Material transfer agreements" (MTAs) are increasingly required before research materials can be shared. Not only are such requirements costly and time consuming, they also include "reach-through" rights such as confidentiality requirements and publication restrictions.[26]

---

21. Sismondo, "Pushing Knowledge in the Drug Industry," pp. 150–51, 162; Sismondo, *Ghost-Managed Medicine*, chs. 3–4; United States Senate Committee on Finance, "Ghostwriting in Medical Literature."

22. Fugh-Berman and Aharii, "Following the Script," pp. 124–29; Applbaum, "Getting to Yes," pp. 143–44; Sismondo, *Ghost-Managed Medicine*, pp. 100–101, ch. 5.

23. Brown, "The Community of Science®," pp. 190–96. On the resistance to and limitations of these reforms, see Greenberg, *Science for Sale*, chs. 13–14.

24. Evans, "Industry Collaboration, Scientific Sharing, and the Dissemination of Knowledge."

25. Biddle, "Can Patents Prohibit Research?," pp. 16–17.

26. Mirowski, *Science-Mart*, pp. 140–42.

With all these cases, the concern about the erosion of prior rules for good conduct in science presupposes that we can talk about what those rules are, or should be, not to mention study how well they constrain behavior in practice. In the long view, the coming to the fore of self-interested and economic incentives represent a change in the perception, if not the behavior, of the scientist. Shapin's natural philosophers were gentlemen who were free and independent by virtue of their class position, a stereotype that certainly obscured the contributions of those who helped make the engines move.[27] A similar situation characterized the man of science of the modern university system that emerged in nineteenth-century Germany. Weber showed that science depended upon a kind of ascetic, self-denying pursuit of truth, as individual researchers needed to specialize to make a lasting contribution but without guarantees that merit will be recognized by a capricious reward system. Only as institutional support for science grew in the twentieth century could most scientists count on secure employment, as an aristocratic hobby or a scholarly calling became a job.[28]

Merton's focus was on the way that the reward system of science directed the scientist's behavior in ways that best served an institutional goal of the production of knowledge. While salaries and faculty chairs were subject to vicissitudes of different bureaucratic contexts within which scientists worked, science's internal reward system depended crucially on the pursuit of achieved status based on contributions to knowledge recognized by one's peers. The crucial "engine of modernity," the institution of science, depended upon a "pre-modern" economy of gift-giving and recognition for its effective functioning.[29]

The sociologist Ferdinand Tönnies had distinguished between traditional societies based upon personal, intimate social interactions and complex societies with impersonal, role-based interactions. Weber described the shift between the two types as the key transition of modernity. The former, small-scale, intimate societies were labeled "*Gemeinschaft*," or community, while complex, impersonal, and role-based societies were labeled "*Gesellschaft*" or "society."[30] The persistence of small *Gemeinschaft*, scientific core sets within larger *Gesellschaft* divisions between fields arguably reflects the need for decomposing scientific problem areas within science in ways that work with our evolved capacities for tracking and evaluating only a small number of intimate acquaintances. What has been called our "Dunbar number" is correlated with brain size among primates and reflects the growing capacity of our

---

27. Shapin, *A Social History of Truth*, chs. 2–3, 8.
28. Weber, "Science as a Vocation"; Shapin, *The Scientific Life*, p. 17; Ben-David, *The Scientist's Role in Society*, pp. 125–29.
29. Ziman, *Real Science*, p. 81.
30. Tönnies, *Community and Society*; Weber, *Economy and Society*, vol. 1, pp. 40–43.

lineage to use gossip and intimate knowledge to manage trust. For humans, Dunbar argued that we are able to track roughly 150 other humans in a way that supports functioning groups dependent upon social grooming.[31]

Arguably, modern, impersonal societies depend upon agglomerations of smaller groups connected by more formal kinds of interactions that have evolved through cultural evolution. Hierarchically nested groups, bureaucratic structures, commercial markets, and larger-scale imagined communities like those we considered in chapter 4 all serve as cultural "work arounds" to direct small group-adapted humans to contribute efficiently to complex, modern societies.[32] The emergence of a dynamic, modern institution of science in the last few centuries, capable of continually generating new discoveries, depends upon the development of methods of decomposing complex conceptual problems and parceling them out to small core sets. Collins's account of experts with tacit knowledge of other experts, including those who can be trusted, depends crucially on this kind of organization. The networks that link different core sets can also serve as conduits for the transmission of symbolic information between different groups that would otherwise be blocked by dominant small-group beliefs.[33]

Where small-group core sets run into problems is when their tacit understanding of the evidence is doubted by outsiders. Quantification as a tool for government scientists to provide "proof" of findings in a politically or legally contested environment represents one technique for credibility when trust in expert judgment erodes. The emergence of a quantitative concept of proof thus served as a political solution to increasing reliance upon scientific experts by governments, with the United States leading the way in the early twentieth century.[34] Shapin's premodern circle of experts—or Collins's self-selecting core sets—while still operating within science, ceased to define the political economy of science as bureaucratic management of science by corporations and the state grew, with quantitative proof designed to manage conflict among experts and wider policy constituencies.

Porter's examination of this transition shows, as one might expect, that the ideal of quantification did not match its practice. As quantification began to be used for political legitimation, it often functioned to suppress minority views or plausible causal claims that lacked quantitative proof at the time. This is why I have argued in chapter 6 that a precautionary principle linked

31. Hill and Dunbar, "Social Network Size in Humans"; Dunbar, *How Many Friends Does One Person Need?*

32. Richerson and Boyd, "Evolution"; Collins, *The Sociology of Philosophies*, ch. 10.

33. Granovetter, "The Strength of Weak Ties"; Hull, *Science as a Process*, pp. 507–9; Sterelny, *The Evolved Apprentice*, pp. 54–56; Lynch, "Cultural Evolution and Social Epistemology"; Galison, *Image and Logic*, ch. 9.

34. Porter, *Trust in Numbers*, chs. 7–9.

to research programs that show some degree of progress is needed to deal with science-based policy challenges, rather than a demand for proof that is epistemologically naïve and politically conservative.

The rise of a compact between science and state began with the growth of scientific experts in government bureaucracy, especially as government agencies and large-scale engineering projects proliferated in the first half of the twentieth century.[35] Government support of research by academic scientists increased dramatically after the experience of government-funded military research during World War II. This led to changes in both the organization of scientific communities and the implied moral community and social epistemology of science. Before the wide employment of scientific experts by government bureaucracies and the expansion of government funding for academic science, small, self-organized core sets subscribing to something like Merton's norms and granted credibility based upon their skilled judgment was the rule. Such *Gemeinschaft* science continued to operate as a premodern world within modernity, a world of person-to-person credit among a small group of scientists who knew each other's work well and could evaluate it well.[36] Even within the Cold War funding regime, increased funding for science across the board ensured continued support of basic science along with applied science and technology.

Only with the growth of "Big Science," as large research teams sharing expensive equipment grew in number, did greater bureaucratization of actual working conditions among scientists begin—and begin to be criticized.[37] The role of formal bureaucracies displaced an emphasis on ineffable expert judgment, especially in the United States. The impact of agonistic legal and policy contexts transformed science used in policy contexts from gentlemanly trust to formal legal briefs. Today, quasi-governmental entities like the IPCC similarly shifted the scientific consensus from the quiet agreement of core sets to formal statements incorporating statistical measures of certainty and mobilizing evidence to fend off criticism. *Gesellschaft* science emphasized top-down management of individual scientific laboratories and larger, collaborative projects, the rise of formal measures of productivity and quality, and a growing external bureaucracy shaping the formation and evaluation of research programs.

It is only with the rise of "neoliberal" or "post-academic" science in the last few decades that *Gemeinschaft* science has truly given way in the heart of the modern university to *Gesellschaft* science. John Ziman has argued that a revo-

35. Owens, "Science in the United States," pp. 824–29; Porter, *Trust in Numbers*, pp. 148–55.
36. Shapin, *A Social History of Truth*; Ziman, *Real Science*, p. 81.
37. Weinberg, "Impact of Large-Scale Science on the United States"; Price, *Little Science, Big Science*; Forman, "Behind Quantum Electronics"; Aronova, "The Congress for Cultural Freedom, Minerva, and the Quest for Instituting 'Science Studies' in the Age of Cold War."

lution is under way that is introducing a new normative ethos within science and a new social role for science in society. To understand these changes, Ziman focuses upon the radically changed basis of science at the level of practice in the university setting, which allows him to see how the ethos of science has changed practically for most scientists. Post-academic science is still carried out in academic settings but in ways that are radically changed as the result of developments external and internal to the institution of science. The growth of university–business partnerships and commercial funding of academic science are key external impacts, emphasized especially by those authors who refer to "neoliberal" science. The key internal changes flow from the limits to growth of the Cold War model, as exponential growth of scientists, publications, and funding levels fell off leaving scientists in the lurch.[38]

The growing employment of more, and more expensive, scientific instruments transformed the division of labor within science and enforced a growing collectivization of scientific research dependent upon more formal management, cooperation, and collaboration. Purpose-built equipment for individual research groups gave way to large-scale scientific infrastructure requiring coordination of different disciplinary groups competing for access. An emphasis on utility and transdisciplinary research teams further altered the autonomy of individual researchers and core sets.[39]

An emerging parasitical bureaucracy imposed formal assessments and accountability on a shrinking tenured faculty, while a growing academic precariat carried out more of the teaching and short-term research contracts. The commercialization of science comes with a greater shift away from Merton's norm of communism and toward the counternorm of secrecy, a form of "epistemic pollution," as corporations look to protect intellectual property and trade secrets. Even the sharing of research materials becomes more difficult as universities look to enforce material transfer agreements, inserting university transfer offices between scientists who had previously sent each other research materials[40]

In understanding the development of post-academic science as "a whole new way of life," Ziman tracks mutual interactions between a more or less formal ethos and the vagaries of concrete practice in a manner similar to chapter 4's call for tracing the multiple, iterative interactions between methodological discourse and concrete scientific practice. Post-academic science

> is the resultant of innumerable improvised solutions to immediate practical problems. It is the product of expediency, not of design. Yet it constitutes a more

38. Ziman, *Real Science*, pp. 60–66, 71–72; Milojević, Radicchi, and Walsh, "Changing Demographics of Scientific Careers."
39. Ziman, *Real Science*, pp. 69–74; Doing, *Velvet Revolution at the Synchrotron*, pp. 80–107.
40. Ziman, *Real Science*, p. 116; Mirowski, *Science-Mart*, ch. 4.

> or less coherent culture, not because it was planned as such but because science is typically a complex, self-organizing social system that adapts opportunistically to changing circumstances.[41]

Therefore it is urgent that scientists and students of science alike look to see the "normalization of deviance" that they may come to regret if current trends continue. This also means that we need to understand how the baseline nature of science may be changing and evaluate that against our best models of how the institution of science functions within different societies, past and present. The key change altering the character of science is the increased subordination of science to the project of spurring national economies through R&D funding, resulting in a greater role for utility over the progress of research programs.[42]

The result is that industrial, post-academic science inverts the norms associated with academic science and associated practices shaping the production of public knowledge dating from the mid-nineteenth century. During this period, even industrial laboratories associated with large corporations, such as the Bell Telephone Laboratories and General Electric Research Laboratory, carved out an important place for publicly shared basic knowledge. While scientists did not follow Merton's norms without qualification, the relative balance of Merton's norms and Mitroff's counternorms has certainly shifted, reflecting real institutional changes in science.[43] Academic science saw market transactions increasingly encroach upon the *Gemeinschaft* system that exchanged recognition for scientific contributions. This system can be described via the apt acronym CUDOS, standing for the norms of *communalism*, *universalism*, *disinterestedness*, *originality*, and *skepticism*. The system resisted purely market mechanisms and resembled the gift-giving *potlatch* system studied by anthropologists. Under post-academic science, dominant norms protect embedded institutions and experts from competition, employing norms from the acronym PLACE, indicating *proprietary*, *local*, *authoritarian*, *commissioned*, and *expert* approaches.[44]

---

41. Ziman, *Real Science*, p. 68.
42. Ibid., pp. 73–76.
43. See Thorpe, "Community and the Market in Michael Polanyi's Philosophy of Science," for a discussion of Polanyi's conflation of science as a traditional community and science as a marketplace of ideas.
44. Ziman, *Real Science*, pp. 25, 31–46, 78–79; Hagstrom, *The Scientific Community*, pp. 12–23; Malinowski, *Argonauts of the Western Pacific*; Mirowski, *The Effortless Economy of Science?*, pp. 383–92. See also the discussion of the extent to which the epistemological health of science and the direction of science by democratic values are challenged by private direction of its research agenda. See Longino, "Science and the Common Good," p. 567; Kitcher, "Reply to Helen Longino," p. 570. Carrier, "Science in the Grip of the Economy," pp. 222–33, argues that commercial science can overcome these limitations as competition between corporate sponsors and the emergence of innovation in applied research tempers the threat posed by commercialization.

ROBERT MERTON AND THE CAPITALIST FRUSTRATION OF SCIENCE

The development of a professionalized field of Science and Technology Studies (STS) has been dominated by a self-serving mythology that it alone has subjected the content of science to sociological scrutiny and in doing so opened science to critical scrutiny that brings experts down to earth. The principle of symmetry is most often seen as the crucial transition from earlier apologetics on behalf of rational science and the liberal social milieus that allegedly sustained it. Viewed this way, the clear division of labor was set between philosophers of science, laying out an internal logic for autonomous science directed along a single path by univocal reason, and sociologists of science, who laid down the basic rules of conduct for the community of science, which testified to the superiority of democratic over totalitarian regimes.

As we saw in chapter 1, Robert Merton's account of the normative ethos of science was rejected by Michael Mulkay as an ideology that put a self-serving view of the behavior of scientists into scientific articles in order to clean up the messy motivations for public view. Merton's assimilation of fascist and Soviet threats to science registered his apologetics for an emerging Cold War view of totalitarianism familiar also in the writings of Karl Popper, Arthur Koestler, and Hannah Arendt.[45] Moreover, his commitment to a functionalist view of society was taken to typify the consensual and organicist view of society that a new generation of social theorists emphasizing conflict and power sought to challenge.[46]

In fact, Merton had a more sophisticated view of the normative structure of science, articulating this theory not only to identify Soviet and Nazi perversion of science but also to challenge the commercial corruption of science in capitalist societies. Merton's early work, and in particular his articles on the ethos of science in 1938 and 1942, articulated a basis for a critical approach to science that was rooted in the concerns about capitalist corruption of science by the Marxist scientists J. B. S. Haldane, J. D. Bernal, Hyman Levy, Joseph Needham, and Lancelot Hogben. These British radical scientists had been influenced by the Soviet delegation to the 1931 Second International Congress of the History of Science, most notably Boris Hessen's analysis of the social and economic roots of Newton's science.[47] To be sure, Merton was

45. In fact, the modern concept of totalitarianism can be traced to disenchanted Marxist Franz Borkenau's appropriation of Vilfredo Pareto's theory of the "circulation of elites," blended with a Marxist theory of economic crisis, as well as George Orwell's wartime writings. See Jones, "Toward a Theory of Totalitarianism," pp. 458–60.

46. Mills, *The Power Elite*; Scimecca, "Paying Homage to the Father," pp. 191–92.

47. Werskey, *The Visible College*; McGucken, *Scientists, Society, and State*, ch. 4; Beddeleem, "Fighting for the Mantle of Science," ch. 1; Hollinger, "The Defense of Democracy and Robert K. Merton's Formulation of the Scientific Ethos," pp. 6–8.

not himself a Marxist, and his postwar work left behind many of these themes of external impacts on science. As Everett Mendelsohn puts it, "Merton was certainly no historical materialist or Marxist in any simple sense; but he definitely was a believer in the impact of societal forces, explicitly including economic ones, on the themes, problem choices, and actions of science."[48]

Moreover, Merton's emphasis on the "functions" of societal institutions are by no means as simplistic as post-functionalist sociology would suggest. Borrowing from anthropologist Bronislaw Malinowski, Merton distinguished between manifest and latent functions of institutions, with manifest functions acting a lot like Mulkay's ideologies while the analyst identifies the latent function of which the participants are unaware. Thus, the manifest function of a rain dance is to produce rain, yet the anthropologist does not believe that the dance actually produces rain. Instead, the dance has the latent function of promoting group solidarity, and its function helps explain why it continues to be practiced despite presumably being an unreliable way to produce rain.[49] Similarly, Mulkay showed, in effect, that the manifest function of belief in scientific method was that it provided a reliable way to produce knowledge, while its latent function was to provide a cleaned-up reconstruction that would justify continued social support of scientists.

The focus by Merton on the conflicts and tensions between different institutions also reflected a more nuanced view than caricatures of structural functionalism would have it. He tended to see functions as simply structural effects that maintained the institution in contrast with those that led to social change. Merton's view differed from Talcott Parsons's account of structural functionalism by rejecting teleological ends defining functions or a focus on the functioning of society conceived as a whole.[50] Moreover, contemporary proponents of cultural group selection like David Sloan Wilson have revived a version of functionalism as those practices that facilitate cooperation within the group are more likely to be passed on through social learning.[51] The wide diffusion of science and technology of the last few centuries, the focus

48. Mendelsohn, "Robert K. Merton," p. 278. See also Turner, "Merton's 'Norms' in Political and Intellectual Context," p. 173: "Although Merton was of course no historical materialist, there was a loose affinity between Merton's Puritanism argument and the general idea, promoted by the British academic Left, that science was propelled by external, social forces."

49. Merton, *Social Theory and Social Structure*, pp. 60–71.

50. For the claim that Merton had a deflationary account of functionalism divorced from teleology, by a philosopher close to Merton, who encouraged her to write the book, see Emmet, *Rules, Roles, and Relations*. For discussion of the difference between Parsons's systems approach to functions and Merton's equivocal one, see Turner, "Robert Merton and Dorothy Emmet," pp. 820–24, 827, 833.

51. Wilson, *Darwin's Cathedral*, pp. 67–69; Campbell, "The Two Distinct Routes beyond Kin Selection to Ultrasociality," pp. 16–17. Ultimately, Campbell's student, Christopher Boehm, showed in *Hierarchy in the Forest*, pp. 199–217, that the policing of moral norms emerges from the reverse dominance hierarchy of modern humans that makes human ultrasociality possible, where the community monitors and enforces desired behavior to suppress within-group selection for traits that erode cooperation.

of Lakatos's reconstructions, may have something to do with its functional characteristics for modern societies, while an erosion of what makes these institutions work may degrade society's ability to reproduce itself materially and culturally. It is precisely on these questions that STS has abdicated its responsibility by seeing the rise of neoliberal science as just another funding regime, equally legitimate as any other.[52]

Merton's norms may then function not as direct constraints on individual behavior but as broadly shared values that scientists internalize as part of their enculturation as scientists. As such, these values can inform structural assessments of changes in the institution of science and policy recommendations for those so committed. In addition, it is the structural characteristics as they relate to individual opportunities that determine whether individuals will follow norms or evade them.

THE SOCIAL PSYCHOLOGY OF SCIENTISTS AND STRAIN THEORY

This structural, rather than purely psychological, approach, was developed by Merton as strain theory, examining the impact of the gap between goals and current status, an approach that has been highly influential in the field of criminology. Merton looked at how the poor responded to societal emphasis on economic success in the United States. Under this theory, the "hopeful poor" and "passive poor" continued to follow societal norms and laws, with the former anticipating eventual conventional success while the latter sought more minimal goals within reach. The "surviving poor," by contrast, sought economic success through disapproved means, such as drug dealing or theft. Finally the "retreating poor" and the "resisting poor" rejected both conventional success as a goal and the accepted means to achieve it, with the former retreating into escapism and the latter working to change them.[53]

The key point is that for different societies, the prevalence of different percentages of unsuccessful individuals and different proportions of responding types varied. Combined with his account of ambivalence and anomie, Merton's structural approach provides the possibility of examining strain

52. Mirowski, *Science-Mart*, pp. 89–92. For an attempt to rehabilitate Merton's norms in light of the indeterminacy of moral rules, using Weber's notion of status groups, see Barnes, "Catching Up with Robert Merton," pp. 180–83. Barnes's argument that "it is not 'norms themselves' but the continuing collective interaction of those who 'follow' them that establishes what norms are taken to imply" is similar to Merton's emphasis, in "The Normative Structure of Science," p. 276, on disinterestedness emerging not from personal or psychological qualities but because "the activities of scientists are subject to rigorous policing, to a degree perhaps unparalleled in any other field of activity."

53. Merton, "Social Structure and Anomie," reprinted in revised form in Merton, *Social Theory and Social Structure*, ch. 4; Deflem, "Anomie, Strain, and Opportunity Structure."

among scientists in today's highly competitive environments and its impact on the ethos of science, integrating structural, cultural, and individual levels of analysis.[54] Merton's formulation of these norms during a critical period when he perceived science to be under threat underscores the point that they are a kind of normative or critical theory applied to the defense of science rather than a purely empirical account of social action.[55] As such, Merton's theory can be understood as a normative social epistemology.

Merton sought to pick out practices and tendencies within the science of his day that failed to live up to the values he identified as crucial to its ability to continue to deliver on its promise. He also sought to incorporate a measure of sociological reflexivity into his understanding of the normative structure of science, including careful attention to the interaction of technical and moral factors in the practice of scientists, as well as an understanding of the destructive interaction between an emphasis on purity by scientists and popular disenchantment with science.

In addition to seeking to ensure that science continued to function well as a knowledge-producing enterprise, he remained attuned to the social responsibility of the scientist for their impact on the wider society, an approach he carried forward in his own work on the social psychology of mass persuasion. While nowhere near as polemical as Feyerabend's complaints about science as a new, oppressive, priestly caste, Merton's early writings identified the dangers of both arrogant experts and anti-scientific publics.

The connection between science and liberal democracy was a fraught one for Merton because developments within science threatened to erode the support for science over time, especially as the public reacted to two trends pulling the threads of the science-society compact in different directions: the commercial corruption of science and the advocacy of "pure science" unconcerned with utility. This version of the basic/applied science distinction, developed by Vannevar Bush, would set the basis for Cold War science funding through the National Science Foundation, albeit after a long delay following the end of World War II's direct commandeering of scientists for the war effort.[56] At Harvard, Merton and James Conant both developed similar ideas

54. Baumer, "Untangling Research Puzzles in Merton's Multilevel Anomie Theory."

55. See Radder, "Mertonian Values, Scientific Norms, and the Commodification of Academic Research," pp. 236–50, for a "deflationary" defense of Merton with a focus on the threat that commodification of scientific knowledge presents to tacit values articulated by Merton in 1942, as well as explicit values articulated in formal codes since the 1970s. More concrete policies can be developed based on reasoning about these values, as Radder does in arguing against allowing patents for academic research. See also Brown, "The Community of Science®," pp. 212–13; Brown, "The Good Old Days," pp. 183–86.

56. Bush, *Science: The Endless Frontier*; Schaffter, *The National Science Foundation*, pp. 17–21; Price, *The Scientific Estate*, pp. 237–59; Greenberg, *The Politics of Pure Science*, ch. 6.

about the scientific ethos, with Conant involved in overseeing the transition from military to civilian control of science after the war.[57]

With the threat to science closer to home than overseas fascists and communists, Merton did not endorse a hands-off, laissez-faire approach to science but one requiring careful intervention and management, though he avoided Bernal's talk of "planning science" that had brought criticism from fellow liberals like Conant and Polanyi.[58] In this sense, Merton's norms for science were no more self-regulating than the market economy was understood to be with the hegemony of Keynesianism.

THE FRUSTRATION OF SCIENCE

In considering the norm of universalism, which is part of the ethos of both science and democracy, Merton rejects any automatic quality to its operation:

> Insofar as lassez-faire democracy permits the accumulation of differential advantages for certain segments of the population, . . . the democratic process leads to increasing regulation by political authority. Under changing conditions, *new technical forms of organization* must be introduced to preserve and extend equality of opportunity. The political apparatus may be required to put democratic values into practice and to maintain universalistic standards.[59]

Mendelsohn traces the origin of Merton's concept of universalism to the "invariance" of Hyman Levy, a Scottish mathematician and communist, active among the British radical scientists who were concerned about the "frustration" of science by capitalism.[60] In the title essay of the 1935 collection, *The Frustration of Science*, Patrick Blackett, physicist and Operations Research pioneer, argued for the necessity of a transition to socialism to prevent science from being artificially stunted in its development.[61] The essay blamed fascism on capitalism's failures to make the most of science's potential to spur industry. In considering the alternatives to fascism, Blackett compared the prospects of capitalist planning and socialist planning; laissez-faire was not even an option up for debate.[62]

57. Turner, "Merton's 'Norms' in Political and Intellectual Context," p. 17; Hershberg, *James B. Conant*, pp. 396–99, 559–77; Dennis, "Historiography of Science," pp. 9–15.

58. Turner, "The Social Study of Science before Kuhn," pp. 43–47.

59. Merton, "Normative Structure of Science," p. 273.

60. Mendelsohn, "Robert K. Merton," p. 281.

61. Budiansky, *Blackett's War*, p. x; Nye, *Blackett*, pp. 1–2, 28–36. Blackett, "The Frustration of Science," pp. 136–40, identifies unemployment among the lower middle classes and the peasantry's opposition to scientific agriculture as the root causes of the rise of fascism and its anti-scientific tendencies.

62. Even Polanyi's defense of an early neoliberal approach to science and economics, that dovetailed with Austrian school economist Friedrich Hayek's approach, took for granted the role of

The idea that science should be invariant or universalistic at this time was not just about thwarting Nazi (and for some, Stalinist) abuse of science. The threat of both types of totalitarian regimes clearly was a central concern of Merton, who turned Left ideas on science to liberal purposes in that regard. He developed a defense of relative autonomy and freedom via science's own normative structure, as opposed to Bernal's vision of science planning.[63] However, the Marxist interpretation of capitalist frustration of science found its parallel in Merton's concern that science's norms could also be corrupted by commercial incentives and governmental exploitation of scientific knowledge for warfare in the United States and England.[64]

Merton explicitly invokes the literature on the frustration of science in the section on the norms of communism, where he considers the conflict between the capitalistic conception of private property and the scientific ethos. According to Mendelsohn, Merton's use of the term "communism" was connected to "his careful reading and general sympathy with the British radical scientists," who had called for science to be directed to socially beneficial ends.[65] As Merton put it,

> The communism of the scientific ethos is incompatible with the definition of technology as "private property" in a capitalist economy. Current writings on the "frustration of science" reflect this conflict.[66]

Merton considered different ways that scientists have sought to resolve the conflict, from taking out patents "as a defensive measure" to ensure its public availability, becoming directly involved in new economic enterprises, and advocating socialism.[67]

In the section on the norm of disinterestedness, Merton evidenced a concern with fraud and the abuse of expert authority. He notes that fraud is more likely as science develops a closer relationship with clients outside the scientific community, like the service professions. The concern here is that the growing entanglement of science with professions tied to a commercial economy make the abuse of expert authority more likely. Science has its authority among the lay public primarily due to its technological fruits. For Merton, the "laity is often in no position to distinguish spurious from genuine claims to such authority." Given the combined erosion of norms within

---

Keynesian stimulus of full employment in a free market. See Thorpe, "Community and the Market in Michael Polanyi's Philosophy of Science," pp. 76–80.

63. Turner, "Merton's 'Norms' in Political and Intellectual Context," pp. 160–66, 172–73, 175.

64. Merton, "Science and the Social Order," pp. 261–62.

65. Mendelsohn, "Robert K. Merton," pp. 284–85.

66. Merton, "The Normative Structure of Science," p. 275.

67. Ibid.

science and the increasing distance between popular and scientific thought, belief in credible science can be eroded by totalitarian thought:

> If anything, the myths will seem more plausible and are certainly more comprehensible to the general public than accredited scientific theories, since they are closer to common-sense experience and to cultural bias. Partly as a result of scientific achievements, therefore, the population at large becomes susceptible to new mysticisms expressed in apparently scientific terms. The borrowed authority of science bestows prestige on the unscientific doctrine.[68]

Contrary to what we might expect given his reputation as an apologist for science, Merton does not call for a greater degree of autonomy to ward off irrationalism.

Instead, Merton argues that the "purity of science" ideal can be dysfunctional. While separating out what would be called basic and applied science in the postwar science-society compact, it can also threaten science's long-term stability:

> It is repeatedly urged that scientists should in their research ignore all considerations other than the advance of knowledge. Attention is to be focused exclusively on the scientific significance of their work with no concern for the practical uses to which it may be put or for its social repercussions generally. The customary justification of this tenet . . . holds that failure to adhere to this injunction will encumber research by increasing the possibility of bias and error. But this *methodological* view overlooks the *social* results of such an attitude. The objective consequences of this attitude have furnished a further basis of revolt against science; an incipient revolt that is found in virtually every society where science has reached a high stage of development.[69]

Sociological reflexivity on the growing gap between a scientific community emphasizing purity and autonomy (even as sections of it violate the principle) and a public increasingly alienated from the knowledge produced leads to a destructive short-sightedness by the scientific community.

Merton blamed this short-sightedness on the "imperious immediacy of interest," where the primary goal of the pursuit of knowledge crowds out consideration of other effects of one's actions, ultimately interfering with the primary goal itself.

> Precisely because scientific research is not conducted in a social vacuum, its effects ramify into other spheres of value and interest. Insofar as these effects are deemed socially undesirable, science is charged with responsibility. The goods

68. Ibid., p. 277.
69. Ibid., p. 261.

> of science are no longer considered an unqualified blessing. Examined from this perspective, the tenet of pure science and disinterestedness has helped to prepare its own epitaph.[70]

A "cult of unintelligibility" emerges as specialized training leads to a growing gap between "esoteric science" and "popular mysticism."

MASS COMMUNICATION AND TECHNICIANS IN SENTIMENT

Merton's concern to avoid the negative effects of rule by experts and scientism took him beyond the British Marxists' tendency to equate socialism with rule by scientists.[71] He urged reflexivity and attention to scientific (and social scientific) responsibility for the unintended and negative effects of modern science and the rise of irrational movements. In this respect, Merton looks closer to the critical theory of Adorno and the Frankfurt school, Marxist-influenced theorists who had sought to rethink Marxist platitudes about the working class in light of the rise of fascism. This resonance is not surprising since Merton's colleague and collaborator, Karl Lazarsfeld, essentially led the empirical wing of the Frankfurt school in exile before it moved its center of gravity to California.

Lazarsfeld's childhood home in Vienna exposed him to members of the Vienna Circle including Otto Neurath and Rudolf Carnap, Otto Bauer and other prominent Austrian socialists, and psychoanalysts Alfred Adler, Karl Bühler, and Charlotta Bühler. His background in science and mathematics, the politics of the Social Democratic Workers' Party (SDAP), and Freudian psychology shaped his interest in the study of mass communication as his political activism shifted to statistical social psychology research, first on behalf of socialist political organizations and eventually as part of a research institute he founded in Vienna in 1931, the Research Center for Economic Psychology. He set up the institute with an eye to the possibilities for funding from business sources, keeping in mind the model of market research on consumer goods then emerging in the United States. His study of the unemployed of Marienthal, Austria, attracted the attention of the Rockefeller Institute and led to a fellowship in the United States from 1933 to 1935. In 1936, with the political situation worsening in Vienna, he settled in Newark, establishing the Newark University Research Center.[72]

70. Ibid., p. 263.
71. Werskey, *The Visible College*, pp. 91–92.
72. Jeřábek, *Paul Lazarsfeld and the Origins of Communications Research*, pp. 6–13, 60–61, 132–33; Jahoda, Lazarsfeld, and Zeisel, *Marienthal: The Sociology of an Unemployed Community* (originally published in 1933); Neurath, "Sixty Years since Marienthal."

Lazarsfeld had begun to collaborate with the Frankfurt Institute from the United States when it was shut down by the Nazi government in 1933; the institute began an affiliation with Columbia University in New York in 1934. In 1938, Adorno joined Lazarsfeld's Radio Research Project in an often-strained collaboration to bring together theoretical and empirical work on mass culture before Adorno moved on to Los Angeles in 1941. Specializing in the study of the new mass communication technologies employing cutting-edge empirical sociology, Lazarsfeld sought to put Theodor Adorno and Max Horkheimer's critical theories about the pathologies of mass culture to the test.[73]

Mass communication was a significant focus of research for Lazarsfeld not only because of the rise of mass media technologies like radio and television but also because the Frankfurt school saw a growing conformism of the working class that challenged the revolutionary role attributed to it by Marxists. This conformity, on Adorno's view, emerged not from the economic or material base of society but from the cultural superstructure. If critical approaches were still possible to challenge fascism and to bring about a more emancipated society, intellectuals would need to find them and to challenge the processes shaping authoritarian personalities. Lazarsfeld looked to test these theories using new empirical methods, including group interviews of test subjects exposed to radio programs. Such research eventually ended up in Adorno's *The Authoritarian Personality*, which fused psychoanalytic theory with empirical research. Key to the research was a personality test, the "F scale," to indicate a psychological propensity to submit to authority that could be manipulated by fascist propaganda.[74]

Lazarsfeld and Merton met shortly after both were hired by Columbia University, their hiring representing a compromise between the sociology department's empirical and theoretical wings.[75] Lazarsfeld's collaboration with Merton was a happier one than his collaboration with Adorno. Merton's endorsement of "middle range theories" facilitated their collaboration, in contrast with theorists like Adorno and Merton's own mentor, Talcott Parsons, who looked to develop systematic theories of the functioning of societies. Lazarsfeld discovered that the effects of media messages were mediated through informal "opinion leaders" in an individual's social networks. This fit well with Merton's own account of the interaction between normative

73. Fleck, *A Transatlantic History of the Social Sciences*, ch. 5; Jay, *The Dialectical Imagination*, ch. 2; Rogers, *A History of Communication Study*, pp. 108–21; Jeřábek, *Paul Lazarsfeld and the Origins of Communications Research*, pp. 60–66; Wiggershaus, *The Frankfurt School*, pp. 236–46.

74. Fleck, *A Transatlantic History of the Social Sciences*, ch. 6; Jay, *The Dialectical Imagination*, chs. 6–7; Adorno, *The Authoritarian Personality*.

75. Rogers, *A History of Communication Study*, pp. 244–45.

structures and reference groups that provided the basis for status comparison for individuals.[76]

They collaborated on a study of the effectiveness of a radio war bond campaign using focused interviews with 100 people.[77] The study provided the occasion to reflect upon the relationship between the technical task of persuading Americans to purchase war bonds and the ethical consequences of such propaganda. If scientists were urged to reflect on the broader consequences of the knowledge they produced by proponents of the social responsibility of science, Merton also advocated a similar reflexivity about the use of the communication research launched by Lazarsfeld to manipulate identification with national identities.[78]

The Frankfurt school attributed the success of fascist propaganda to its capacity to latch onto preexisting personality structures stabilized as the result of past environmental influence that preempted rational calculation of economic self-interest. In countering that threat, they sought to undermine the individual's susceptibility to manipulation, rather than simply employing manipulative techniques for democratic purposes. Antifascist programs "should not be limited to devices for manipulating people in such a way that they will behave more democratically, but they should be devoted to increasing the kind of self-awareness and self-determination that makes any kind of manipulation impossible."[79] For Merton, a similar kind of manipulation was evident in the war bond drive that appealed to the listener's patriotism in ways that simulated organic community, what he called "*pseudo-Gemeinschaft*."

In weaponizing applied social science research for competing clients representing distinct interest groups, social scientists risked eroding their objectivity and opening themselves to calls for outside regulation. Reflexive studies, or "applied social research on applied social research," were needed to assess the image of the social scientist held by the wider public and to develop proposals for self-regulation of the profession along the lines of what medical or legal professionals had already done. In this context, the use of social science research contracts to "lend support to [a client's] predetermined courses of action" compromised the objectivity of social scientists. To remedy this situation, the social scientists needed to evaluate the "value framework of the policy maker." Without such reflexivity about the societal functions of their research, social scientists were more likely to seek psychological techniques to enhance morale among African American workers at an industrial plant

76. Merton, *Social Theory and Social Structure*, pp. 326–32; Katz and Lazarsfeld, *Personal Influence*.

77. Rogers, *A History of Communication Study*, pp. 245–46.

78. See Brown, *The Social Responsibility of the Scientist*, especially pp. 3–15, on the "Faustian bargain" that comes with state funding of science.

79. Adorno, *The Authoritarian Personality*, pp. 5–6, 10.

than suggest that segregation in the workplace should be eliminated. This narrow focus was facilitated by the technical expediency of measuring psychological attitudes compared to a more wide-ranging examination of social structures and policies. For Merton, this was the difference between a *technician* and a *socially oriented scientist*.[80]

The anxiety about mass culture that Merton and Adorno shared in their different ways had to do with the extent to which mass media had created the means to manipulate society as real, organic community was dismantled and replaced with a manufactured sense of community produced through the production of cultural meaning exploiting emotion rather than reason. Merton and his collaborators' study of the bond drive was the result of an opportunistic "natural experiment" in mass persuasion that had inadvertently revealed the significance of emotional appeals to patriotism. A War Bond Day on CBS radio led by radio star Kate Smith had raised $39 million, far exceeding two previous drives that had only raised $1 million and $2 million, respectively. A content analysis coupled with the new technique of focused interviews of listeners and non-listeners revealed that appeal to the listener's patriotism and sacrifice for the war effort was far more effective than pragmatic appeals to the anti-inflation benefits of war bond purchases.[81]

Key to the effectiveness of the exercise was that Smith's perceived trustworthiness, sincerity, and lack of ulterior motives disarmed the listener's readiness to spot signs of manipulation and self-interest on the part of mass media messages. It was not that listeners accepted whatever radio hosts told them. Kate's self-presentation was deliberately manipulated to maximize pledges over the course of the drives. Merton discussed the tendency of the market to instrumentalize human relationships, leading to a tendency to distrust or suspect the motives of others, making the (instrumental) appeal to non-instrumental values all the more urgent: "The very same society that produces this sense of alienation and estrangement generates in many a craving for reassurance, an acute need to believe, a flight to faith." As a result, the successful salesman will seek to feign or simulate genuine friendliness, sincerity, or other commonly valued qualities. "In place of a sense of *Gemeinschaft*—genuine community of values—there intrudes *pseudo-Gemeinschaft*—the feigning of personal concern with the other fellow in order to manipulate him the better."[82]

It was precisely this loss of organic community with the rise in commodification that underwrote the Frankfurt school's account of the rise of fascism.

80. Merton, *The Sociology of Science*, pp. 77, 80, 83.
81. Morgan, "Nature's Experiments and Natural Experiments in the Social Sciences," pp. 347–49.
82. Merton, *Mass Persuasion*, pp. 143, 142.

On this view, it is the perceived loss of community, allegedly stolen by another group, that fueled fascism. As Jay puts it,

> Liberalism, Horkheimer and Adorno pointed out, had promised happiness without power to both the Jews and the masses. But the masses, denied both happiness and power, turned their fury on the Jews, out of the mistaken belief that what had been withheld from them had been given to the Jews.[83]

The retreat of Adorno and Horkheimer to the cultural alternative of aesthetic modernism represented their failed attempt to find a point of resistance to the rise of mass, consumer culture with its corrosive effects on oppositional political agency. Any sense of either proletarian or organic folk culture had given way to the sway of the ad man and the ideologue. As Jameson developed the point, "the historically unique tendential effect of late capitalism on all such groups has been to dissolve and to fragment or atomize them into agglomerations (*Gesellschaften*) of isolated and equivalent private individuals, by way of the corrosive action of universal commodification and the market system."[84]

In his treatment of the alien, ethnic community portrayed in the *Godfather* movies, Jameson's own account of this double-edged, pseudo-*Gemeinschaft* striving, fueled by anomie and commodification, reflect well this mid-twentieth-century confluence of neo-Marxist critical theory and structural-functionalist sociology:

> In the United States, . . . ethnic groups are not only the objects of prejudice, they are also the object of envy; and these two impulses are deeply intermingled and reinforce each other mutually. The dominant white middle-class groups—already given over to *anomie* and social fragmentation and atomization—find in the ethnic and racial groups which are the object of their social repression and status contempt at one and the same time the image of some older collective ghetto or ethnic neighborhood solidarity; they feel the envy and *ressentiment* of the *Gesellschaft* for the older *Gemeinschaft* which it is simultaneously exploiting and liquidating.[85]

Merton's concerns with the erosion of the ethos of science as the result of the external pressures of authoritarian political structures and anti-intellectual populism reflected a similar concern for the erosion of community for society, *Gemeinschaft* for *Gesellschaft*, in science. Neoliberal science increasingly substitutes bureaucratic management for autonomous core sets, which

---

83. Jay, *The Dialectical Imagination*, p. 231.
84. Jameson, *The Jameson Reader*, p. 129.
85. Ibid., p. 145. Hobsbawm, *Globalisation, Democracy and Terrorism*, pp. 88–94, develops a similar analysis of xenophobia and national football (soccer) teams.

can lead to the same kind of cynicism about manipulation and control that Merton was concerned about.

Merton also used this case as a springboard for ethical reflection on the social responsibility of the social scientist. For Merton, his own study of the social psychology of mass persuasion raised some of the same concerns as he had about the "technicians in sentiment" who had tinkered with Smith's script to maximize contributions, as both were "a means of social control."[86] In thinking about this case, Merton introduced consideration of the now fashionable topic of social values and science. The social scientist

> may, in short, affirm an occupational philosophy which appears to absolve him of any responsibility for the use to which his discoveries in methods of mass persuasion may be put. With its specious and delusory distinction between "ends" and "means" and its insistence that the intrusion of social values into the work of scientists makes for special pleading, this philosophy fails to note that the investigator's social values do influence his choice and definition of problems. The investigator may naively suppose that he is engaged in the value-free activity of research, whereas in fact he may simply have so defined his research problems that the results will be of use to one group in the society, and not to others. His very choice and definition of a problem reflect his tacit values.[87]

Merton then asks what an investigation of propaganda would look like that was oriented toward democratic values like "respect for the dignity of the individual."[88]

Here sociological attention to the remote consequences of purposive social action clarifies the ethical choices facing the researcher. Such a reflective practitioner

> would not only have asked which techniques of persuasion produce the *immediate result* of moving a given proportion of people to action, but also, what are the *further, more remote* but not necessarily less significant, *effects* of these techniques upon the individual personality and the society? He would be, in short, sensitized to certain questions stemming from his democratic values which would otherwise be readily overlooked. For example, he would ask, Does the unelaborated appeal to sentiment which displaces the information pertinent to assessing this sentiment blunt the critical capacity of listeners? What are the effects upon the personality of being subjected to virtual terrorization by advertisements which threaten the individual with social ostracism unless he uses the

86. Merton, "Technical and Moral Dimensions of Policy Research," p. 84.

87. Ibid., p. 86. The role of plural social values in directing science is developed in Longino, *The Fate of Knowledge* and Kitcher, *Science, Truth, and Democracy*, albeit on the model of the invisible hand arising from individual preferences. See Mirowski, *The Effortless Economy of Science?*, pp. 21–24.

88. Merton, "Technical and Moral Dimensions of Policy Research," p. 87.

> advertised defense against halitosis or B.O.? Or, more relevantly, what are the effects, in addition to increasing the sale of bonds, of terrorizing the parents of boys in military service by the threat that only through their purchase of war bonds can they ensure the safety of their sons and their ultimate return home? . . . No single advertising or propaganda campaign may significantly affect the psychological stability of those subjected to it. But a society subjected to a flow of "effective" half-truths and the exploitation of mass anxieties may all the sooner lose that mutuality of confidence and reciprocal trust so essential to a stable social structure. A "morally neutral" investigation of propaganda will be less likely than an inquiry stemming from democratic values to address itself to such questions.[89]

Merton's account of the social responsibility of the scientist and the impact of values on the selection of scientific problems and the knowledge produced provide an initial sociological framework for addressing the kind of questions asked by the early Lakatos in the epigraph to this chapter as well as the later Feyerabend's criticism of the growing entanglement of science and the state.

TOWARD A CRITICAL THEORY OF SCIENCE

Ultimately, for Lakatos and Feyerabend, knowledge has a dialectical character because they rejected the notion of a static characterization of knowledge, seeing knowledge as split between its explanatory role at a moment in time and its function of generating new knowledge in a way that would surpass and exhaust its role as a mirror of nature. Lakatos rejected a correspondence view of knowledge, as our best theories did not just give us an approximate description of the world but were part of a crucial historical dynamic, where scientists produced successive theories within a research tradition with a limited lifespan, that would give way in turn to new research programs. Feyerabend took the point further by emphasizing that research programs were not self-contained and that, potentially, any research program could be all research programs, meaning that there were no essential boundaries between different paradigms or research programs and no fixed way to extend meaning to new circumstances. Incommensurability played the role of breaking down self-contained research programs that led one falsely to believe that the correct and final observation language had been found.

At the same time, while seeing knowledge as rooted in essentially historical traditions, Lakatos and Feyerabend denied that knowledge should be relativized to self-contained communities, that scientists possessed the knowledge in their bones, shaped by a shared culture or practice. One needs

89. Ibid., p. 87.

to rationally reconstruct research traditions because their *telos*, their direction and thrust over time, was not evident to a contemporary participant. Static social determination was no better than static correspondence with nature. Feyerabend's criticism of Wittgenstein's "monistic" perspective was part of a similar attempt to show that the experts have no monopoly on the meaning of their own work, something that Feyerabend explored in every stage of his career, ultimately rejecting relativism for its misleading picture of self-contained and autonomous communities. Feyerabend even entertained the notion that modern science was a peculiar and temporary practice that would die off as new ways of being in the world emerged.[90]

A dialectical approach to knowledge arises most fundamentally out of an attempt to come to terms with the idea that humans are simultaneously objects of scientific explanation and authors of those very same scientific explanations, a theme explored in post-Kantian philosophy.[91] In what sense is science the expression of culture or cognition, and in what sense are we commanded by nature? Is science merely nature speaking without equivocation, or are we free to impose our own fancies upon the world? In the latter case, are we just creatures of society, or is there some meaningful sense of freedom in our collective representations of the world? Key to Hegel's philosophy was the attempt to reconcile these competing views in a picture where the conflicts are moments of a larger movement rather than logical contradictions of the ordinary kind.

As Charles Taylor has argued, Hegel developed his dialectical philosophy out of a reaction to the German Romantic philosophers of the 1790s like Johann Gottfried von Herder, who in turn were reacting to the Enlightenment idea that a scientific understanding of society would involve treating humans on analogy with Newton's physics. For the French philosophes, or in a different way the Scottish enlightenment figures like David Hume or Adam Smith, humans are social atoms subject to natural forces as thoroughly as falling objects. Their interests and desires are individual in kind and are to be reconciled by the application of reason to society, remaking social institutions and traditions to maximize human happiness.[92]

On this view, different ways of organizing society are conventions designed to maximize utility, or they are superstitious traditions standing in the way of such happiness. For Herder, individuals are not a bundle of desires,

90. Feyerabend, *Conquest of Abundance*; Feyerabend, *Tyranny of Science*; Brown, "The Abundant World."

91. Considered from a naturalistic point of view, reconciling lawlike determinism and the experience of will or autonomy involves recognizing that conscious will is an illusion accompanying our actions that facilitates the attribution of responsibility for actions to ourselves and others, albeit one that can be manipulated in kind. See Wegner, *The Illusion of Conscious Will*, pp. 1–25.

92. The following discussion follows Taylor, *Hegel and Modern Society*, ch. 1.

creatures of nature; they are ruled, or ought to be, by an expressive unity, an aesthetic whole that should be allowed to come to its fullest expression. Key to that expressive unity was not only that each individual should have their own autonomy, their own life project or form, but that each individual should also share in the expressive unity of their own national cultures, and ultimately the expression of life in the universe. Humans are not social atoms but part of cultures and ultimately part of an unfolding universe in which autonomy and connection are linked. Individuals are integral forms linked with others as part on an emerging whole.

At the same time, the generation of the 1790s was obsessed with Kant's idea of radical freedom. For Kant, if we could explain our behavior entirely through external natural laws, we would explain away all freedom and morality. Only by rejecting a utilitarian definition of morality as maximizing happiness, with the human subject understood as a bundle of desires, could a conception of morality as the free moral choice of an autonomous subject become possible. As Taylor puts it, "[i]nstead of being dispersed throughout his diverse desires and inclinations the morally free subject must be able to gather himself together, as it were, and make a decision about his total commitment."[93] Morality as a categorical imperative is rationally binding as the result of the exercise of will independent of any individual desires or goals. Humans are not just reflections of natural processes that determine us; we are products of an autonomous moral will that we impose upon ourselves. Like Herder's expressive theory of freedom, Kant struggled with the conflict between our animal nature, or determination by natural law, and our capacity for creating and imposing an order upon the world.

The generation of German philosophers reacting to the triumph and tragedy of the French Revolution sought to combine these two influences, "combining the fullest rational autonomy with the greatest expressive unity."[94] Hegel distinguished himself from his contemporaries by seeking to synthesize these trends through reason rather than intuition, imagination, or artistic expression. To do so, he had to develop a new logic that combined a sense of how oppositions between freedom and nature, the individual and society, or the knower and the world could be reconciled, but only after we had come to terms with their oppositional natures. The logic of contradiction, his account of how identity and opposition, sameness and difference, emerge in time from spiral processes of essential change sought to make sense of how human identity emerged as a process of self-realization.

Where Descartes's dualism encouraged us to think about mind and body as distinct, with pure thought and will standing outside material causation,

93. Ibid., p. 3.
94. Ibid., p. 12.

Herder brought them together in an expressive view of the self. On this view, language is the medium of our thought that also mediates between the individual self and others. Herder rejected a view of language as simply a set of labels for preexisting things; he saw it as a medium for the expression of consciousness, just as art expresses the experience of its creator within a particular medium. Hegel's conception of *Geist*, which Taylor translates as cosmic spirit, does not reflect a "ghostly" spirit behind the scenes guiding history but is instantiated in the first instance by the finite, material reality of the world as it exists. In this sense, before Marx's materialist inversion of Hegel, Hegel already saw that there was no separate spiritual realm or God standing apart from the reality of the universe. For Hegel, the "spirit is a bone."[95]

The difference with a crudely materialistic conception of the world is that Hegel combined an Aristotelian conception of forms with a historical picture of their emergence through time. The essences that make up the world emerge over time; they are "something that could not be known in advance" but emerge through a process of self-realization in engagement with the world.[96] Taylor calls this Hegel's "principle of necessary embodiment":

> Thus where the Aristotelian conception of the relation of matter and form, or hylomorphism, as it has been called, gives a notion of living beings in which soul is inseparable from body, so this theory of expression gives us a view of thinking beings in which thought is inseparable from its medium. And hence it takes just those functions—pure thought, reflection, deliberation—which one would be most tempted to attribute to a disembodied mind, and reclaims them for embodied existence as necessarily couched in an external medium.[97]

At the same time, this embodiment conflicts with the tendency toward change as the subject struggles against these conditions of its own existence.

> It is this inner complexity which makes possible the subject's relation to its self/other. In order to be at all a conscious being, the subject must be embedded in life; but in order to realize the perfection of consciousness it must fight and overcome the natural bent of life as a limit. The conditions of its existence are in conflict with the demands of its perfection; and yet for it to exist is to seek perfection.[98]

Marx's own Hegelian recognition of the self-activity of the working class emerging in the material practice of the proletariat similarly reflected a view

---

95. Hegel, *Phenomenology of Spirit*, pp. 208–11; Žižek, *The Sublime Object of Ideology*, pp. 206–9.
96. Taylor, *Hegel and Modern Society*, p. 16.
97. Ibid., p. 18.
98. Ibid., pp. 21–22.

in which essential forms emerge in time only via the conflict between the subject and historical forms handed down from the past.

The relevance of Hegel for understanding Marx's mature thinking was long neglected as his earliest writings engaging Hegel were only published by the 1920s and 1930s, and Stalin's dialectical materialism obscured earlier explorations of this connection in Lenin and others. Marx objected above all to Hegel's picture of the state as the vehicle for liberating humanity, with his own view holding that the state stood in the way of the proletariat's self-realization.[99] György Lukács famously revived a Hegelian reading of Marx in 1923, a reading which influenced Lakatos's early work, Karl Mannheim, and the Frankfurt school among others. Key to this interpretation was reviving Hegel's "dialectical relation between subject and object in the historical process" without which "dialectics ceases to be revolutionary" and "thought remains contemplative and fails to become practical."[100]

Echoing his fellow Hungarian's dialectical reading of Marx, Lakatos developed a thoroughgoing sociology of scientific knowledge that reconciled the socio-material determination of knowledge and the dialectic between social causes and scientific theories. His second dissertation at Cambridge reflected a dialectical reading of the history of mathematics adapted to Popper's talk of conjectures and refutations.[101] Lakatos and Feyerabend explored their shared interest in a dialectical reading of the history of science in their correspondence and in the way they staged a synthesis of Popper and Kuhn in *Criticism and the Growth of Knowledge*.

Popper had already rejected a picture of science in which a single theory should be adopted in any area of science based upon the empirical confirmation of its predictions. Kuhn's idea of a paradigm blocked the development of alternative theories according to Popper, which was the lifeblood of his emphasis on critical thinking. However, Popper's emphasis on falsification and his picture of knowledge as a third world apart from the world of physical and mental events lacked the dialectical emphasis found in Lakatos and Feyerabend's work.[102] Lakatos's early work on Lenin's "inexhaustible atoms," whereby science's continually changing engagement with the natural world generated ever-changing ontologies, was carried forward into his mature philosophy of science as a picture of the ever-changing representations of natural processes through research programs that would flourish and die like any living thing, though the insights that they had brought about would be preserved in changed form in the totality of science. Feyerabend appealed to

99. Avineri, "The Hegelian Origins of Marx's Political Thought," pp. 33–37.
100. Lukács, *History and Class Consciousness*, p. 3.
101. Lakatos, *Proofs and Refutations*; Kadvany, *Imre Lakatos and the Guises of Reason*; Lynch, "Imre Lakatos and the Inexhaustible Atom."
102. Popper, *Objective Knowledge*, ch. 3.

Trotsky's concept of uneven and combined development in his contribution to *Criticism and the Growth of Knowledge*, which similarly rejected a notion of knowledge as a static correspondence with nature rather than a shifting set of relationships between different scientific theories.[103]

In 1981, Hacking pointed to Lakatos's "thoroughly Hegelian and somewhat Hungarian conception of the events of modern philosophy," an interpretation worked up into book-length treatments by Larvor and Kadvany in 1998 and 2001.[104] Before this time, the hints of Lakatos's continuing Hegelianism were ignored or suppressed, just as with Marx. At the same time, Lakatos confined his dialectical approach to understanding knowledge rather than to the world as a whole, and emphasized the fallible and divided nature of knowledge in real time rather than the emergence of a unitary reason at the end of history. At the same time, Lakatos's treatment of mathematics as inherently dialectical went further than Hegel or Popper in introducing fallibility into the realm of logical necessity, an accomplishment praised by Feyerabend.[105]

For Hacking, "Lakatos, educated in Hungary in an Hegelian and Marxist tradition, took for granted the post-Kantian, Hegelian, demolition of correspondence theories." Lakatos sought "to provide a theory of objectivity without a representational theory of truth." What makes Lakatos's theory "internal" is that he identifies the mechanisms behind the *growth* of knowledge that are internal to a research program rather than identifying a theory's correspondence to an *external* reality.[106] What counts as internal is judged by his account of dynamic research programs that capture most, but not all of, scientific knowledge, as a process unfolding from research programs whose historical emergence and struggle are in turn related to the society that produces—and sometimes thwarts—them. Thus, Stalin's suppression of Mendelian genetics reflected not merely one scientific research program's partisan view of the state of science but an external assault on the scientific community that altered the state of knowledge in a destructive way. Reason was not guaranteed to prevail as it was for Hegel.

The replacement of one research program by another represents an *Aufhebung* on Hegel's terms, an overcoming that preserves itself in altered form, a "dialectical transition in which a lower stage is both annulled and preserved in a higher one."[107] To see this, note that research programs are fruitful of new knowledge for a time before their own demise sets the stage for the rise

103. Lakatos, "Falsification and the Methodology of Scientific Research Programmes"; Feyerabend, "Consolations for the Specialist."

104. Hacking, "Lakatos's Philosophy of Science," p. 129; Larvor, *Lakatos*; Kadvany, *Imre Lakatos and the Guises of Reason.*

105. Musgrave and Pigden, "Imre Lakatos"; Bağçe and Başkent, "An Examination of Counterexamples in *Proofs and Refutations*," p. 19.

106. Hacking, "Lakatos's Philosophy of Science," pp. 118, 130.

107. Taylor, *Hegel and Modern Society*, p. 49.

of new programs that reframe their contributions to knowledge. Einstein's relativity is not logically compatible with Newton's account of motion, but the latter is preserved as a limited perspective within science that still has value for certain purposes and is preserved within the pedagogical structure of science in order to demonstrate how current science now sees its past.[108] Logical falsifications are framed as anomalies for shelving or further work in time; only after the triumph of the new program is an anomaly redefined as a "crucial experiment," "an honorific title" applied "*long after the event.*"[109]

In his philosophy of mathematics, lemma incorporation plays the role of Hegelian contradictions, "the motor force of the method of proofs and refutations" resulting in mathematical concepts whose meanings were changed retrospectively.[110] Monster-barring, by contrast, narrows the content of the mathematical theorem to preserve the logical truth of the theorem in light of a newly discovered exception, thereby maintaining continuity of logical structure at the cost of blocking new discovery. A similar idea explains why Lakatos focused on research programs rather than theories. Only by examining a series of theories produced according to a common template can we tell if progress is being made. Feyerabend accepted that historical trajectories were the best way to evaluate scientific progress, but he denied that lineages respected neat distinctions between separate research programs. Feyerabend's concept of incommensurabilty facilitated criticism between different schools of thought and were capable of leading to hybrid perspectives, in opposition to Kuhn's interpretation. Qualitative change in scientific concepts could be elicited by employing a new observation language as an external framework that would reinterpret established scientific concepts.[111]

Both Lakatos and Feyerabend rejected Kuhn and Polanyi's "elitist authoritarianism," whereby the expertise and interpretations of scientific communities had to be respected by outsiders.[112] The simplest way to understand this is to see that scientists are in the midst of the movement of history and cannot claim to know with authority what the future significance of their own work will be, let alone the future direction of knowledge as a whole. As a retrospective reconstruction of how the scientific present emerged from its pasts, Lakatos's internal history is not a snapshot of scientific representations of nature or beliefs shaped by the existing forms of scientific life. Scientific facts and theories are conventional in that they are shaped historically by societal forces and the integral forms imposed on them by scientific research

108. Zahar, *Einstein's Revolution*, pp. 18–19; Brush, "Should the History of Science Be Rated X?"
109. Lakatos, "History of Science and Its Rational Reconstructions," p. 100.
110. Kadvany, *Imre Lakatos and the Guises of Reason*, p. 48.
111. Feyerabend, "Consoloations for the Specialist."
112. Lakatos and Feyerabend, *For and Against Method*, pp. 27–31, 95–97.

programs. Knowledge is not about shared beliefs, justified or otherwise, but is an alienated product of scientific labor.[113]

When Lakatos argued that a properly reconstructed history should banish "real" history to the footnotes, this reflected his view that scientists' psychological attitudes to their theories may be irrelevant to its place in a dialectical succession of theories driven by a positive heuristic. Lakatos's Methodology of Scientific Research Programs (MSRP) represents not an internalism opposed to the sociological determination of thought but a sociology of knowledge that is dialectical in being driven by the contradictions between existing scientific concepts and the current state of knowledge.

If Kuhn looks like a sleepwalker, mixing reference to conceptual schemes from his mentor, James Conant, Polanyi's concept of tacit knowledge, and fortuitous discovery of Fleck's early work on thought collectives, Lakatos had a long familiarity with Hegel, Marx, Lukács, and the classic sociologists of knowledge that prepared him to meet the challenge of the new history and philosophy of science.[114] A synthesis of critical rationalism and historicism was prefigured already in Lakatos's 1947 Hungarian dissertation, as discussed in chapter 2. The dissertation, now lost, possibly at Lakatos's hand by way of dissembling his past upon leaving Hungary,[115] has been reconstructed from the reports of his dissertation readers and from articles Lakatos published at the time. Sándor Karácsony reported that Lakatos had argued

> that scientific concept-building is not objective, not the alphabet of nature, but it is a relation and a function of the prevailing social structure. Therefore, natural science, during the course of its progress, occasionally outgrows its proper system of concepts, as long as socialist society has not yet reached its static rest state, at which it will force scientific concept-building into a form more independent of time and facts.[116]

One can see the reference to the emergence of static truth with the arrival of socialist society as itself a limit on Lakatos's thought at the time imposed by the Hungarian communist state at the time.

At the same time, a surviving article from the dissertation cites Lukács's *History and Class Consciousness*, and Karácsony notes that the dissertation was "based on dialectic Marxism, but in its modern and not its orthodox form."[117] Lakatos was pushing the bounds of Marxism available to him, as Lukács himself had by then disavowed the book to conform to Stalinist

113. Kutrovátz, "Imre Lakatos' Hungarian Dissertation," pp. 125–29. Compare Feyerabend, "Consolations for a Specialist"; Bartley, "Alienation Alienated."
114. Fuller, *Thomas Kuhn*.
115. Kadvany, *Imre Lakatos and the Guises of Reason*, p. 287.
116. Quoted in Kutrovátz, "Lakatos' Philosophical Work in Hungary," p. 125.
117. Kutrovátz, "Imre Lakatos' Hungarian Dissertation," pp. 369, n.13, 372, 374, n. 21.

orthodoxy. At the same time, Lakatos's participation as an intellectual hit man in the destruction of Eötvös College for its deviations from orthodoxy ironically informed his own dissatisfaction with student radicals in the 1960s pushing politically correct theories. Before he left Hungary, he became involved with reformers and pushed for freedom of inquiry.[118]

Lakatos's concepts of lemma incorporation and progressive problem-shifts were anticipated by his emphasis in 1947 on confronting "contradictions." The other dissertation reader, Ottó Varga, described Lakatos's thesis to be that "the dissolution of contradictions is made possible only when we view any scientific statement as a creation of the society which created that science—that is, truth is always a function of history."[119] Based on the surviving evidence, Lakatos's Hungarian dissertation has to be seen as one of the forerunners of a fully general sociology of scientific knowledge, like that of Ludwik Fleck, which did not exempt the exact sciences from social determination as had Mannheim.[120] Lakatos came to reject the dogmatism of Marxism but continued to accept the dialectical approach that "unmasks the seemingly unchanging and eternal things as historical categories revealing their birth, flourishing, and demise."[121]

Lakatos and Feyerabend's development of a dialectical approach to science offers a useful model for integrating the sociology of scientific knowledge with normative, critical accounts of the development of science. Lakatos and Feyerabend set themselves up as opposites, planning a book "for and against method," or defending reason versus saying farewell to it. At the same time, they both developed criteria for when outside criticism of established perspectives was warranted. Their polemical conflict over reason and method obscures a more concrete, but still significant, difference in their assessment of when minority positions should gain a greater hearing.

Lakatos allowed that degenerating programs could recover and become progressive again so that it would not be irrational to knowingly pursue "risky" strategies with the promise of high rewards. His emphasis was on the existence of multiple perspectives within established science of the past few centuries, and he was admittedly more critical of some established programs such as sociology or environmentalist approaches to intelligence. By contrast, Feyerabend was willing to consider more unorthodox perspectives deserving consideration and support, from the evolving Marxism of Lenin, Trotsky, and

---

118. Long, "Lakatos in Hungary," pp. 260–70; Martin, "'The Battle Is On.'" Long notes that Lakatos had attended lectures by Lukács, though he thought Lukács was "too bourgeois" and "silly," possibly influencing Party criticism of his work (p. 272).

119. Kutrovátz, "Lakatos' Philosophical Work in Hungary," p. 126; Kutrovátz, "Imre Lakatos' Hungarian Dissertation," p. 373.

120. Bloor, *Knowledge and Social Imagery*. But see Kaiser, "A Mannheim for All Seasons."

121. Kutrovátz, "Lakatos' Philosophical Work in Hungary," p. 117.

Luxemburg to parascientific programs like witchcraft, alchemy, or Aristotelian physics that departed from a strictly materialistic worldview associated with modern science.[122]

In *Method and Appraisal in the Physical Sciences*, Lakatos and Feyerabend came closest to their planned debate, *For and Against Method*. In the opening essay, Lakatos outlines his Methodology of Scientific Research Programs and his general approach emphasizing rational reconstructions of the trajectories of theories in the history of science. Historiography starts with "basic judgments" of significant scientific achievements and incorporates as much of the "external" sociology of knowledge internally within a theory of knowledge as possible.[123] This amounts to case-based reasoning now common in science and engineering ethics, as we saw in chapter 6.[124]

Feyerabend, in the concluding essay, denies that reconstructions on this basis can distinguish rational and irrational theory appraisal *in a moment of time* and also observes that one already has to be convinced that science of the last few centuries is epistemologically superior to other traditions. If one accepts this point, as Hume accepts induction as a custom internal to forms of life in which one participates, then the difference between the two views diminishes.[125] For Feyerabend, Lakatos's internalist perspective, properly developed, does not escape from the "sociology of knowledge"; it but merely "provides standards that aid the scientist in evaluating the historical situation in which he makes his decisions; it does not contain *rules* that tell him what to do."[126]

Thus, Feyerabend recognized that if one set aside Lakatos's rhetorical flourishes on behalf of rationality, we are left with a dialectical account of the history of science that brings the sociology of knowledge and accounts of competition and development of research programs together. Kasavin argues that Lakatos's MSRP "comes pretty close to a descriptive sociology of the scientific knowledge," which is precisely Feyerabend's argument in this volume. At the same time, Lakatos rejects the view that his internal/external distinction corresponds to an intellectual/social history distinction.[127]

Shipovalova similarly shows how the turn to material practice in recent science studies indicates how the constructed, material objects of science can

122. Feyerabend, "On the Critique of Scientific Reason," pp. 315, 318–19. On Marxism as a science, see also Lakatos and Feyerabend, *For and Against Method*, pp. 106–7.

123. Lakatos, "History of Science and Its Rational Reconstruction" (1976), pp. 22, 33–34.

124. This is close to Toulmin's approach despite Lakatos's criticisms of Toulmin. See Jonsen and Toulmin, *The Abuse of Casuistry*; Porus, "What Do the Marxist 'Dialectics of Cognition' and Lakatos's 'Sophisticated Falsificationism' Have in Common?," pp. 38–39.

125. Feyerabend, "On the Critique of Scientific Reason," p. 323, n. 44.

126. Ibid., pp. 328, 323.

127. Kasavin, "Uniting the Cognitive and the Social." See Lakatos, "History of Science and Its Rational Reconstruction" (1976), p. 2, n. 1, for his rejection of the view that the internal/external distinction corresponds to an intellectual/social history distinction.

continue to evolve with the advance of laboratory practice, providing a demonstration that "reality is necessarily conceived as indefinite and inexhaustible" in science.[128] Shipovalova argues that Feyerabend's greater emphasis on hybridity and proliferation of theories in science further challenges the dominance of single paradigms. The dominance of paradigms is not only the result of entrenched observation languages but can result when these paradigms make common cause with dominant social ideologies. Feyerabend's analysis here resonates with Lakatos's Hungarian dissertation, with its emphasis on how capitalism's idealist ideologies and the silos of distinct, national traditions in science impede the interaction and progress of scientific programs.[129]

For Feyerabend, Lakatos's approach never makes possible decisive, normative judgments but does transform the history of science into a dialectical account of the emergence of reason in human history. The result is that "history has been transformed to such an extent that a slight change in our standards, say from research programme standards to Hegelian standards, enables us to read it as a history of reason itself."[130] As a historical point, Lakatos and Feyerabend's correspondence on these resonances with Hegelian and Marxist dialectics and their decision to frame their public debate in this way suggest to me a stronger sense of the influence of dialectics than simply rhetorical flourishes.[131] Thus, in 1968, after receiving a letter from Feyerabend endorsing the application of "dialectical materialism" to science, Lakatos sends Feyerabend a copy of the first page of his Hungarian dissertation and follows it up with a reference to a footnote in his *Proofs and Refutations* endorsing dialectics.[132]

Both Lakatos and Feyerabend maintained a commitment to "Popperian" critical thinking, while accepting Kuhn's point that scientists generally neglect (and generally should neglect) the "ocean of anomalies" surrounding scientific theories. Where Kuhn saw this as grounds for deference to established expertise, Lakatos and Feyerabend saw this as implying that one has more freedom to pursue positive projects rather than just fighting off negative objections.[133] This is the expressive theory of reason that Hegel developed, something that Feyerabend pushed to its limits in his final writings.[134]

For Lakatos, his background developing a Marxist sociology of scientific knowledge meant that he had no objections to the "conventional" character of

128. Shipovalova, "Contemporary Science Studies with or without Hidden Marxist Roots?"

129. Feyerabend, "How to Defend Society from Science"; Kutrovátz, "Imre Lakatos' Hungarian Dissertation."

130. Feyerabend, "On the Critique of Scientific Reason," p. 330.

131. Porus, "What Do the Marxist 'Dialectics of Cognition' and Lakatos's 'Sophisticated Falsificationism' Have in Common?"; Musgrave and Pigden, "Imre Lakatos."

132. Lakatos and Feyerabend, *For and Against Method*, pp. 148–51.

133. Feyerabend, "Consolations for the Specialist"; Lakatos, "History of Science and Its Rational Reconstruction" (1976), p. 10.

134. Feyerabend, *The Tyranny of Science*.

knowledge espoused by Popper, but sought only to trace how long a positive heuristic could last before being exhausted while encouraging the creative development of multiple perspectives that would provide a ready pool of tools available for when the occasion demanded. What I would suggest is that the reception of Lakatos in Anglo-American philosophy was distorted by a lack of familiarity with the concepts Lakatos was trading in, and the result was a sharp, ideological split between rationalist or realist philosophy of science and the new sociology of scientific knowledge that waited a generation to be reconsidered.[135] The result is that we are poorly positioned to see how we should respond to linked scientific and societal developments today.

135. Fuller, *Social Epistemology*; Rouse, *Knowledge and Power*; Harding, *Whose Science? Whose Knowledge?*; Longino, *Science as Social Knowledge*.

Bibliography

Abbot, Patrick, et al. "Inclusive Fitness Theory and Eusociality." *Nature* 471, no. 7339 (2011): E1–E4.

Adorno, Theodor W. *The Authoritarian Personality*. New York: Norton, 1982.

Ahmed, Moinuddin, et al. "Continental-Scale Temperature Variability during the Past Two Millennia." *Nature Geoscience* 6 (2013): 339–46.

Akrich, Madeleine, and Bruno Latour. "A Summary of a Convenient Vocabulary for the Semiotics of Human and Nonhuman Assemblies." In *Shaping Technology/ Building Society: Studies in Sociotechnical Change*, edited by Wiebe E. Bijker and John Law, 259–64. Cambridge, MA: MIT Press, 1992.

Alevizaki, Maria, Calliope C. Alevizaki, and Dimitri A. Koutras. "Author's Reply." *Wiener klinische Wochenschrift* 118, no. 11 (2006): 366–67.

Alevizaki, Maria, Emily Mantzou, Adriana T. Cimponeriu, Calliope C. Alevizaki, and Dimitri A. Koutras. "TSH May Not Be a Good Marker for Adequate Thyroid Hormone Replacement Therapy." *Wiener klinische Wochenschrift* 117, no. 18 (2005): 636–40.

Allen, David T., and David R. Shonnard. *Green Engineering: Environmentally Conscious Design of Chemical Processes*. Upper Saddle River, NJ: Prentice Hall, 2002.

Alvard, Michael. "Kinship and Cooperation." *Human Nature: An Interdisciplinary Biosocial Perspective* 20, no. 4 (2009): 394–416.

Alvard, Michael S. "Kinship, Lineage, and an Evolutionary Perspective on Cooperative Hunting Groups in Indonesia." *Human Nature* 14, no. 2 (2003): 129–63.

Anastas, Paul T., and John C. Warner. *Green Chemistry: Theory and Practice*. Oxford: Oxford University Press, 1998.

Anastas, P. T., and J. B. Zimmerman. "Design through the 12 Principles of Green Engineering." *Environmental Science and Engineering* 37, no. 5 (2003): 94A–101A.

Anderson, Benedict. *Imagined Communities: Reflections on the Origin and Spread of Nationalism*. London: Verso, 1983.

Anstey, Erica Hesch. "A Feminist Perspective on the Precautionary Principle and the Problem of Endocrine Disruption under Neoliberal Globalization Policies." PhD dissertation. University of South Florida, 2006.

Applbaum, Kalman. "Getting to Yes: Corporate Power and the Creation of a Psychopharmeceutical Blockbuster." In *The Pharmaceutical Studies Reader*, edited by Sergio Sismondo and Jeremy A. Greene, 133–49. Chichester, England: Wiley Blackwell, 2015.

Apple, Sam. "John Arnold Made a Fortune at Enron. Now He's Declared War on Bad Science." *Wired*, January 22, 2017.

Aronova, Elena. "The Congress for Cultural Freedom, Minerva, and the Quest for Instituting 'Science Studies' in the Age of Cold War." *Minerva* 50, no. 3 (2012): 307–37.

Asimov, Isaac. *The Foundation Trilogy*. New York: Del Rey, 2014.

Audette, Ray V. *Neanderthin: Eat Like a Caveman to Achieve a Lean, Strong, Healthy Body*. New York: St. Martin's, 1999.

Auerbach, Charlotte. *Genetics in the Atomic Age*. New York: Oxford University Press, 1965.

Avineri, Shlomo. "The Hegelian Origins of Marx's Political Thought." *The Review of Metaphysics* 21, no. 1 (1967): 33–56.

Avineri, Shlomo. *The Social and Political Thought of Karl Marx*. London: Cambridge University Press, 1990.

Avital, Eytan, and Eva Jablonka. *Animal Traditions: Behavioural Inheritance in Evolution*. Cambridge: Cambridge University Press, 2000.

Azapagic, Adisa. "Life-Cycle Assessment: A Tool for Identification of More Sustainable Products and Processes." In *Handbook of Green Chemistry and Technology*, edited by James H. Clark and Duncan J. Macquarrie, 62–85. Oxford: Blackwell Science, 2002.

Bachelard, Gaston. *The Formation of the Scientific Mind: A Contribution to a Psychoanalysis of Objective Knowledge*. Manchester, England: Clinamen, 2002.

Bağçe, Samet, and Can Başkent. "An Examination of Counterexamples in *Proofs and Refutations*." *Philosophia scientiae* 13, no. 2 (2009): 3–20.

Baird, Davis. *Thing Knowledge: A Philosophy of Scientific Instruments*. Berkeley: University of California Press, 2004.

Baker, G. P., and P. M. S. Hacker. *Skepticism, Rules and Language*. Oxford: Blackwell, 1985.

Baldwin, Melina. "Scientific Autonomy, Public Accountability, and the Rise of 'Peer Review' in the Cold War United States." *Isis* 109, no. 3 (2018): 538–58.

Ball, Kylie, David Crawford, Anna Timperio, and Jo Salmon. "Eating Behaviors and the Food Environment." In *Obesogenic Environments: Complexities, Perceptions, and Objective Measures*, edited by Amelia Lake, Tim G. Townshend, and Seraphim Alvanides. Chichester, England: Wiley Blackwell, 2010.

Bandy, Alex. *Chocolate and Chess (Unlocking Lakatos)*. Budapest: Akadémiai Kiadó, 2009.

Barash, David. "Genes Are Selfish; People Are Not." *Wall Street Journal*, June 12, 2015.

Bar-Hillel, Yehoshua. "Indexical Expressions." *Mind* 63 (1967): 359–79.

Barker, Gillian, Eric Desjardins, and Trevor Pearce, eds. *Entangled Life: Organism and Environment in the Biological and Social Sciences*. Dordrecht: Springer, 2013.

Barnes, Barry. "Catching Up with Robert Merton: Scientific Collectives as Status Groups." *Journal of Classical Sociology* 7, no. 2 (2007): 179–92.

Barnes, Barry. *Interests and the Growth of Knowledge*. London: Routledge and Kegan Paul, 1977.

Barnes, Barry. "Social Life as Bootstrapped Induction." *Sociology* 17, no. 4 (1983): 524–45.

Barnes, Barry, and David Bloor. "Relativism, Rationalism and the Sociology of Knowledge." In *Rationality and Relativism*, edited by Martin Hollis and Steven Lukes, 21–47. Oxford: Blackwell, 1982.

Barnes, Barry, David Bloor, and John Henry. *Scientific Knowledge: A Sociological Analysis*. London: Athlone, 1996.

Barrett, Deirdre. *Supernormal Stimuli: How Primal Urges Overran Their Evolutionary Purpose*. New York: Norton, 2010.

Barrett, Deirdre. *Waistland: The (R)Evolutionary Science behind Our Weight and Fitness Crisis*. New York: Norton, 2007.

Barrett, Jeff, and P. Kyle Stanford. "Prediction." In *The Philosophy of Science: An Encyclopedia*, II, edited by Sahotra Sarkar and Jessica Pfeifer, 624–38. New York: Routledge, 2006.

Barthes, Roland. *Mythologies*. London: Jonathan Cape, 1973.

Bartley, W. W. "Alienation Alienated: The Economics of Knowledge versus the Psychology and Sociology of Knowledge." In *Evolutionary Epistemology, Rationality and the Sociology of Knowledge*, edited by Gerard Radnitzky and W. W. Bartley, 423–52. Lasalle, IL: Open Court, 1987.

Bartley, W. W. *Unfathomed Knowledge, Unmeasured Wealth: On Universities and the Wealth of Nations*. Lasalle, IL: Open Court, 1990.

Bartley, William Warren, III. *Wittgenstein*. LaSalle, IL: Open Court, 1985.

Basart, Josep M., and Montse Serra. "Engineering Ethics beyond Engineers' Ethics." *Science and Engineering Ethics* 19, no. 1 (2013): 179–87.

Bassett, Ross. "Aligning India in the Cold War Era: Indian Technical Elites, the Indian Institute of Technology at Kanpur, and Computing in India and the United States." *Technology and Culture* 50, no. 4 (2009): 783–810.

Bauman, Z. "On the Philosophical Status of Ethnomethodology." *Sociological Review* 21, no. 1 (1973): 5–23.

Baumer, Eric P. "Untangling Research Puzzles in Merton's Multilevel Anomie Theory." *Theoretical Criminology* 11, no. 1 (2007): 63–93.

Bazerman, Charles. *Shaping Written Knowledge: The Genre and Activity of the Experimental Article in Science*. Madison: University of Wisconsin Press, 1988.

Beardsley, Tim. "Profile: Thereza Imanishi-Kari—Starting with a Clean Slate." *Scientific American* 275, no. 5 (1996): 50–52.

Beddeleem, Martin. "Fighting for the Mantle of Science: The Epistemological Foundations of Neoliberalism, 1931–1951." PhD dissertation. University of Montreal, 2017.

Begley, C. Glenn, and Lee M. Ellis. "Raise Standards for Preclinical Cancer Research." *Nature* 483, no. 7391 (2012): 531–33.

Ben-David, Joseph. *The Scientist's Role in Society: A Comparative Study*. Chicago: University of Chicago Press, 1984.

Berreby, David. "And Now, Overcoming All Binary Oppositions, It's . . . That Damned Elusive Bruno Latour." *Lingua Franca* 4, no. 6 (1994).

Bhaskar, Roy. *A Realist Theory of Science*. Hassocks: Harvester Press, 1978.

Bhaskar, Roy. *Reclaiming Reality: A Critical Introduction to Contemporary Philosophy*. London: Verso, 1989.

Biddle, Justin B. "Can Patents Prohibit Research? On the Social Epistemology of Patenting and Licensing in Science." *Studies in History and Philosophy of Science* 45 (2014): 14–23.

Biondi, Bernadette, and David S. Cooper. "Thyroid Hormone Therapy for Hypothyroidism." *Endocrine* 66, no. 1 (2019): 18–26.

Birch, Jonathan, and Okasha Samir. "Kin Selection and Its Critics." *Bioscience* 65, no. 1 (2015): 22–33.

Blackburn, Henry, David Jacobs, Daan Kromhout, and Alessandro Menotti. "Review of *Big Fat Surprise* Should Have Questioned Author's Claims." *Lancet* 392, no. 10152 (2018): 1014.

Blackburn, Robin. "Marxism: Theory of Proletarian Revolution." *New Left Review*, no. 97 (1976): 3–35.

Blackett, P. M. S. "The Frustration of Science." In *The Frustration of Science*, by Sir Daniel Hall et al., 129–41. New York: Norton, 1935.

Blain, J. Craig, and Jack W. Szostak. "Progress toward Synthetic Cells." *Annual Review of Biochemistry* 83 (2014): 11.1–11.26.

Bloor, David. "Institutions and Rule-Scepticism: A Reply to Martin Kusch." *Social Studies of Science* 34, no. 4 (2004): 593–601.

Bloor, David. *Knowledge and Social Imagery*. London: Routledge and Kegan Paul, 1976.

Bloor, David. "Left and Right Wittgensteinians." In *Science as Practice and Culture*, edited by Andrew Pickering, 266–82. Chicago: University of Chicago Press, 1992.

Bloor, David. "The Living Foundations of Mathematics." *Social Studies of Science* 17 (1987): 337–58.

Bloor, David. *Wittgenstein: A Social Theory of Knowledge*. New York: Columbia University Press, 1983.

Bloor, David. *Wittgenstein, Rules and Institutions*. London: Routledge, 2002.

Boas, Franz. "Museums of Ethnology and Their Classification." *Science* 9, no. 228 (1887): 587–89.

Boehm, Christopher. "Emergency Decisions, Cultural-Selection Mechanics, and Group Selection." *Current Anthropology* 37, no. 5 (1996): 763–93.

Boehm, Christopher. *Hierarchy in the Forest: The Evolution of Egalitarian Behavior*. Cambridge, MA: Harvard University Press, 1999.

Boivin, N. L., et al. "Ecological Consequences of Human Niche Construction: Examining Long-Term Anthropogenic Shaping of Global Species Distributions." *Proceedings of the National Academy of Sciences* 113, no. 23 (2016): 6388–96.

Bologh, Roslyn Wallach. "The Promise and Failure of Ethnomethodology from a Feminist Perspective: Comment on Rogers." *Gender and Society* 6, no. 2 (1992): 199–206.

Booth, William. "A Clash of Cultures at Meeting on Misconduct." *Science* 243, no. 4891 (1989): 598.

Borofsky, Robert. *Yanomami: The Fierce Controversy and What We Can Learn from It*. Berkeley: University of California Press, 2005.

Bowthorpe, Janie E. *Stop the Thyroid Madness: A Patient Revolution against Decades of Inferior Thyroid Treatment*. 2nd ed. Fredericksburg, TX: Laughing Grape Publishing, 2011/2012.

Bowthorpe, Janie E., ed. *Stop the Thyroid Madness II: How Thyroid Experts Are Challenging Ineffective Treatments and Improving the Lives of Patients*. Dolores, CO: Laughing Grape Publishing, 2014.

Bourke, Andrew F. G. *Principles of Social Evolution*. Oxford: Oxford University Press, 2011.

Bowles, Samuel. "Did Warfare among Ancestral Hunter-Gatherers Affect the Evolution of Human Social Behaviors?" *Science* 324, no. 5932 (2009): 1293–98.

Boyd, Richard. "Constructivism, Realism, and Philosophical Method." In *Inference, Explanation, and Other Frustrations: Essays in the Philosophy of Science*, edited by John Earman, 131–98. Berkeley: University of California Press, 1992.

Boyd, Richard. "The Current Status of Scientific Realism." In *Scientific Realism*, edited by Jarrett Leplin. Berkeley: University of California Press, 1984.

Boyd, Robert, and Peter J. Richerson. *Culture and the Evolutionary Process*. Chicago: University of Chicago Press, 1985.

Brissette, Christy. "Is Adrenal Fatigue Real? Forget the Label and Treat the Stress." *Washington Post*, April 5, 2017.

Broad, William J. "Esoteric Wedge of Academia Is Roiled by Hunt for Bomber." *New York Times*, August 5, 1995.

Broad, William, and Nicholas Wade. *Betrayers of the Truth: Fraud and Deceit in the Halls of Science*. New York: Simon & Schuster, 1982.

Brown, James Robert. "The Community of Science®." In *The Challenge of the Social and the Pressure of Practice: Science and Values Revisited*, edited by Martin Carrier, Don Howard, and Janet Kourany, 189–216. Pittsburgh, PA: University of Pittsburgh Press, 2008.

Brown, James Robert. "The Good Old Days: Medical Research Then and Now." In *Science Transformed? Debating Claims of an Epochal Break*, edited by Gregor Schiemann, Hans Radder, and Alfred Nordmann, 177–88. Pittsburgh, PA: University of Pittsburgh Press, 2011.

Brown, John K. "A Different Counterfactual Perspective on the Eads Bridge." *Technology's Stories*, August 2014. http://www.technologystories.org.

Brown, John K. "Not the Eads Bridge: An Exploration of Counterfactual History of Technology." *Technology and Culture* 55, no. 3 (2014): 521–59.

Brown, John K., Gary Lee Downey, and Maria Paula Diogo. "The Normativities of Engineers: Engineering Education and History of Technology." *Technology and Culture* 50, no. 4 (2009): 737–52.

Brown, Martin, ed. *The Social Responsibility of the Scientist*. London: Collier-Macmillan, 1971.

Brown, Matthew J. "The Abundant World: Paul Feyerabend's Metaphysics of Science." *Studies in History and Philosophy of Science Part A* 57 (2016): 142–54.

Brown, Matthew J. "The Functional Complexity of Scientific Evidence." *Metaphilosophy* 46, no. 1 (2015): 65–83.

Brown, Phil, Sabrina McCormick, Brian Mayer, Stephen Zavestoski, Rachel Morello-Frosch, Rebecca Gasior Altman, and Laura Senier. "'A Lab of Our Own': Environmental Causation of Breast Cancer and Challenges to the Dominant Epidemiological Paradigm." *Science, Technology, & Human Values* 31, no. 5 (2006): 499–536.

Brown, Phil, Rachel Morello-Frosch, and Stephen Zavestoski. *Contested Illnesses: Citizens, Science, and Health Social Movements*. Berkeley: University of California Press, 2011.

Browning, Gary K. "Marx's Doctoral Dissertation: The Development of a Hegelian Thesis." In *The Hegel-Marx Connection*, edited by Tony Burns and Ian Fraser, 131–45. London: Palgrave Macmillan, 2000.

Brucker-Davis, F. "Effects of Environmental Synthetic Chemicals on Thyroid Function." *Thyroid* 8, no. 9 (1998): 827–56.

Brumfiel, Geoff. "Academy Affirms Hockey-Stick Graph." *Nature* 441 (2006): 1032–33.

Brush, Stephen G. "Should the History of Science Be Rated X?" *Science* 183, no. 4130 (1974): 1164–72.

Bschir, Karim. "Feyerabend and Popper on Theory Proliferation and Anomaly Import: On the Compatibility of Theoretical Pluralism and Critical Rationalism." *HOPOS: The Journal of the International Society for the History of Philosophy of Science* 5, no. 1 (2015): 24–55.

Budiansky, Stephen. *Blackett's War: The Men Who Defeated the Nazi U-Boats and Brought Science to the Art of Warfare*. New York: Knopf, 2013.

Bunge, Mario. "The Inexhaustible Electron." *Science and Society* 14, no. 2 (1950): 115–21.

Buranyi, Stephen. "Is the Staggeringly Profitable Business of Scientific Publishing Bad for Science?" *Guardian*, June 27, 2017.

Burghardt, Gordon M., and Kevin Laland. "Deconstructing Niche Construction: A Conversation between Gordon Burghardt and Kevin Laland." *This View of Life*, June 19, 2017.

Bush, Vannevar. *Science: The Endless Frontier*. Washington, DC: United States Government Printing Office, 1945.

Cadegiani, Flavio A., and Claudio E. Kater. "Adrenal Fatigue Does Not Exist: A Systematic Review." *BMC Endocrine Disorders* 16, no. 48 (2016): 1–16.

Callon, Michel. "Some Elements of a Sociology of Translation: Domestication of the Scallops and the Fishermen of St Brieuc Bay." *Sociological Review* 32, no. 1 (1984): 196–233.

Callon, Michel, and Bruno Latour. "Unscrewing the Big Leviathan: How Actors Macro-Structure Reality and How Sociologists Help Them to Do So." In *Advances in Social Theory and Methodology: Toward an Integration of Micro- and Macro-*

Sociologies, edited by K. Knorr-Centina and A. V. Cicourel, 276–303. Boston: Routledge and Kegan Paul, 1981.

Callon, Michel, and John Law. "On Interests and Their Transformation: Enrolment and Counter-Enrolment." *Social Studies of Science* 12 (1982): 615–26.

Callon, Michel, John Law, and Arie Rip, eds. *Mapping the Dynamics of Science and Technology: Sociology of Science in the Real World*. London: Macmillan, 1986.

Calvert-Minor, Chris. "Blurry Humanism: A Reply to Michael Lynch." *Human Studies* 37, no. 1 (2014): 147–52.

Campbell, Donald T. "The Two Distinct Routes beyond Kin Selection to Ultrasociality: Implications for the Humanities and Social Sciences." In *The Nature of Prosocial Development: Interdisciplinary Theories and Strategies*, edited by Diane L. Bridgeman, 11–41. New York: Academic Press, 1983.

Campbell, Donald T. "Variation and Selective Retention in Sociocultural Systems." In *Social Change in Developing Areas: A Reinterpretation of Evolutionary Theory*, edited by Herbert R. Barringer, George I. Blanksten, and Raymond W. Mack, 19–49. Cambridge, MA: Schenkman, 1965.

Carafoli, Ernesto. "Scientific Misconduct: The Dark Side of Science." *Rendiconti Lincei* 26, no. 3 (2015): 369–82.

Carlson, Elof Axel. *Genes, Radiation, and Society: The Life and Work of H. J. Muller*. Ithaca, NY: Cornell University Press, 1981.

Carnap, Rudolf. *The Logical Structure of the World: Pseudoproblems in Philosophy*. Berkeley: University of California Press, 1967.

Carnap, Rudolf. *The Logical Syntax of Language*. London: Routledge & Kegan, 1959.

Carrier, Martin. "Science in the Grip of the Economy: On the Epistemic Impact of the Commercialization of Research." In *The Challenge of the Social and the Pressure of Practice: Science and Values Revisited*, edited by Martin Carrier, Don Howard, and Janet Kourany, 217–34. Pittsburgh, PA: University of Pittsburgh Press, 2008.

Carson, Rachel. *Silent Spring*. New York: Fawcett Crest, 1962.

Cartwright, Nancy, Jordi Cat, Lola Fleck, and Thomas E. Uebel. *Otto Neurath: Philosophy between Science and Politics*. Cambridge: Cambridge University Press, 1996.

Castleman, Barry I., and Grace E. Ziem. "Toxic Pollutants, Science, and Corporate Influence." *Archives of Environmental Health* 44, no. 2 (1989): 68–127.

Cat, Jordi, "Political Economy." Supplement to "Otto Neurath." *The Stanford Encyclopedia of Philosophy*, edited by Edward N. Zalta. Fall 2019, https://plato.stanford.edu/archives/fall2019/entries/neurath.

Cavalli-Sforza, L. L., and M. W. Feldman. *Cultural Transmission and Evolution: A Quantitative Approach*. Princeton, NJ: Princeton University Press, 1981.

Cavers, David F. "The Food, Drug, and Cosmetic Act of 1938: Its Legislative History and Its Substantive Provisions." *Law and Contemporary Problems* 6, no. 1 (1939): 2–42.

Chagnon, N., and P. Bugos. "Kin Selection and Conflict: An Analysis of a Yanomamö Ax Fight." In *Evolutionary Biology and Human Social Behavior: An Anthropological Perspective*, edited by N. Chagnon and W. Irons, 213–38. North Scituate: Duxbury.

Chagnon, Napoleon A. "Kin-Selection Theory, Kinship, Marriage and Fitness among the Yanomamö Indians." In *Sociobiology: Beyond Nature/Nurture?*, edited by George W. Barlow and James Silverberg, 545–87. Boulder, CO: Westview Press, 1980.

Chase, Alston. *Harvard and the Unabomber: The Education of an American Terrorist*. New York: Norton, 2003.

Chen, Irene A., and Mattanjah S. de Vries. "From Underwear to Non-Equilibrium Thermodynamics: Physical Chemistry Informs the Origin of Life." *Physical Chemistry Chemical Physics* 18, no. 30 (2016): 20005–6.

Chomsky, Noam. *Necessary Illusions: Thought Control in Democratic Societies*. Boston: South End Press, 1989.

Chubin, Daryl E. "Collins's Programme and the Hardest Possible Case." *Social Studies of Science* 12 (1982):136–39.

Chubin, Daryl E., and Sal Restivo. "The 'Mooting' of Science Studies: Research Programmes and Science Policy." In *Science Observed: Perspectives on the Social Study of Science*, edited by Karin Knorr-Cetina and Michael Mulkay, 55–83. London: Sage, 1983.

Clarke, Adele E. "A Social Worlds Research Adventure: The Case of Reproductive Science." In *Theories of Science in Society*, edited by S. E. Cozzens and T. F. Gieryn, 15–42. Bloomington: Indiana University Press, 1990.

Cleaver, Harry. *Reading Capital Politically*. Austin: University of Texas Press, 1979.

Close, Frank. *Too Hot to Handle: The Race for Cold Fusion*. Princeton, NJ: Princeton University Press, 2014.

Cokol, Murat, et al. "How Many Scientific Papers Should Be Retracted?" *EMBO Reports* 8, no. 5 (2007): 422–23.

Colborn, Theo, Dianne Dumanoski, and John Peterson Myers. *Our Stolen Future: Are We Threatening Our Fertility, Intelligence, and Survival? A Scientific Detective Story with a New Epilogue by the Authors*. New York: Penguin Group, 1997.

Collier, Andrew. *Scientific Realism and Socialist Thought*. Hertfordshire, England: Harvester Wheatsheaf, 1989.

Collin, Finn. *Science Studies as Naturalized Philosophy*. Dordrecht: Springer, 2010.

Collins, Harry. *Are We All Scientific Experts Now?* Cambridge: Polity, 2014.

Collins, Harry. *Artificial Experts: Social Knowledge and Intelligent Machines*. Cambridge, MA: MIT Press, 1990.

Collins, H. M. *Changing Order: Replication and Induction in Scientific Practice*. London: Sage, 1985.

Collins, H. M. "An Empirical Relativist Programme in the Sociology of Scientific Knowledge." In *Science Observed: Perspectives on the Social Study of Science*, edited by Karin Knorr-Cetina and Michael Mulkay, 85–113. London: Sage, 1983.

Collins, Harry. *Gravity's Ghost: Scientific Discovery in the Twenty-First Century*. Chicago: University of Chicago Press, 2010.

Collins, H. M. "Public Experiments and Displays of Virtuosity: The Core-Set Revisited." *Social Studies of Science* 18, no. 4 (1988): 725–48.

Collins, H. M. "Son of Seven Sexes: The Social Destruction of a Physical Phenomenon." *Social Studies of Science* 11, no. 1 (1981): 33–62.

Collins, H. M. "Stages in the Empirical Programme of Relativism." *Social Studies of Science* 11, no. 1 (1981): 3–10.

Collins, Harry. "Three Dimensions of Expertise." *Phenomenology and the Cognitive Sciences* 12, no. 2 (2013): 253–73.

Collins, H. M., and Robert Evans. *Rethinking Expertise*. Chicago: University of Chicago Press, 2007.

Collins, Harry, and Martin Kusch. *The Shape of Actions: What Humans and Machines Can Do*. Cambridge, MA: MIT Press, 1998.

Collins, H. M., and T. J. Pinch. "The Construction of the Paranormal: Nothing Unscientific Is Happening." *Sociological Review* 27 (1979): 237–70.

Collins, Harry, and Trevor Pinch. *The Golem at Large: What You Should Know about Technology*. Cambridge: Cambridge University Press, 1998.

Collins, Harry, and Trevor Pinch. *The Golem: What You Should Know about Science*. 2nd ed. Cambridge: Cambridge University Press, 1998.

Collins, H. M., and Steven Yearly. "Epistemological Chicken." In *Science as Practice and Culture*, edited by Andrew Pickering, 301–26. Chicago: University of Chicago Press, 1992.

Collins, Randall. *Conflict Sociology: Toward an Explanatory Science*. New York: Academic Press, 1975.

Collins, Randall. *The Credential Society: An Historical Sociology of Education and Stratification*. New York: Academic Press, 1979.

Collins, Randall. *The Sociology of Philosophies: A Global Theory of Intellectual Change*. Cambridge: Belknap Press of Harvard University Press, 1998.

Collison, Patrick, and Michael Nielsen. "Science Is Getting Less Bang for Its Buck." *Atlantic*, November 16, 2018.

Conlon, Eddie, and Henk Zandvoort. "Broadening Ethics Teaching in Engineering: Beyond the Individualistic Approach." *Science and Engineering Ethics* 17, no. 2 (2011): 217–32.

Connolly, William E. *Identity\Difference: Democratic Negotiations of Political Paradox*. Ithaca, NY: Cornell University Press, 1991.

Cordain, Loren, S. Boyd Eaton, Anthony Sebastian, and Neil Mann. "Origins and Evolution of the Western Diet: Health Implications for the 21st Century." *American Journal of Clinical Nutrition* 81, no. 2 (2005): 341–54.

Cordain, L., J. B. Miller, S. B. Eaton, and N. Mann. "Plant-Animal Subsistence Ratios and Macronutrient Energy Estimations in Worldwide Hunter-Gatherer Diets." *American Journal of Clinical Nutrition* 71, no. 3 (2000): 682–92.

Costa, James T. "Hamiltonian Inclusive Fitness: A Fitter Fitness Concept." *Biology Letters* 9, no. 6 (2013): 20130335.

Coyne, Jerry. "Big Dust-Up about Kin Selection." March 24, 2011. https://whyevolutionistrue.wordpress.com/2011/03/24/big-dust-up-about-kin-selection/.

Cremo, Michael A. "Forbidden Archaeology." In *You Are Still Being Lied To: The New Disinformation Guide to Media Distortion, Historical Whitewashes and Cultural Myths*, edited by Russ Kick, 390–95. New York: Disinformation Books, 1999. http://www.mcremo.com/YASBLT_forbiddenarchaeology.pdf.

Cremo, Michael, and Richard L. Thompson. *The Hidden History of the Human Race*. Los Angeles: Bhaktivedanta Book Publishing, 1996.

da Cunha, Ivan Ferreira. "The Utopia of Unified Science: The Political Struggle of Otto Neurath and the Vienna Circle." *Principia: An International Journal of Epistemology* 17, no. 2 (2013): 319–29.

Daemmrich, Arthur, and Jeremy Greene. "From Visible Harm to Relative Risk: Centralization and Fragmentation of Pharmacovigilance." In *The Fragmentation of U.S. Healthcare: Causes and Solutions*, edited by Einer Elhauge, 301–22. Oxford: Oxford University Press, 2010.

D'Agostino, Fred. "Verballed? Incommensurability 50 Years On." *Synthese* 191, no. 3 (2014): 517–38.

D'Amico, Robert D. *Historicism and Knowledge*. New York: Routledge, 1989.

Darwin, Charles. *The Life and Letters of Charles Darwin: Including an Autobiographical Chapter*, edited by Francis Darwin, 3 vols. London: John Murray, 1887.

Daston, Lorraine, and Peter Galison. *Objectivity*. New York: Zone Books, 2007.

Davies, Kate. *The Rise of the U.S. Environmental Health Movement*. Lanham, MD: Rowman & Littlefield, 2013.

Davis, Michael. "Engineering Ethics, Individuals, and Organizations." *Science and Engineering Ethics* 12, no. 2 (2006): 223–31.

Davis, Michael. *Thinking Like an Engineer: Studies in the Ethics of a Profession*. New York: Oxford University Press, 1998.

Dawkins, Richard. "The Descent of Edward Wilson." *Prospect*, June 2012. http://www.prospectmagazine.co.uk/magazine/edward-wilson-social-conquest-earth-evolutionary-errors-origin-species#.UnQHvxAXWAo.

Dawkins, Richard. *The God Delusion*. Boston: Houghton Mifflin, 2006.

Dawkins, Richard. "Obscurantism to the Rescue." *Quarterly Review of Biology* 72, no. 4 (1997): 397–99.

Dawkins, Richard. *The Selfish Gene*. Oxford: Oxford University Press, 1976.

de Punder, Karin, and Leo Pruimboom. "Stress Induces Endotoxemia and Low-Grade Inflammation by Increasing Barrier Permeability." *Frontiers in Immunology* 6, no. 223 (2015).

Deakin, Michael A. B. "On Urbach's Analysis of the 'IQ Debate.'" *British Journal for the Philosophy of Science* 37, no. 1 (1986): 60–65.

Dear, Peter. *Discipline and Experience: The Mathematical Way in the Scientific Revolution*. Chicago: University of Chicago Press, 1995.

Dear, Peter. "Narratives, Anecdotes, and Experiments: Turning Experience into Science in the Seventeenth Century." In *The Literary Structure of Scientific Argument: Historical Studies*, edited by Peter Dear, 135–63. Philadelphia: University of Pennsylvania Press, 2015.

Deflem, Mathieu. "Anomie, Strain, and Opportunity Structure: Robert K. Merton's Paradigm of Deviant Behavior." In *The Handbook of the History and Philosophy of Criminology*, edited by Ruth A. Triplett, 140–55. Malden, MA: Wiley Blackwell, 2018.

Dennis, Michael Aaron. "Historiography of Science: An American Perspective." In *Companion Encyclopedia of Science in the Twentieth Century*, edited by John Krige and Dominique Pestre, 1–26. London: Routledge, 2003.

Descombes, Vincent. *The Institutions of Meaning: A Defense of Anthropological Holism*. Cambridge, MA: Harvard University Press, 2014.

Doing, Park A. "Applying Ethnographic Insight to Engineering Ethics: Epistemography and Accountability in the Space Shuttle *Challenger* Failure and the Macondo Well Blowout." *Engineering Studies* 4, no. 3 (2012): 233–48.

Doing, Park. "Give Me a Laboratory and I Will Raise a Discipline: The Past, Present, and Future Politics of Laboratory Studies." In *The Handbook of Science and Technology Studies*, edited by Edward J. Hackett, Olga Amsterdamska, Michael E. Lynch, and Judy Wajcman, 279–95. Cambridge, MA: MIT Press, 2007.

Doing, Park. *Velvet Revolution at the Synchrotron: Biology, Physics, and Change in Science*. Cambridge, MA: MIT Press, 2009.

Doorn, Neelke. "Responsibility Ascriptions in Technology Development and Engineering: Three Perspectives." *Science and Engineering Ethics* 18, no. 1 (2012): 69–90.

Dörries, Matthias. "Politics, Geological Past, and the Future of the Earth." *Historical Social Research/Historische Sozialforschung* 40, no. 2 (2015): 22–36.

Downey, Gary Lee. "Low Cost, Mass Use: American Engineers and the Metrics of Progress." *History and Technology* 22, no. 3 (2007): 289–308.

Downey, Gary Lee, and Juan C. Lucena. "Knowledge and Professional Identity in Engineering." *History and Technology* 20, no. 4 (2004): 393–420.

Downey, Gary Lee, Juan C. Lucena, and Carl Mitcham. "Engineering Ethics and Identity: Emerging Initiatives in Comparative Perspective." *Science and Engineering Ethics* 13 (2007): 463–87.

Dreger, Alice. *Galileo's Middle Finger: Heretics, Activists, and the Search for Justice in Science*. New York: Penguin Press, 2015.

Duhem, Pierre. *The Aim and Structure of Physical Theory*. Princeton, NJ: Princeton University Press, 1954.

Dunbar, Robin. *How Many Friends Does One Person Need? Dunbar's Number and Other Evolutionary Quirks*. Cambridge, MA: Harvard University Press, 2010.

Durham, William H. *Coevolution: Genes, Culture, and Human Diversity*. Stanford: Stanford University Press, 1991.

Dusek, Val. "Lakatos between Marxism and the Hungarian Heuristic Tradition." *Studies in East European Thought* 67, no. 1–2 (2015): 61–73.

Eaton, S. B., and Melvin Konner. "Paleolithic Nutrition: A Consideration of Its Nature and Current Implications." *New England Journal of Medicine* 312, no. 5 (1985): 283–89.

Edwards, Marc A., and Siddhartha Roy. "Academic Research in the 21st Century: Maintaining Scientific Integrity in a Climate of Perverse Incentives and Hypercompetition." *Environmental Engineering Science* 34, no. 1 (2017): 51–61.

Edwards, Paul N. *The Closed World: Computers and the Politics of Discourse in Cold War America*. Cambridge, MA: MIT Press, 1996.

Edwards, Paul N. *A Vast Machine: Computer Models, Climate Data, and the Politics of Global Warming*. Cambridge, MA: MIT Press, 2010.
Elster, Jon. *Logic and Society: Contradictions and Possible Worlds*. New York: Wiley, 1979.
Emmet, Dorothy. *Rules, Roles, and Relations*. London: Macmillan, 1966.
Engels, Friedrich. *The Condition of the Working Class in England*. Oxford: Oxford University Press, 1993.
Engels, Frederick. *Ludwig Feuerbach and the Outcome of Classical German Philosophy*. New York: International Publishers, 1941.
Epstein, Steven. "The Construction of Lay Expertise: AIDS Activism and the Forging of Credibility in the Reform of Clinical Trials." *Science, Technology, & Human Values* 20 (1995): 408–37.
Epstein, Steven. *Impure Science: AIDS, Activism, and the Politics of Pure Knowledge*. Berkeley: University of California Press, 1996.
Etter, Martin A. "Sahlins and Sociobiology." *American Ethnologist* 5, no. 1 (1978): 160–69.
Evans, James A. "Industry Collaboration, Scientific Sharing, and the Dissemination of Knowledge." *Social Studies of Science* 40, no. 5 (2010): 757–91.
Fang, Ferric C., and Arturo Casadevall. "Retracted Science and the Retraction Index." *Infection and Immunity* 79, no. 10 (2011): 3855–59.
Feinman, Richard D., and Jeff S. Volek. "Low Carbohydrate Diets Improve Atherogenic Dyslipidemia Even in the Absence of Weight Loss." *Nutrition & Metabolism* 3, no. 1 (2006): 24.
Felt, Ulrike, Rayvon Fouché, Clark A. Miller, and Laurel Smith-Doerr, eds. *The Handbook of Science and Technology Studies*. 4th ed. Cambridge, MA: MIT Press, 2017.
Feuerbach, Ludwig. *The Essence of Christianity*. New York: Harper and Brothers, 1957.
Feyerabend, Paul. *Against Method: Outline of an Anarchistic Theory of Knowledge*. London: Verso, 1978.
Feyerabend, Paul. *Against Method: Outline of an Anarchistic Theory of Knowledge*. 3rd ed. London: Verso, 1993.
Feyerabend, Paul K. "Art as a Product of Nature as a Work of Art." *World Futures* 40, no. 1–3 (1994): 87–100.
Feyerabend, P. K. "An Attempt at a Realistic Interpretation of Experience." *Proceedings of the Aristotelian Society* 58 (1957–1958): 143–70.
Feyerabend, Paul. *Conquest of Abundance: A Tale of Abstraction versus the Richness of Being*. Chicago: University of Chicago Press, 1999.
Feyerabend, Paul K. "Consolations for the Specialist." In *Criticism and the Growth of Knowledge*, edited by Alan Musgrave and Imre Lakatos, 197–230. Cambridge: Cambridge University Press, 1970.
Feyerabend, Paul K. "Dialectical Materialism and the Quantum Theory." *Slavic Review* 25, no. 3 (1966): 414–17.
Feyerabend, P. K. "Explanation, Reduction, and Empiricism." *Minnesota Studies in the Philosophy of Science* 3 (1962): 23–97.

Feyerabend, Paul K. "How to Defend Society from Science." In *Scientific Revolutions*, edited by Ian Hacking, 156–67. Oxford: Oxford University Press, 1981.

Feyerabend, Paul. *Killing Time: The Autobiography of Paul Feyerabend*. Chicago: University of Chicago Press, 1995.

Feyerabend, Paul. "More Clothes from the Emperor's Bargain Basement: A Review of Laudan's *Progress and Its Problems*." In *Philosophical Papers*, vol. 2, *Problems of Empiricism*, 231–46. Cambridge: Cambridge University Press, 1981.

Feyerabend, Paul. "On the Critique of Scientific Reason." In *Method and Appraisal in the Physical Sciences*, edited by Colin Howson, 309–39. 2 vols. Cambridge: Cambridge University Press, 1976.

Feyerabend, Paul. *Philosophical Papers*. 2 vols. Cambridge: Cambridge University Press, 1981.

Feyerabend, Paul. *Philosophy of Nature*, translated by Dorothea Lotter, edited by Helmut Heit and Eric Oberheim. Cambridge: Polity, 2016.

Feyerabend, Paul. "Problems of Empiricism." In *Beyond the Edge of Certainty: Essays in Contemporary Science and Philosophy*, edited by Robert G. Colodny, 145–260. Pittsburgh, PA: University of Pittsburgh Press, 1965.

Feyerabend, Paul. *Science in a Free Society*. London: NLB, 1978.

Feyerabend, Paul. *The Tyranny of Science*, edited by Eric Oberheim. London: Polity Press, 2011.

Feyerabend, Paul. "Wittgenstein's Philosophical Investigations." *Philosophical Review* 64, no. 3 (1955): 449–83.

Fine, Arthur. "Science Made Up: Constructivist Sociology of Scientific Knowledge." In *The Disunity of Science: Constructivist Sociology of Scientific Knowledge*, edited by Peter Galison and David Stump, 231–54. Stanford, CA: Stanford University Press, 1996.

Fine, Arthur. *The Shaky Game: Einstein, Realism and the Quantum Theory*. Chicago: University of Chicago Press, 1996.

Fine, Arthur. "Unnatural Attitudes: Realist and Instrumentalist Attachments to Science." *Mind* 95 (1986): 149–79.

Fish, Stanley. *Doing What Comes Naturally: Change, Rhetoric, and the Practice of Theory in Literary and Legal Studies*. Durham, NC: Duke University Press, 1989.

Fish, Stanley. "Professor Sokal's Bad Joke." *New York Times*, May 21, 1996.

Fisher, Erik. "Editorial Overview: Public Science and Technology Scholars: Engaging Whom?" *Science and Engineering Ethics* 17, no. 4 (2011): 607–20.

Fleck, Christian. *A Transatlantic History of the Social Sciences: Robber Barons, the Third Reich and the Invention of Empirical Social Research*. London: Bloomsbury Publishing, 2011.

Fleck, Ludwik. *Genesis and Development of a Scientific Fact*. Chicago: University of Chicago Press, 1981.

Forman, Paul. "Behind Quantum Electronics: National Security as Basis for Physical Research in the United States, 1940–1960." *Historical Studies in the Physical and Biological Sciences* 18, no. 1 (1987): 149–229.

Frickel, Scott, Sahra Gibbon, Jeff Howard, Joanna Kempner, Gwen Ottinger, and David J. Hess. "Undone Science: Charting Social Movement and Civil Society

Challenges to Research Agenda Setting." *Science, Technology, & Human Values* 35, no. 4 (2010): 444–73.
Frickel, Scott, and Kelly Moore. *The New Political Sociology of Science: Institutions, Networks, and Power*. Madison: University of Wisconsin Press, 2006.
Friedman, Michael. "Carnap's *Aufbau* Reconsidered." *Nous* 21 (1987): 521–45.
Fry, Iris. "The Emergence of Life on Earth and the Darwinian Revolution." In *The Cambridge Encyclopedia of Darwin and Evolutionary Thought*, edited by Michael Ruse, 322–29. New York: Cambridge University Press, 2013.
Fry, Iris. "Philosophical Aspects of the Origin-of-Life Problem: The Emergence of Life and the Nature of Science." In *Exploring the Origin, Extent, and Future of Life: Philosophical, Ethical and Theological Perspectives*, edited by Constance M. Bertka, 61–79. Cambridge: Cambridge University Press, 2009.
Fuchs, Stephan. *Against Essentialism: A Theory of Culture and Society*. Cambridge, MA: Harvard University Press, 2001.
Fuchs, Stephan. *The Professional Quest for Truth: A Social Theory of Science and Knowledge*. Albany: State University of New York Press, 1992.
Fugh-Berman, Adriane, and Shahram Aharii. "Following the Script: How Drug Reps Make Friends and Influence Doctors." In *The Pharmaceutical Studies Reader*, edited by Sergio Sismondo and Jeremy A. Greene, 123–32. Chichester, England: Wiley Blackwell, 2015.
Fujimura, Joan H. "Crafting Science: Standardized Packages, Boundary Objects, and 'Translation.'" In *Science as Practice and Culture*, edited by Andrew Pickering, 168–211. Chicago: University of Chicago Press, 1992.
Fuller, Steve. "Bounded Rationality in Law and Science." PhD dissertation. University of Pittsburgh, 1985.
Fuller, Steve, ed. *The Cognitive Turn: Sociological and Psychological Perspectives on Science*. Dordrecht: Kluwer, 1989.
Fuller, Steve. *Dissent over Descent: Intelligent Design's Challenge to Darwinism*. Cambridge: Icon Books, 2008.
Fuller, Steve. "The Dissent over Dissent over Descent." *Philosophy of the Social Sciences* 40 (2010): 478–503.
Fuller, Steve. *Knowledge: The Philosophical Quest in History*. London: Routledge, 2015.
Fuller, Steve. "Naturalized Epistemology Sublimated: Rapprochement without the Ruts." *Studies in History and Philosophy of Science Part A* 22, no. 2 (1991): 277–93.
Fuller, Steve. *The Philosophy of Science and Its Discontents*. Boulder, CO: Westview Press, 1989.
Fuller, Steve. *Philosophy, Rhetoric, and the End of Knowledge: The Coming of Science and Technology Studies*. Madison: University of Wisconsin Press, 1993.
Fuller, Steve. "Precautionary and Proactionary as the New Right and the New Left of the Twenty-First Century Ideological Spectrum." *International Journal of Politics, Culture, and Society* 25, no. 4 (2012): 157–74.
Fuller, Steve. *Social Epistemology*. Bloomington, IN: Indiana University Press, 1988.

Fuller, Steve. "Social Epistemology for Theodicy without Deference: Response to William Lynch." *Symposium: Theoretical and Applied Inquiries in Philosophy and Social Sciences* 3, no. 2 (2016): 207–18.

Fuller, Steve. *Thomas Kuhn: A Philosophical History for Our Times*. Chicago: University of Chicago Press, 2000.

Fuller, Steve. "What Has Atheism Ever Done for Science?" In *Religion and the New Atheism: A Critical Appraisal*, edited by Amarnath Amarasingam, 57–77. Leiden: Brill, 2009.

Fuller, Steve, and James H. Collier. *Philosophy, Rhetoric, and the End of Knowledge: A New Beginning for Science and Technology Studies*. 2nd ed. Mahwah, NJ: Lawrence Erlbaum Associates, 2004.

Fuller, Steve, and Veronika Lipińska. *The Proactionary Imperative: A Foundation for Transhumanism*. New York: Palgrave Macmillan, 2014.

Fuller, Steve, and Sujatha Raman, eds. *Teaching Science and Technology Studies: A Guide for Curricular Planners*. Blacksburg, VA: Science Studies Center, Virginia Polytechnic Institute and State University, 1991.

Fumento, Michael. "Big Fat Fake: The Atkins Diet Controversy and the Sorry State of Science Journalism." *Reason*, March 2003. https://reason.com/2003/03/01/big-fat-fake-2/.

Fuss, Diana. *Essentially Speaking: Feminism, Nature and Difference*. New York: Routledge, 1989.

Galison, Peter. "Aufbau/Bauhaus: Logical Positivism and Architectural Modernism." *Critical Inquiry* 16 (1990): 709–52.

Galison, Peter. "History, Philosophy, and the Central Metaphor." *Science in Context* 2 (1988): 197–211.

Galison, Peter. *How Experiments End*. Chicago: University of Chicago Press, 1987.

Galison, Peter. *Image and Logic: A Material Culture of Microphysics*. Chicago: University of Chicago Press, 1997.

Galison, Peter, and David J. Stump, eds. *The Disunity of Science: Boundaries, Contexts, and Power*. Stanford, CA: Stanford University Press, 1996.

Gardner, Andy. "The Price Equation." *Current Biology* 18, no. 5 (2008): R198–R202.

Gardner, Christopher D., John F. Trepanowski, Liana C. Del Gobbo, Michelle E. Hauser, Joseph Rigdon, John P. A. Ioannidis, Manisha Desai, and Abby C. King. "Effect of Low-Fat vs Low-Carbohydrate Diet on 12-Month Weight Loss in Overweight Adults and the Association with Genotype Pattern or Insulin Secretion: The Dietfits Randomized Clinical Trial." *JAMA* 319, no. 7 (2018): 667–79.

Gardner, Howard. *Frames of Mind: The Theory of Multiple Intelligences*. New York: Basic Books, 2011.

Garfinkel, Harold. *Studies in Ethnomethodology*. Cambridge: Polity Press, 1967.

Garver, Newton, and Seung-Chong Lee. *Derrida and Wittgenstein*. Philadelphia: Temple University Press, 1994.

Geary, Daniel. *Beyond Civil Rights: The Moynihan Report and Its Legacy*. Philadelphia: University of Pennsylvania Press, 2015.

Giddens, Anthony. *New Rules of Sociological Method: A Positive Critique of Interpretive Sociologies*. 2nd ed. Stanford, CA: Stanford University Press, 1993.

Giddens, Anthony. *Studies in Social and Political Theory*. New York: Basic Books, 1977.

Gieryn, Thomas. "Boundary Work and the Demarcation of Science from Non-Science: Strains and Interests in Professional Ideologies of Scientists." *American Sociological Review* 48 (1983): 781–95.

Gieryn, Thomas F., and Anne E. Figert. "Ingredients for a Theory of Science in Society: O-Rings, Ice Water, C-Clamp, Richard Feynman, and the Press." In *Theories of Science in Society*, edited by Susan E. Cozzens and Thomas F. Gieryn. Bloomington: Indiana University Press, 1990.

Gilbert, G. Nigel, and Michael Mulkay. *Opening Pandora's Box: A Sociological Analysis of Scientists' Discourse*. Cambridge: Cambridge University Press, 1984.

Giskes, K., F. van Lenthe, M. Avendano-Pabon, and J. Brug. "A Systematic Review of Environmental Factors and Obesogenic Dietary Intakes among Adults: Are We Getting Closer to Understanding Obesogenic Environments?" *Obesity Reviews* 12, no. 5 (2011): e95–e106.

Glas, Eduard. "The 'Popperian Programme' and Mathematics. Part I: The Fallibilist Logic of Mathematical Discovery." *Studies in History and Philosophy of Science Part A* 32, no. 1 (2001): 119–37.

Glas, Eduard. "The 'Popperian Programme' and Mathematics. Part II: From Quasi-Empiricism to Mathematical Research Programmes." *Studies in History and Philosophy of Science Part A* 32, no. 2 (2001): 355–76.

Gleeson, Denis, and Michael Erben. "Meaning in Context: Notes towards a Critique of Ethnomethodology." *British Journal of Sociology* 27, no. 4 (1976): 474–83.

Gluckman, Peter, and Mark Hanson. *Mismatch: Why Our World No Longer Fits Our Bodies*. Oxford: Oxford University Press, 2006.

Goldman, Alvin I. "Foundations of Social Epistemics." *Synthese* 73, no. 1 (1987): 109–44.

Goldman, Alvin, and Thomas Blanchard. "Social Epistemology." In *The Stanford Encyclopedia of Philosophy*, edited by Edward N. Zalta. Summer 2018, https://plato.stanford.edu/archives/sum2018/entries/epistemology-social.

Gooding, David. "Putting Agency Back into Experiment." In *Science as Practice and Culture*, edited by Andrew Pickering, 65–112. Chicago: University of Chicago Press, 1992.

Gottlieb, Martin. "Pattern Emerges in Bomber's Tract." *New York Times*, August 2, 1995.

Gottweis, Herbert. "German Politics of Genetic Engineering and Its Deconstruction." *Social Studies of Science* 25 (1995): 195–235.

Gottweis, Herbert. *Governing Molecules: The Discursive Politics of Genetic Engineering in Europe and the United States*. Cambridge, MA: MIT Press, 1998.

Gould, Stephen Jay. *The Mismeasure of Man*. New York: Norton, 1981.

Gould, Stephen Jay. "Nonoverlapping Magisteria." *Natural History* 106, no. 2 (1997): 16–22.

Gould, Stephen Jay. *Punctuated Equilibrium*. Cambridge, MA: Belknap Press of Harvard University Press, 2007.

Graedel, Thomas E. "Green Chemistry and Sustainable Development." In *Handbook of Green Chemistry and Technology*, edited by James H. Clark and Duncan J. Macquarrie, 56–61. Oxford: Blackwell Science, 2002.

Graham, Loren R. *Lysenko's Ghost: Epigenetics and Russia*. Cambridge, MA: Harvard University Press, 2016.

Graham, Loren R. "The Socio-Political Roots of Boris Hessen: Soviet Marxism and the History of Science." *Social Studies of Science* 15, no. 4 (1985): 705–22.

Gramsci, Antonio. *Prison Notebooks*. 2 vols. New York: Columbia University Press, 1992.

Granovetter, Mark S. "The Strength of Weak Ties." *American Journal of Sociology* 78, no. 6 (1973): 1360–80.

Gray, Asa. "Darwin on the *Origin of Species*." *Atlantic*, 1860, http://www.theatlantic.com/magazine/archive/1860/07/darwin-on-the-origin-of-species/304152/.

Greenberg, Daniel S. *The Politics of Pure Science*. New York: World Publishing Company, 1967.

Greenberg, Daniel S. *Science for Sale: The Perils, Rewards, and Delusions of Campus Capitalism*. Chicago: University of Chicago Press, 2007.

Greimas, Algirdas Julien. *The Social Sciences: A Semiotic View*. Minneapolis: University of Minnesota Press, 1990.

Gremillion, Kristen J. *Ancestral Appetites: Food in Prehistory*. New York: Cambridge University Press, 2011.

Griffiths, Paul E., and Russell D. Gray. "The Developmental Systems Perspective: Organism-Environment Systems as Units of Development and Evolution." In *Phenotypic Integration: Studying the Ecology and Evolution of Complex Phenotypes*, edited by Katherine Preston and Massimo Pigliucci, 409–31. Oxford: Oxford University Press, 2004.

Grinnell, Frederick. *The Scientific Attitude*. 2nd ed. New York: Guilford Press, 1992.

Gross, Paul R., and Norman Levitt. *Higher Superstition: The Academic Left and Its Quarrels with Science*. Baltimore, MD: Johns Hopkins University Press, 1994.

Gruber, Helmut. *Red Vienna: Experiment in Working-Class Culture, 1919–1934*. New York: Oxford University Press, 1991.

Guston, David H. "The Pumpkin or the Tiger? Michael Polanyi, Frederick Soddy, and Anticipating Emerging Technologies." *Minerva* 50, no. 3 (2012): 363–79.

Guyenet, Stephan J. "Nutrition Science Initiative (Nusi) in Retrospect." https://www.stephanguyenet.com/nutrition-science-initiative-nusi-in-retrospect/.

Guyenet, Stephan J., and Michael W. Schwartz. "Regulation of Food Intake, Energy Balance, and Body Fat Mass: Implications for the Pathogenesis and Treatment of Obesity." *Journal of Clinical Endocrinology and Metabolism* 97, no. 3 (2012): 745–55.

Hacker, Barton C. *The Dragon's Tail: Radiation Safety in the Manhattan Project, 1942–1946*. Berkeley: University of California Press, 1987.

Hacking, Ian. "Lakatos's Philosophy of Science." In *Scientific Revolutions*, edited by Ian Hacking, 128–43. Oxford: Oxford University Press, 1981.

Hacking, Ian. "Language, Truth, and Reason." In *Rationality and Relativism*, edited by Martin Hollis and Steven Lukes, 48–66. Oxford: Blackwell, 1982.

Hacking, Ian. "Making up People." *London Review of Books* 28 (August 17, 2006).

Hacking, Ian. *Representing and Intervening: Introductory Topics in the Philosophy of Natural Science*. Cambridge: Cambridge University Press, 1983.

Hacking, Ian. "The Self-Vindication of the Laboratory Sciences." In *Science as Practice and Culture*, edited by Andrew Pickering, 29–64. Chicago: University of Chicago Press, 1992.

Hacohen, Malachi Haim. *Karl Popper, the Formative Years, 1902–1945: Politics and Philosophy in Interwar Vienna*. Cambridge: Cambridge University Press, 2000.

Hagerman, Paul J., and Randi J. Hagerman. "The Fragile-X Premutation: A Maturing Perspective." *American Journal of Human Genetics* 74, no. 5 (2004): 805–16.

Hagerman, R. J., M. Leehey, W. Heinrichs, F. Tassone, R. Wilson, J. Hills, J. Grigsby, B. Gage, and P. J. Hagerman. "Intention Tremor, Parkinsonism, and Generalized Brain Atrophy in Male Carriers of Fragile X." *Neurology* 57, no. 1 (2001): 127–130.

Hagstrom, Warren O. *The Scientific Community*. New York: Basic Books, 1965.

Hall, Kevin D., et al. "Energy Expenditure and Body Composition Changes after an Isocaloric Ketogenic Diet in Overweight and Obese Men." *American Journal of Clinical Nutrition* 104, no. 2 (2016): 324–33.

Hall, Kevin D., Stephan J. Guyenet, and Rudolph L. Leibel. "The Carbohydrate-Insulin Model of Obesity Is Difficult to Reconcile with Current Evidence." *JAMA Internal Medicine* 178, no. 8 (2018): 1103–5.

Hamilton, W. D. "The Evolution of Altruistic Behavior." *American Naturalist* 97, no. 896 (1963): 354–56.

Hamilton, W. D. "The Genetical Evolution of Social Behaviour. I." *Journal of Theoretical Biology* 7, no. 1 (1964): 1–16.

Hamilton, W. D. "The Genetical Evolution of Social Behaviour. II." *Journal of Theoretical Biology* 7, no. 1 (1964): 17–52.

Hamilton, W. D. "Innate Social Aptitudes of Man: An Approach from Evolutionary Genetics." In *Biosocial Anthropology*, edited by Robin Fox, 133–55. New York: Wiley, 1975.

Hamlin, Christopher. "Third Wave Science Studies: Toward a History and Philosophy of Expertise." In *The Challenge of the Social and the Pressure of Practice: Science and Values Revisited*, edited by Martin Carrier, Don Howard, and Janet Kourany, 160–85. Pittsburgh, PA: University of Pittsburgh Press, 2008.

Han, Kyonghee, and Gary Lee Downey. *Engineers for Korea*. Synthesis Lectures on Global Engineering. San Rafael, CA: Morgan and Claypool, 2014.

Hands, John. *Cosmosapiens: Human Evolution from the Origin of the Universe*. New York: Overlook Duckworth, 2015.

Hanekamp, Jaap. "Neither Acceptable nor Certain—Cold War Antics for 21st Century Precautionary Culture." *Erasmus Law Review* 2, no. 2 (2009): 221–57.

Hanekamp, Jaap C., and Aalt Bast. "Hormesis in Precautionary Regulatory Culture: Models Preferences and the Advancement of Science." *Human and Experimental Toxicology* 26, no. 11 (2007): 855–73.

Hannaway, Owen. *The Chemists and the Word: The Didactic Origins of Chemistry*. Baltimore, MD: Johns Hopkins University Press, 1975.

Haraway, Donna. "A Giant Bumptious Litter: Donna Haraway on Truth, Technology, and Resisting Extinction." *Logic*, December 7, 2019.

Haraway, Donna. "A Manifesto for Cyborgs: Science, Technology, and Socialist Feminism in the 1980s." *Socialist Review* 15 (1985): 64–107.

Haraway, Donna. *Primate Visions: Gender, Race, and Nature in the World of Modern Science*. New York: Routledge, 1989.

Haraway, Donna. "The Promises of Monsters: A Regenerative Politics for Inappropriate/d Others." In *Cultural Studies*, edited by Lawrence Grossberg, Cary Nelson, and Paula A. Treichler, 295–337. New York: Routledge, 1992.

Haraway, Donna. *Simians, Cyborgs, and Women: The Reinvention of Nature*. London: Free Association, 1990.

Haraway, Donna. "Situated Knowledges: The Science Question in Feminism and the Privilege of Partial Perspective." *Feminist Studies* 14 (1988): 575–99.

Harding, Sandra G., ed. *The "Racial" Economy of Science: Toward a Democratic Future*. Bloomington: Indiana University Press, 1993.

Harding, Sandra G. *The Science Question in Feminism*. Ithaca, NY: Cornell University Press, 1986.

Harding, Sandra. *Whose Science? Whose Knowledge? Thinking from Women's Lives*. Ithaca, NY: Cornell University Press, 1991.

Harding, Sandra, and Jean F. O'Barr, eds. *Sex and Scientific Inquiry*. Chicago: University of Chicago Press, 1987.

Harman, Oren, and Michael R. Dietrich, eds. *Rebels, Mavericks, and Heretics in Biology*. New Haven, CT: Yale University Press, 2008.

Harré, Rom, and Edward H. Madden. *Causal Powers: A Theory of Natural Necessity*. Totowa, NJ: Rowman & Littlefield, 1975.

Harris, Charles E., Michael S. Pritchard, and Michael J. Rabins. *Engineering Ethics: Concepts and Cases*. Belmont, CA: Wadsworth, 1995.

Harvey, David. *The Condition of Postmodernity: An Enquiry into the Origins of Cultural Change*. Oxford: Basil Blackwell, 1989.

Harvey, Frank P. *Explaining the Iraq War: Counterfactual Theory, Logic, and Evidence*. Cambridge: Cambridge University Press, 2012.

Harvey, Van A. *Feuerbach and the Interpretation of Religion*. Cambridge: Cambridge University Press, 1997.

Hazen, Robert M. "Emergence and the Experimental Pursuit of the Origin of Life." In *Exploring the Origin, Extent, and Future of Life: Philosophical, Ethical and Theological Perspectives*, edited by Constance M. Bertka, 21–46. Cambridge: Cambridge University Press, 2009.

Hegel, Georg Wilhelm Friedrich. *Encyclopedia of the Philosophical Sciences in Basic Outline, Part 1, Science of Logic*. Cambridge: Cambridge University Press, 2010.

Hegel, G. W. F. *Phenomenology of Spirit*. Oxford: Oxford University Press, 1977.

Heit, Helmut, and Eric Oberheim. "Paul Feyerabend, an Historical Philosopher of Nature." In *Philosophy of Nature* by Paul Feyerabend, vii–xxvii. Cambridge: Polity, 2016.

Henderson, W. O. *The Life of Friedrich Engels*. 2 vols. London: Frank Cass, 1976.

Henrich, Natalie, and Joseph Henrich. *Why Humans Cooperate: A Cultural and Evolutionary Explanation*. Oxford: Oxford University Press, 2007.

Herkert, Joseph R. "Future Directions in Engineering Ethics Research: Microethics, Macroethics and the Role of Professional Societies." *Science and Engineering Ethics* 7, no. 3 (2001): 403–14.

Herkert, Joseph R. "Professional Societies, Microethics, and Macroethics: Product Liability as an Ethical Issue in Engineering Design." *International Journal of Engineering Education* 19, no. 1 (2003): 163–67.

Herkert, Joseph R. "Ways of Thinking about and Teaching Ethical Problem Solving: Microethics and Macroethics in Engineering." *Science and Engineering Ethics* 11, no. 3 (2005): 373–85.

Hershberg, James G. *James B. Conant: Harvard to Hiroshima and the Making of the Nuclear Age*. New York: Knopf, 1993.

Hessen, Boris. "The Social and Economic Roots of Newton's *Principia*." In *Science at the Crossroads*, edited by N. Bukharin, 1–62. London: Kniga, 1931.

Hill, R. A., and R. I. M. Dunbar. "Social Network Size in Humans." *Human Nature* 14, no. 1 (2003): 53–72.

Hobsbawm, Eric. *Globalisation, Democracy and Terrorism*. London: Little, Brown, 2007.

Holcomb, Harmon R. *Sociobiology, Sex, and Science*. Albany: State University of New York Press, 1993.

Hollander, Samuel. *Friedrich Engels and Marxian Political Economy*. New York: Cambridge University Press, 2011.

Hollinger, David A. "The Defense of Democracy and Robert K. Merton's Formulation of the Scientific Ethos." *Knowledge and Society: Studies in the Sociology of Culture Past and Present* 4 (1983): 1–15.

Hollis, Martin, and Steven Lukes, eds. *Rationality and Relativism*. Oxford: Blackwell, 1982.

Holton, Gerald. "Subelectrons, Presuppositions, and the Millikan–Ehrenhaft Dispute." *Historical Studies in the Physical Sciences* 9 (1978): 161–224.

Horgan, John. *The End of Science: Facing the Limits of Knowledge in the Twilight of the Scientific Age*. Boulder, CO: Basic Books, 2015.

Hormone Health Network. "Adrenal Fatigue." https://www.hormone.org/diseases-and-conditions/adrenal-fatigue.

Howard-Grenville, Jennifer, Andrew J. Nelson, Andrew G. Earle, and Julie A. Haack. "'If Chemists Don't Do It, Who Is Going To?': Peer-Driven Occupational Change and the Emergence of Green Chemistry." *Administrative Science Quarterly* 62, no. 3 (2017): 524–60.

Hoyningen-Huene, Paul. "Paul Feyerabend and Thomas Kuhn." In *The Worst Enemy of Science? Essays in Memory of Paul Feyerabend*, edited by John Preston, Gonzalo Munévar, and David Lamb, 102–14. New York: Oxford University Press, 2000.

Hoyningen-Huene, Paul. "Two Letters of Paul Feyerabend to Thomas S. Kühn on a Draft of the Structure of Scientific Revolutions." *Studies in History and Philosophy of Science Part A* 26, no. 3 (1995): 353–87.

Hoyningen-Huene, Paul, Eric Oberheim, and Hanne Andersen. "On Incommensurability." *Studies in History and Philosophy of Science* 27, no. 1 (1996): 131–41.

Hu, Tian, Katherine T. Mills, Lu Yao, Kathryn Demanelis, Mohamed Eloustaz, William S. Yancy, Jr., Tanika N. Kelly, Jiang He, and Lydia A. Bazzano. "Effects of Low-Carbohydrate Diets versus Low-Fat Diets on Metabolic Risk Factors: A Meta-Analysis of Randomized Controlled Clinical Trials." *American Journal of Epidemiology* 176, no. 7 (2012): S44–S54.

Hughes, William O. H., Benjamin P. Oldroyd, Madeleine Beekman, and Francis L. W. Ratnieks. "Ancestral Monogamy Shows Kin Selection Is Key to the Evolution of Eusociality." *Science* 320, no. 5880 (2008): 1213–16.

Huizenga, John R. *Cold Fusion: The Scientific Fiasco of the Century*. New York: University of Rochester Press, 1992.

Hull, David L. *Science as a Process: An Evolutionary Account of the Social and Conceptual Development of Science*. Chicago: University of Chicago Press, 1988.

Hunter, Michael. *Science and Society in Restoration England*. Cambridge: Cambridge University Press, 1981.

Hvistendahl, Mara. "China's Publication Bazaar." *Science* 342, no. 6162 (2013): 1035–39.

Inhofe, James M. "The Science of Climate Change Senate Floor Statement." U.S. Senate Committee on Environment and Public Works, July 28, 2003. http://www.epw.senate.gov/speechitem.cfm?party=rep&id=230594.

Ioannidis, John P. A. "Why Most Published Research Findings Are False." *PLoS Medicine* 2, no. 8 (2005): e124.

Jablokow, Kathryn W. "Engineers as Problem-Solving Leaders: Embracing the Humanities." *IEEE Technology and Society Magazine* (Winter, 2007): 29–35.

Jablonka, Eva, and Marion J. Lamb. *Epigenetic Inheritance and Evolution: The Lamarckian Dimension*. Oxford: Oxford University Press, 1995.

Jablonka, Eva, and Marion J. Lamb. *Evolution in Four Dimensions: Genetic, Epigenetic, Behavioral, and Symbolic Variation in the History of Life*. Cambridge, MA: MIT Press, 2005.

Jahoda, Marie, Paul F. Lazarsfeld, and Hans Zeisel. *Marienthal: The Sociology of an Unemployed Community*. Abingdon, England: Routledge, 2017.

Jameson, Fredric. *The Jameson Reader*. Oxford: Blackwell, 2000.

Jamison, Dale. *The Making of Green Engineers: Sustainable Development and the Hybrid Imagination*. San Rafael, CA: Morgan and Claypool, 2013.

Jasanoff, Sheila. "Acceptable Evidence in a Pluralistic Society." In *Acceptable Evidence: Science and Values in Risk Management*, edited by Deborah G. Mayo and Rachelle D. Hollander, 29–47. New York: Oxford University Press, 1991.

Jasanoff, Sheila S. "Contested Boundaries in Policy-Relevant Science." *Social Studies of Science* 17 (1987): 195–230.

Jasanoff, Sheila. "Future Imperfect: Science, Technology, and the Imaginations of Modernity." In *Dreamscapes of Modernity: Sociotechnical Imaginaries and the Fabrication of Power*, edited by Sheila Jasanoff and Sang-Hyun Kim, 1–33. Chicago: University of Chicago Press, 2015.

Jasanoff, Sheila, ed. *States of Knowledge: The Co-Production of Science and the Social Order*. London: Routledge, 2004.

Jasanoff, Sheila, and Sang-Hyun Kim, eds. *Dreamscapes of Modernity: Sociotechnical Imaginaries and the Fabrication of Power*. Chicago: University of Chicago Press, 2015.

Jasanoff, Sheila, Gerald E. Markle, James C. Petersen, and Trevor Pinch, eds. *Handbook of Science and Technology Studies*. Thousand Oaks, CA: Sage, 1995.

Jay, Martin. *The Dialectical Imagination: A History of the Frankfurt School and the Institute of Social Research, 1923–1950*. Boston: Little, Brown, 1973.

Jeřábek, Hynek. *Paul Lazarsfeld and the Origins of Communications Research*. London: Routledge, 2017.

John Paul II. "Message to the Pontifical Academy of Sciences." *Quarterly Review of Biology* 72, no. 4 (1997): 381–83.

Jolly, J. Christopher. "Thresholds of Uncertainty: Radiation and Responsibility in the Fallout Controversy." PhD dissertation. Oregon State University, 2003.

Jones, Geoffrey, and Christina Lubinski. "Historical Origins of Environmental Sustainability in the German Chemical Industry, 1950s–1980s." Working Paper. Harvard Business School, August 30, 2013.

Jones, William David. "Toward a Theory of Totalitarianism: Franz Borkenau's Pareto." *Journal of the History of Ideas* 53, no. 3 (1992): 455–66.

Jonsen, Albert R., and Stephen Toulmin. *The Abuse of Casuistry: A History of Moral Reasoning*. Berkeley: University of California Press, 1988.

Joppke, Christian, and Steven Lukes. *Multicultural Questions*. Oxford: Oxford University Press, 1999.

Josephson-Storm, Jason Ā. *The Myth of Disenchantment: Magic, Modernity, and the Birth of the Human Sciences*. Chicago: University of Chicago Press, 2017.

Jung, Chang Hee, and Min-Seon Kim. "Molecular Mechanisms of Central Leptin Resistance in Obesity." *Archives of Pharmacal Research* 36, no. 2 (2013): 201–7.

Kaczynski, David. *Every Last Tie: The Story of the Unabomber and His Family*. Durham, NC: Duke University Press, 2016.

Kadvany, John. *Imre Lakatos and the Guises of Reason*. Durham, NC: Duke University Press, 2001.

Kainz, Howard P. *Hegel's Philosophy of Right, with Marx's Commentary: A Handbook for Students*. The Hague: Martinus Nijhoff, 1974.

Kaiser, David. "A Mannheim for All Seasons: Bloor, Merton, and the Roots of the Sociology of Scientific Knowledge." *Science in Context* 11, no. 1 (2008): 51–87.

Kalra, Sanjay, and Sachin Kumar Khandelwal. "Why Are Our Hypothyroid Patients Unhappy? Is Tissue Hypothyroidism the Answer?" *Indian Journal of Endocrinology and Metabolism* 15, Supplement 2 (2011): S95–S98.

Kasavin, Ilya T. "Uniting the Cognitive and the Social: Lakatos Unmasked?" *Epistemology & Philosophy of Science* 55, no. 3 (2018): 41–46.

Kathren, Ronald L. "Historical Development of the Linear Nonthreshold Dose-Response Model as Applied to Radiation." *Pierce Law Review* 1, no. 1/2 (2003): 5–30.

Kathren, Ronald L. "Pathway to a Paradigm: The Linear Nonthreshold Dose-Response Model in Historical Context." *Health Physics* 70 (1996): 376–90.

Katz, Elihu, and Paul F. Lazarsfeld. *Personal Influence; the Part Played by People in the Flow of Mass Communications*. Glencoe, IL: Free Press, 1955.

Kearns, Cristin E., Laura A. Schmidt, and Stanton A. Glantz. "Sugar Industry and Coronary Heart Disease Research: A Historical Analysis of Internal Industry Documents." *JAMA Internal Medicine* 176, no. 11 (2016): 1680–85.

Keim, Brandon. "E. O. Wilson Proposes New Theory of Social Evolution." *Wired*, August 26, 2010. http://www.wired.com/2010/08/kin-selection-challenged/. Accessed August 14, 2015.

Keller, Evelyn Fox. *A Feeling for the Organism: The Life and Work of Barbara McClintock*. San Francisco: W. H. Freeman, 1983.

Keller, Evelyn Fox. *Reflections on Gender and Science*. New Haven, CT: Yale University Press, 1985.

Keohane, Kieran. "Central Problems in the Philosophy of the Social Sciences after Postmodernism: Reconciling Consensus and Hegemonic Theories of Epistemology and Political Ethics." *Philosophy & Social Criticism* 19 (1993): 145–69.

Keys, Ancel. *Coronary Heart Disease in Seven Countries*. New York: American Heart Association, 1970.

Keys, Ancel. "Coronary Heart Disease—the Global Picture." *Atherosclerosis* 22, no. 2 (1975): 149–92.

Keys, Ancel. "Sucrose in the Diet and Coronary Heart Disease." *Atherosclerosis* 14, no. 2 (1971): 193–202.

Khan, Tauseef A., and John L. Sievenpiper. "Controversies about Sugars: Results from Systematic Reviews and Meta-Analyses on Obesity, Cardiometabolic Disease and Diabetes." *European Journal of Nutrition* 55, Supplement 2 (2016): 25–43.

Khan, Tauseef A., and John L. Sievenpiper. "Metabolic Improvement with Fructose Restriction: Is It the Fructose or the Weight Loss?" *Obesity* 24, no. 3 (2016): 549.

Kidd, Ian James. "Reawakening to Wonder: Wittgenstein, Feyerabend, and Scientism." In *Wittgenstein and Scientism*, edited by Jonathan Beale and Ian James Kidd, 101–15. London: Routledge, 2017.

Kim, Kyung-Man. "Critical Theory Criticized: Giddens's Double Hermeneutic and the Problem of Language Game Change." *Cultural Studies ↔ Critical Methodologies* 4, no. 1 (2004): 28–44.

Kirsch, David A. *The Electric Vehicle and the Burden of History*. East Brunswick, NJ: Rutgers University Press, 2000.

Kitcher, Philip. "Reply to Helen Longino." *Philosophy of Science* 69, no. 4 (2002): 569–72.

Kitcher, Philip. *Science, Truth, and Democracy*. New York: Oxford University Press, 2001.

Kitcher, Philip. *Vaulting Ambition: Sociobiology and the Quest for Human Nature*. Cambridge, MA: MIT Press, 1985.

Kitcher, Philip, and Evelyn Fox Keller. *The Seasons Alter: How to Save Our Planet in Six Acts*. New York: Liveright, 2017.

Kitzmiller v. Dover Area School District. 400 F. Supp. 2d 707, United States District Court for the Middle District of Pennsylvania, 2005.

Klein, Julie Thompson. *Interdisciplinarity: History, Theory, and Practice*. Detroit, MI: Wayne State University Press, 1990.

Klein, Richard G. *The Human Career: Human Biological and Cultural Origins*. 3rd ed. Chicago: University of Chicago Press, 2009.

Kline, Ronald R. "From Progressivism to Engineering Studies: Edwin T. Layton's *The Revolt of the Engineers*." *Technology and Culture* 49, no. 4 (2008): 1018–24.

Kline, Ronald R. "Using History and Sociology to Teach Engineering Ethics." *IEEE Technology and Society Magazine* (Winter 2001/2002): 13–20.

Knorr-Cetina, Karin D. "The Couch, the Cathedral, and the Laboratory: On the Relationship between Experiment and Laboratory in Science." In *Science as Practice and Culture*, edited by Andrew Pickering, 113–38. Chicago: University of Chicago Press, 1992.

Knorr-Cetina, Karin D. "Introduction: The Micro-Sociological Challenge of Macro-Sociology: Towards a Reconstruction of Social Theory and Methodology." In *Advances in Social Theory and Methodology: Toward an Integration of Micro- and Macro-Sociologies*, edited by K. Knorr-Cetina and A. V. Cicourel, 1–47. Boston: Routledge Kegan Paul, 1981.

Knorr-Cetina, Karin D. "Scientific Communities or Transepistemic Arenas of Research? A Critique of Quasi-Economic Models of Science." *Social Studies of Science* 12 (1982): 101–30.

Knorr-Cetina, Karin, and Michael Mulkay. "Introduction: Emerging Principles in Social Studies of Science." In *Science Observed: Perspectives on the Social Study of Science*, edited by Karin Knorr-Cetina and Michael Mulkay, 1–17. London: Sage, 1983.

Knorr-Cetina, Karin, and Michael Mulkay, eds. *Science Observed: Perspectives on the Social Study of Science*. London: Sage.

Koertge, Noretta, ed. *A House Built on Sand: Exposing Postmodernist Myths about Science*. New York: Oxford University Press, 1998.

Koetsier, Tuen. *Lakatos' Philosophy of Mathematics*. Amsterdam: Elsevier, 1991.

Kofman, Ava. "Bruno Latour, the Post-Truth Philosopher, Mounts a Defense of Science." *New York Times*, October 25, 2018.

Kornblith, Hilary, ed. *Naturalizing Epistemology*. 2nd ed. Cambridge, MA: MIT Press, 1994.

Krauss, Ronald M., Patricia J. Blanche, Robin S. Rawlings, Harriett S. Fernstrom, and Paul T. Williams. "Separate Effects of Reduced Carbohydrate Intake and Weight Loss on Atherogenic Dyslipidemia." *American Journal of Clinical Nutrition* 83, no. 5 (2006): 1025–31.

Kripke, Saul. *Wittgenstein on Rules and Private Language*. Cambridge, MA: Harvard University Press, 1982.

Kuhn, Thomas. "Dubbing and Redubbing: The Vulnerability of Rigid Designation." *Minnesota Studies in the Philosophy of Science* 14 (1990): 298–318.

Kuhn, Thomas S. "The Function of Dogma in Science." In *Scientific Change: Historical Studies in the Intellectual, Social, and Technical Conditions for Scientific*

Discovery and Technical Invention, from Antiquity to the Present, edited by A. C. Crombie, 347–69. New York: Basic Books, 1963.

Kuhn, Thomas S. "Possible Worlds in History of Science." In *Possible Worlds in Humanities, Arts and Sciences*, edited by Sture Allen. Berlin: Walter de Gruyter, 1989.

Kuhn, Thomas S. *The Structure of Scientific Revolutions.* 2nd ed. Chicago: University of Chicago Press, 1970.

Kuipers, Remko S., Martine F. Luxwolda, D. A. Janneke Dijck-Brouwer, S. Boyd Eaton, Michael A. Crawford, Loren Cordain, and Frits A. J. Muskiet. "Estimated Macronutrient and Fatty Acid Intakes from an East African Paleolithic Diet." *British Journal of Nutrition* 104, no. 11 (2010): 1666–87.

Kumar, Ajay. "Historical Development of Green Chemistry." *Chemical Educator* 18 (2013): 50–53.

Kusch, Martin. "Rule-Scepticism and the Sociology of Scientific Knowledge." *Social Studies of Science* 34, no. 4 (2004): 571–91.

Kutrovátz, Gábor. "Imre Lakatos' Hungarian Dissertation." In *Appraising Lakatos: Mathematics, Methodology, and the Man*, edited by George Kampis, Ladislav Kvasz, and Michael Stöltzner, 353–74. Dordrecht: Kluwer Academic Publishers, 2002.

Kutrovátz, Gábor. "Lakatos' Philosophical Work in Hungary." *Studies in East European Thought* 60, no. 1/2 (2008): 113–33.

La Berge, Ann F. "How the Ideology of Low Fat Conquered America." *Journal of the History of Medicine and Allied Sciences* 63, no. 2 (2008): 139–77.

LaCapra, Dominick. *Rethinking Intellectual History: Texts, Contexts, Language.* Ithaca, NY: Cornell University Press, 1983.

Laclau, Ernesto. *New Reflections on the Revolution of Our Time*. London: Verso, 1990.

Laclau, Ernesto, and Chantal Moufee. *Hegemony and Socialist Strategy: Towards a Radical Democratic Politics*. London: Verso, 1985.

Lahsen, Myanna. "Anatomy of Dissent: A Cultural Analysis of Climate Skepticism." *American Behavioral Scientist* 57, no. 6 (2013): 732–53.

Lahsen, Myanna. "Seductive Simulations? Uncertainty Distribution around Climate Models." *Social Studies of Science* 35, no. 6 (2005): 895–922.

Lakatos, Imre. "Anomalies versus 'Crucial Experiments': A Rejoinder to Professor Grünbaum." In *Mathematics, Science and Epistemology*. Vol. 2 of *Philosophical Papers*, edited by John Worrall and Gregory Currie, 211–23. Cambridge: Cambridge University Press, 1980.

Lakatos, Imre. "Criticism and the Methodology of Scientific Research Programmes." *Proceedings of the Aristotelian Society* 69 (1968): 149–86.

Lakatos, Imre. "Falsification and the Methodology of Scientific Research Programmes." In *Criticism and the Growth of Knowledge*, edited by Imre Lakatos and Alan Musgrave, 91–196. Cambridge: Cambridge University Press, 1970.

Lakatos, Imre. "History of Science and Its Rational Reconstruction." In *Method and Appraisal in the Physical Sciences*, edited by Colin Howson, 1–40. Cambridge: Cambridge University Press, 1976.

Lakatos, Imre. "History of Science and Its Rational Reconstructions." *PSA: Proceedings of the Biennial Meeting of the Philosophy of Science Association* (1970): 91–136.

Lakatos, Imre. "Lectures on Scientific Method" (1973). In *For and Against Method: Including Lakatos' Lectures on Scientific Method and the Lakatos-Feyerabend Correspondence*, by Imre Lakatos and Paul Feyerabend, edited by Matteo Motterlini, 19–112. Chicago: University of Chicago Press, 1999.

Lakatos, Imre. "A Letter to the Director of the London School of Economics." In *Mathematics, Science and Epistemology. Philosophical Papers*, Vol. 2, edited by John Worrall and Gregory Currie, 247–53. Cambridge: Cambridge University Press, 1980.

Lakatos, Imre. *The Methodology of Scientific Research Programmes*. Vol. 1 of *Philosophical Papers*, edited by John Worrall and Gregory Currie. Cambridge: Cambridge University Press, 1978.

Lakatos, Imre. "On Rearing Scholars" (1956). In Imre Lakatos and Paul Feyerabend, *For and Against Method: Including Lakatos' Lectures on Scientific Method and the Lakatos-Feyerabend Correspondence*, by Imre Lakatos and Paul Feyerabend, edited by Matteo Motterlini, 375–82. Chicago: University of Chicago Press, 1999.

Lakatos, Imre. *Proofs and Refutations: The Logic of Mathematical Discovery*. Cambridge: Cambridge University Press, 1976.

Lakatos, Imre, and Paul Feyerabend. *For and Against Method: Including Lakatos' Lectures on Scientific Method and the Lakatos-Feyerabend Correspondence*. Edited by Matteo Motterlini. Chicago: University of Chicago Press, 1999.

Lakatos, Imre, and Alan Musgrave, eds. *Criticism and the Growth of Knowledge*. Cambridge: Cambridge University Press, 1970.

Laland, Kevin N., and Gillian R. Brown. "Niche Construction, Human Behavior, and the Adaptive-Lag Hypothesis." *Evolutionary Anthropology: Issues, News, and Reviews* 15, no. 3 (2006): 95–104.

Laland, Kevin N., and Gillian R. Brown. *Sense and Nonsense: Evolutionary Perspectives on Human Behavior*. 2nd ed. Oxford: Oxford University Press, 2011.

Lappé, Frances Moore. *Diet for a Small Planet*. New York: Ballantine Books, 1971.

Larvor, Brendan. *Lakatos: An Introduction*. London: Routledge, 1998.

Latour, Bruno. *Aramis, or The Love of Technology*. Cambridge, MA: Harvard University Press, 1996.

Latour, Bruno. "Don't Throw the Baby Out with the Bath School! A Reply to Collins and Yearly." In *Science as Practice and Culture*, edited by Andrew Pickering, 343–68. Chicago: University of Chicago Press, 1992.

Latour, Bruno. *Down to Earth: Politics in the New Climatic Regime*. Cambridge: Polity, 2018.

Latour, Bruno. "Give Me a Laboratory and I Will Raise the World." In *Science Observed: Perspectives on the Social Study of Science*, edited by Karin Knorr-Cetina and Michael Mulkay, 141–70. London: Sage, 1983.

Latour, Bruno. *An Inquiry into Modes of Existence: An Anthropology of the Moderns*. Cambridge, MA: Harvard University Press, 2013.

Latour, Bruno. *Pandora's Hope: Essays on the Reality of Science Studies*. Cambridge, MA: Harvard University Press, 1999.

Latour, Bruno. *The Pasteurization of France*. Cambridge: Harvard University Press, 1988.

Latour, Bruno. "The Politics of Explanation: An Alternative." In *Knowledge and Reflexivity*, edited by Steve Woolgar, 155–76. Newbury Park, CA: Sage, 1988.

Latour, Bruno. "Postmodern? No, Simply *A*modern! Steps toward an Anthropology of Science." *Studies in History and Philosophy of Science* 21 (1990): 145–71.

Latour, Bruno. *Reassembling the Social: An Introduction to Actor-Network-Theory*. Oxford: Oxford University Press, 2005.

Latour, Bruno. *Science in Action: How to Follow Scientists and Engineers through Society*. Cambridge: Harvard University Press, 1987.

Latour, Bruno. "Technology Is Society Made Durable." In *A Sociology of Monsters: Essays on Power, Technology and Domination*, edited by John Law, 103–31. London: Routledge, 1991.

Latour, Bruno. *We Have Never Been Modern*. Cambridge, MA: Harvard University Press, 1993.

Latour, Bruno. "Where Are the Missing Masses? The Sociology of a Few Mundane Artifacts." In *Shaping Technology/Building Society: Studies in Sociotechnical Change*, edited by Wiebe E. Bijker and John Law, 225–58. Cambridge, MA: MIT Press, 1992.

Latour, Bruno, and Steve Woolgar. *Laboratory Life: The Construction of Scientific Facts*. Princeton, NJ: Princeton University Press, 1986.

Law, John. "STS as Method." In *The Handbook of Science and Technology Studies*, edited by Ulrike Felt, Rayvon Fouché, Clark A. Miller, and Laurel Smith-Doerr, 31–57. Cambridge, MA: MIT Press, 2017.

Laymon, Ronald. "Feyerabend, Brownian Motion, and the Hiddenness of Refuting Facts." *Philosophy of Science* 44, no. 2 (1977): 225–47.

Layton, Edwin T. *The Revolt of the Engineers: Social Responsibility and the American Engineering Profession*. Cleveland, OH: Press of Case Western Reserve University, 1971.

Leahey, Erin. "Not by Productivity Alone: How Visibility and Specialization Contribute to Academic Earnings." *American Sociological Review* 72, no. 4 (2007): 533–61.

Lebow, Richard Ned. *Forbidden Fruit: Counterfactuals and International Relations*. Princeton, NJ: Princeton University Press, 2010.

Lebow, Richard Ned, and Janice Gross Stein. *We All Lost the Cold War*. Princeton, NJ: Princeton University Press, 1994.

Lecourt, Dominique. *Marxism and Epistemology*. London: NLB, 1975.

Lenin, V. I. *Materialism and Empirio-Criticism: Critical Comments on a Reactionary Philosophy*. London: Lawrence & Wishart, 1948.

Lenoir, Timothy. "Was That Last Turn a Right Turn? The Semiotic Turn and A. J. Greimas." *Configurations* 2 (1994): 119–36.

Levins, Richard, and Richard Lewontin. *The Dialectical Biologist*. Cambridge: Harvard University Press, 1985.

Lewens, Tim. *The Meaning of Science: An Introduction to the Philosophy of Science*. New York: Basic Books, 2016.

Lewontin, R. C. "The Apportionment of Human Diversity." *Evolutionary Biology* 6 (1972): 381–98.

Lewontin, R. C. "Gene, Organism and Environment." In *Evolution from Molecules to Men*, edited by D. S. Bendall, 273–85. Cambridge: Cambridge University Press, 1983.

Lewontin, R. C. "Organism and Environment." In *Learning, Development, and Culture: Essays in Evolutionary Epistemology*, edited by H. C. Plotkin, 151–70. Chichester, England: Wiley, 1982.

Lewontin, R. C. "The Organism as the Subject and Object of Evolution." *Scientia* 77, no. 118 (1983): 65–82.

Lewontin, Richard C. "Race and Intelligence." *Bulletin of the Atomic Scientists* 26, no. 3 (1970): 2–8.

Lewontin, Richard. *The Triple Helix: Gene, Organism, and Environment*. Cambridge: Harvard University Press, 2002.

Li, Jessica, and Shengli Fu. "A Systematic Approach to Engineering Ethics Education." *Science and Engineering Ethics* 18, no. 2 (2012): 339–49.

Li, Victor. "Marshall Sahlins and the Apotheosis of Culture." *CR: The New Centennial Review* 1, no. 3 (2001): 201–87.

Liao, Xiaoyun, Stephen Rong, and David C. Queller. "Relatedness, Conflict, and the Evolution of Eusociality." *PLOS Biology* 13, no. 3 (2015): e1002098, 1–14.

Linthorst, J. A. "An Overview: Origins and Development of Green Chemistry." *Foundations of Chemistry* 12, no. 1 (2010): 55–68.

Livergood, Norman D. *Activity in Marx's Philosophy*. The Hague: Martinus Nijhoff, 1967.

Livingston, Eric. *The Ethnomethodological Foundations of Mathematics*. London: Routledge and Kegan Paul, 1986.

Long, Jancis. "Lakatos in Hungary." *Philosophy of the Social Sciences* 28, no. 2 (1998): 244–311.

Longino, Helen E. *The Fate of Knowledge*. Princeton, NJ: Princeton University Press, 2001.

Longino, Helen E. "Science and the Common Good: Thoughts on Philip Kitcher's Science, Truth, and Democracy." *Philosophy of Science* 69, no. 4 (2002): 560.

Longino, Helen E. *Science as Social Knowledge: Values and Objectivity in Scientific Inquiry*. Princeton, NJ: Princeton University Press, 1990.

Love, Alan C., and Rudolf A. Raff. "Knowing Your Ancestors: Themes in the History of Evo-Devo." *Evolution & Development* 5, no. 4 (2003): 327–30.

Lukács, György. *History and Class Consciousness: Studies in Marxist Dialectics*. Cambridge, MA: MIT Press, 1971.

Lustig, Robert H. "Prophecy and Propaganda." In *Pure, White, and Deadly: How Sugar Is Killing Us and What We Can Do to Stop It*, by John Yudkin, xi–xv. New York: Penguin Books, 2012.

Lustig, Robert H. "Response to 'Metabolic Improvement with Fructose Restriction: Is It the Fructose or the Weight Loss?'" *Obesity* 24, no. 3 (2016): 550.

Lustig, Robert H., Laura A. Schmidt, and Claire D. Brindis. "The Toxic Truth about Sugar." *Nature* 482, no. 7383 (2012): 27–29.

Lutts, Ralph H. "Chemical Fallout: Rachel Carson's *Silent Spring*, Radioactive Fallout, and the Environmental Movement." *Environmental Review* 9, no. 3 (1985): 210–25.

Lynch, Michael. *Art and Artifact in Laboratory Science: A Study of Shop Work and Shop Talk in a Research Laboratory*. London: Routledge and Kegan Paul, 1985.

Lynch, Michael. "Extending Wittgenstein: The Pivotal Move from Epistemology to the Sociology of Science." In *Science as Practice and Culture*, edited by Andrew Pickering, 215–65. Chicago: University of Chicago Press, 1992.

Lynch, Michael. "From Normative to Descriptive and Back: Science and Technology Studies and the Practice Turn." In *Science after the Practice Turn in the Philosophy, History, and Social Studies of Science*, edited by Léna Soler, Sjoerd Zwart, Michael Lynch, and Vincent Israel-Jost, 93–139. London: Routledge, 2014.

Lynch, Michael. "From the 'Will to Theory' to the Discursive Collage: A Reply to Bloor's 'Left and Right Wittgensteinians.'" In *Science as Practice and Culture*, edited by Andrew Pickering, 283–300. Chicago: University of Chicago Press, 1992.

Lynch, Michael. "Going Full Circle in the Sociology of Knowledge: Comment on Lynch and Fuhrman." *Science, Technology, & Human Values* 17 (1992): 228–32.

Lynch, Michael. *Scientific Practice and Ordinary Action: Ethnomethodology and Social Studies of Science*. Cambridge: Cambridge University Press, 1993.

Lynch, Michael, and David Bogen. *The Spectacle of History: Speech, Text, and Memory at the Iran-Contra Hearings*. Durham, NC: Duke University Press, 1996.

Lynch, William T. "After the Gold Rush: Cleaning up after Steve Fuller's Theosis." *Philosophy of the Social Sciences* 48, no. 5 (2018): 505–23.

Lynch, William T. "Arguments for a Non-Whiggish Hindsight: Counterfactuals and the Sociology of Knowledge." *Social Epistemology* 3, no. 4 (1989): 361–65.

Lynch, William T. "Between Kin Selection and Cultural Relativism: Cultural Evolution and the Origin of Inequality." *Perspectives on Science* 24, no. 2 (2019): 278–315.

Lynch, William T. "The Challenge to Consensus: The Relevance of the Lakatos-Feyerabend Debate for Contemporary Science and Technology Studies." *Epistemology & Philosophy of Science* 55, no. 3 (2018): 57–61.

Lynch, William T. "Complexity, Natural Selection, and Cultural Evolution." *Social Epistemology Review and Reply Collective* 5, no. 3 (2016): 64–72. http://wp.me/p1Bfg0-2Ln.

Lynch, William T. "Cultural Evolution and Social Epistemology: A Darwinian Alternative to Steve Fuller's Theodicy of Science." *Social Epistemology* 31, no. 2 (2017): 224–34.

Lynch, William T. "Darwinian Social Epistemology: Science and Religion as Evolutionary Byproducts Subject to Cultural Evolution." *Social Epistemology Review and Reply Collective* 5, no. 2 (2016): 26–68. http://wp.me/p1Bfg0-2Ci.

Lynch, William T. "Ideology and the Sociology of Scientific Knowledge." *Social Studies of Science* 24, no. 2 (1994): 197–227.

Lynch, William T. "Imre Lakatos and the Inexhaustible Atom: The Hidden Marxist Roots of History and Philosophy of Science." *Epistemology & Philosophy of Science* 55, no. 3 (2018): 25–34.

Lynch, William T. "King's Evidence." *New Scientist*, August 20, 2005, 39–40.

Lynch, William T. "Social Epistemology Transformed: Steve Fuller's Account of Knowledge as a Divine Spark for Human Domination." *Symposium: Theoretical and Applied Inquiries in Philosophy and Social Sciences* 3, no. 2 (2016): 191–205.

Lynch, William T. *Solomon's Child: Method in the Early Royal Society of London*. Stanford, CA: Stanford University Press, 2001.

Lynch, William T. "The Utility of the Present in Reconstructing Science's Past: Historical Counterfactuals and Contemporary Possibilities." *Scientia Poetica* 8 (2004): 241–50.

Lynch, William T. "What Does the Double Hermeneutic Explain/Justify?" *Social Epistemology* 7, no. 2 (1993): 193–204.

Lynch, William T., and Ellsworth Fuhrman. "Ethnomethodology as Technocratic Ideology: Policing Epistemic Boundaries." *Science, Technology, & Human Values* 17, no. 2 (1992): 234–36.

Lynch, William T., and Ellswsorth R. Fuhrman. "Recovering and Expanding the Normative: Marx and the New Sociology of Scientific Knowledge." *Science, Technology, & Human Values* 16, no. 2 (1991): 233–48.

Lynch, William T., and Ronald Kline. "Engineering Practice and Engineering Ethics." *Science, Technology, & Human Values* 25, no. 2 (2000): 195–225.

Lyotard, Jean-François. *The Differend: Phrases in Dispute*. Minneapolis: University of Minnesota Press, 1988.

Machrafi, Hatim. *Extended Non-Equilibrium Thermodynamics: From Principles to Applications in Nanosystems*. Milton, UK: CRC Press, 2019.

Mackenzie, Donald. *Inventing Accuracy: An Historical Sociology of Nuclear Missile Guidance*. Cambridge, MA: MIT Press, 1990.

Maddox, Brenda. *Rosalind Franklin: The Dark Lady of DNA*. New York: HarperCollins, 2002.

Mah, Harold. *The End of Philosophy, the Origin of "Ideology": Karl Marx and the Crisis of the Young Hegelians*. Berkeley: University of California Press, 1987.

Maio, Marcos Chor, and Ricardo Ventura Santos. "Antiracism and the Uses of Science in the Post–World War II: An Analysis of UNESCO's First Statements on Race (1950 and 1951)." *Vibrant: Virtual Brazilian Anthropology* 12, no. 2 (2015): 1–26.

Malinowski, Bronislaw. *Argonauts of the Western Pacific: An Account of Native Enterprise and Adventure in the Archipelagoes of Melanesian New Guinea*. London: Routledge, 1922.

Mann, Michael E., Raymond S. Bradley, and Malcolm K. Hughes. "Northern Hemisphere Temperatures during the Past Millennium: Inferences, Uncertainties, and Limitations." *Geophysical Research Letters* 26, no. 6 (1999): 759–62.

Margolis, Joseph. *Pragmatism without Foundations: Reconciling Realism and Relativism*. Southampton, England: Basil Blackwell, 1986.

Margulis, Lynn, and Dorion Sagan. *Acquiring Genomes: A Theory of the Origin of Species*. New York: Basic Books, 2002.
Markman, Ellen. "Natural Kinds." In *Naturalizing Epistemology*, 2nd ed., edited by Hilary Kornblith, 77–103. Cambridge, MA: MIT Press, 1994.
Marks, Jonathan. *What It Means to Be 98% Chimpanzee: Apes, People, and Their Genes*. Berkeley: University of California Press, 2002.
Marteel-Parrish, Anne E., and Martin A. Abraham. *Green Chemistry and Engineering: A Pathway to Sustainability*. Hoboken, NJ: Wiley, 2014.
Martin, Eric C. "'The Battle Is On': Lakatos, Feyerabend, and the Student Protests." *European Journal for Philosophy of Science* 9, no. 28 (2019): 1–33.
Martin, Keith. "Green Catalysts for Industry." In *Handbook of Green Chemistry and Technology*, edited by James H. Clark and Duncan J. Macquarrie, 321–77. Oxford: Blackwell Science, 2002.
Marx, Karl. *Capital*. Vol. 1. London: Electric Book Company, 2000.
Marx, Karl. "The Civil War in France." In *The Civil War in France: The Paris Commune*, 2nd ed., Karl Marx and V. I. Lenin, 9–87. New York: International Publishers, 1988.
Marx, Karl. "Economic and Philosophical Manuscripts of 1844." In *Collected Works*, vol. 3, 229–346. New York: International Publishers, 1975.
Marx, Karl. *German Ideology*. London: Electric Book Company, 2001.
Marx, Karl. "Theses on Feuerbach." In *The Marx-Engels Reader*, 2nd ed., edited by Robert C. Tucker, 143–45. New York: Norton, 1978.
Marx, Karl. *Wage-Labour and Capital*. New York: International Publishers, 1933.
Marx, Karl, and Friedrich Engels. *The Cologne Communist Trial*. New York: International Publishers, 1971.
Maynard Smith, J. "Group Selection and Kin Selection." *Nature* 201, no. 4924 (1964): 1145–47.
Maynard Smith, John, and Eörs Szathmáry. *The Major Transitions in Evolution*. Oxford: Oxford University Press, 1995.
McLellan, David. *The Young Hegelians and Karl Marx*. New York: F.A. Praeger, 1969.
McCormick, Sabrina. "Democratizing Science Movements: A New Framework for Mobilization and Contestation." *Social Studies of Science* 37, no. 4 (2007): 609–23.
McCormick, Sabrina. "From 'Politico-Scientists' to Democratizing Science Movements: The Changing Climate of Citizens and Science." *Organization and Environment* 22, no. 1 (2009): 34–51.
McCormick, Sabrina. *Mobilizing Science: Movements, Participation, and the Remaking of Knowledge*. Philadelphia: Temple University Press, 2009.
McGucken, William. *Scientists, Society, and State: The Social Relations of Science Movement in Great Britain, 1931–1947*. Columbus, OH: Ohio State University Press, 1984.
McNeil, Maureen, Michael Arribas-Ayllon, Joan Haran, Adrian Mackenzie, and Richard Tutton. "Conceptualizing Imaginaries of Science, Technology, and Society." In *The Handbook of Science and Technology Studies*, edited by Ulrike Felt,

Rayvon Fouché, Clark A. Miller, and Laurel Smith-Doerr, 435–63. Cambridge, MA: MIT Press, 2017.

Meheus, Joke, ed. *Inconsistency in Science*. Dordrecht: Springer, 2002.

Meiksins, Peter. "The 'Revolt of the Engineers' Reconsidered." *Technology and Culture* 29, no. 2 (1988): 219–46.

Meja, Volker, and Nico Stehr, eds. *Knowledge and Politics: The Sociology of Knowledge Dispute*. London: Routledge, 1990.

Mendelsohn, Everett. "Robert K. Merton: The Celebration and Defense of Science." *Science in Context* 3, no. 1 (2008): 269–89.

Merton, Robert K. *Mass Persuasion: The Social Psychology of a War Bond Drive*. New York: Harper and Brothers, 1946.

Merton, Robert K. "The Normative Structure of Science." In *The Sociology of Science: Theoretical and Empirical Investigations*, by Robert K. Merton, 267–78. Chicago: University of Chicago Press, 1973.

Merton, Robert K. "Science and the Social Order." In *The Sociology of Science: Theoretical and Empirical Investigations*, by Robert K. Merton, 254–66. Chicago: University of Chicago Press, 1973.

Merton, Robert King. *Science, Technology and Society in Seventeenth-Century England*. New York: H. Fertig, 1970.

Merton, Robert K. "Social Structure and Anomie." *American Sociological Review* 3, no. 5 (1938): 672–82.

Merton, Robert King. *Social Theory and Social Structure*. Glencoe, IL: Free Press, 1957.

Merton, Robert K. *Sociological Ambivalence, and Other Essays*. New York: Free Press, 1976.

Merton, Robert K. *The Sociology of Science: Theoretical and Empirical Investigations*. Chicago: University of Chicago Press, 1973.

Merton, Robert K. "Technical and Moral Dimensions of Policy Research." In *The Sociology of Science: Theoretical and Empirical Investigations*, by Robert K. Merton, 70–98. Chicago: University of Chicago Press, 1973.

Merton, Robert K. "The Unanticipated Consequences of Purposive Social Action." *American Sociological Review* 1, no. 6 (1936): 894–904.

Merton, Robert K., and Elinor Barber. *The Travels and Adventures of Serendipity: A Study in Sociological Semantics and the Sociology of Science*. Princeton, NJ: Princeton University Press, 2011.

Meyer, Stephen C. "Of Clues and Causes: A Methodological Interpretation of Origin of Life Studies." PhD dissertation. St. Catharine's College, University of Cambridge, 1990.

Meyer, Stephen C. *Signature in the Cell: DNA and the Evidence for Intelligent Design*. New York: HarperOne, 2009.

Michael, Mike, and Lynda Birke. "Enrolling the Core Set: The Case of the Animal Experimentation Controversy." *Social Studies of Science* 24 (1994): 81–95.

Michaelsson, Luba Freja, Bjarke Borregaard Medici, Jeppe Lerche la Cour, Christian Selmer, Michael Røder, Hans Perrild, Nils Knudsen, Jens Faber, and Birte Nygaard. "Treating Hypothyroidism with Thyroxine/Triiodothyronine Combination Therapy in Denmark: Following Guidelines or Following Trends?" *European Thyroid Journal* 4, no. 3 (2015): 174–80.

Miller, Kenneth R. "The Flagellum Unspun: The Collapse of 'Irreducible Complexity.'" In *Debating Design: From Darwin to DNA*, edited by William A. Dembski and Michael Ruse, 81–97. New York: Cambridge University Press, 2004.

Milliken, P. Jane, and Herbert C. Northcott. "Seeking Validation: Hypothyroidism and the Chronic Illness Trajectory." *Qualitative Health Research* 6, no. 2 (1996): 202–23.

Mills, C. Wright. *The Power Elite*. New York: Oxford University Press, 1959.

Milojević, Staša, Filippo Radicchi, and John P. Walsh. "Changing Demographics of Scientific Careers: The Rise of the Temporary Workforce." *Proceedings of the National Academy of Sciences of the United States of America* 115, no. 50 (2018): 12616–23.

Mins, Henry F. "Marx's Doctoral Dissertation." *Science and Society* 12, no. 1 (1948): 157–69.

Mirowski, Philip. *The Effortless Economy of Science?* Durham, NC: Duke University Press, 2004.

Mirowski, Philip. "The Future(s) of Open Science." *Social Studies of Science* 48, no. 2 (2018): 171–203.

Mirowski, Philip. *Science-Mart: Privatizing American Science*. Cambridge, MA: Harvard University Press, 2011.

Mirowski, Philip, and Dieter Plehwe, eds. *The Road from Mont Pèlerin: The Making of the Neoliberal Thought Collective*. Cambridge, MA: Harvard University Press, 2015.

Mitcham, Carl. "A Historico-Ethical Perspective on Engineering Education: From Use and Convenience to Policy Engagement." *Engineering Studies* 1, no. 1 (2009): 35–53.

Mitroff, Ian I. "Norms and Counter-Norms in a Select Group of the Apollo Moon Scientists: A Case Study of the Ambivalence of Scientists." *American Sociological Review* 39, no. 4 (1974): 579–95.

Moleski, Martin X. "Polanyi vs. Kuhn: Worldviews Apart." *Tradition and Discovery: The Polanyi Society Periodical* 33, no. 2 (2006): 8–24.

Molteni, Megan. "The Struggles of a $40 Million Nutrition Science Crusade." *Wired*, June 18, 2018.

Moore, David S. *The Developing Genome: An Introduction to Behavioral Epigenetics*. Oxford: Oxford University Press, 2015.

Moore, Kelly. *Disrupting Science: Social Movements, American Scientists, and the Politics of the Military, 1945–1975*. Princeton, NJ: Princeton University Press, 2008.

Morgan, Mary S. "Nature's Experiments and Natural Experiments in the Social Sciences." *Philosophy of the Social Sciences* 43, no. 3 (2013): 341–57.

Motterlini, Matteo. "Reconstructing Lakatos: A Reassessment of Lakatos' Epistemological Project in the Light of the Lakatos Archive." *Studies in History and Philosophy of Science* 33 (2002): 487–509.

Moynihan, Daniel P. *The Negro Family: The Case for National Action*. Washington, DC: Office of Policy Planning and Research, U.S. Department of Labor, 1965.

Muhlhausler, Beverly S., and Gérard P. Ailhaud. "Omega-6 Polyunsaturated Fatty Acids and the Early Origins of Obesity." *Current Opinion in Endocrinology, Diabetes, and Obesity* 20, no. 1 (2013): 56–61.

Mulkay, Michael. "Knowledge and Utility: Implications for the Sociology of Knowledge." *Social Studies of Science* 6 (1976): 445–70.

Mulkay, Michael J. "Norms and Ideology in Science." *Social Science Information* 15, no. 4–5 (1976): 637–56.

Murray, Patrick. *Marx's Theory of Scientific Knowledge*. Atlantic Highlands, NJ: Humanities Press International, 1988.

Musgrave, Alan, and Charles Pigden. "Imre Lakatos." In *The Stanford Encyclopedia of Philosophy*, edited by Edward N. Zalta. Winter 2016. https://plato.stanford.edu/archives/win2016/entries/lakatos/.

Nash, Linda. "Furthering the Environmental Turn." *Journal of American History* 100, no. 1 (2013): 131–35.

Nelson, Lynn Hankinson. *Who Knows: From Quine to a Feminist Empiricism*. Philadelphia: Temple University Press, 1990.

Neurath, Otto. *Empiricism and Sociology*. Dordrecht: D. Reidel, 1973.

Neurath, Otto. *Philosophical Papers, 1913–1946*. Dordrecht: D. Reidel, 1983.

Neurath, Otto. "Pseudorationalism of Falsification." In *Philosophical Papers, 1913–1946*, edited by Robert S. Cohen and Marie Neurath, 121–31. Dordrecht: D. Reidel, 1983.

Neurath, Paul. "Sixty Years since Marienthal." *Canadian Journal of Sociology/Cahiers canadiens de sociologie* 20, no. 1 (1995): 91–105.

Niaz, Mansoor. "The Oil Drop Experiment: A Rational Reconstruction of the Millikan–Ehrenhaft Controversy and Its Implications for Chemistry Textbooks." *Journal of Research in Science Teaching* 37, no. 5 (2000): 480–508.

Nickles, Thomas. "Justification and Experiment." In *The Uses of Experiment: Studies in the Natural Sciences*, edited by D. Gooding, T. Pinch, and S. Schaffer, 299–333. Cambridge: Cambridge University Press, 1989.

Nickles, Thomas. "Reconstructing Science: Discovery and Experiment." In *Theory and Experiment: Recent Insights and New Perspectives on Their Relation*, edited by Diderik Batens and Jean Paul Van Bendegem, 33–53. Dordrecht: Springer Netherlands, 1988.

Noble, David F. *America by Design: Science, Technology, and the Rise of Corporate Capitalism*. Oxford: Oxford University Press, 1979.

Nowak, Martin, and Benjamin Allen. "Inclusive Fitness Theorizing Invokes Phenomena That Are Not Relevant for the Evolution of Eusociality." *PLOS Biology* 13, no. 4 (2015): 1–5.

Nowak, Martin A., Corina E. Tarnita, and Edward O. Wilson. "The Evolution of Eusociality." *Nature* 466, no. 7310 (2010): 1057–62.

Nye, Mary Jo. *Blackett: Physics, War, and Politics in the Twentieth Century*. Cambridge, MA: Harvard University Press, 2004.

Nye, Mary Jo. *Michael Polanyi and His Generation: Origins of the Social Construction of Science*. Chicago: University of Chicago Press, 2011.

Obasogie, Osagie, Julie N. Harris-Wai, Katherine Darling, and Carolyn Keagy. "Race in the Life Sciences: An Empirical Assessment, 1950–2000." *Fordham Law Review* 83, no. 6 (2015): 3089–114.

Oberheim, Eric. *Feyerabend's Philosophy*. Berlin: Walter de Gruyter, 2006.

Oberheim, Eric. "On the Historical Origins of the Contemporary Notion of Incommensurability: Paul Feyerabend's Assault on Conceptual Conservativism." *Studies in History and Philosophy of Science Part A* 36, no. 2 (2005): 363–90.

Oddie, Graham. "Truthlikeness." In *The Stanford Encyclopedia of Philosophy*, edited by Edward N. Zalta. Winter 2016. https://plato.stanford.edu/archives/win2016/entries/truthlikeness/.

Odling-Smee, F. John, Kevin N. Laland, and Marcus W. Feldman. *Niche Construction: The Neglected Process in Evolution*. Princeton, NJ: Princeton University Press, 2003.

Ollman, Bertell. *Alienation: Marx's Conception of Man in Capitalist Society*. Cambridge: Cambridge University Press, 1976.

Olson, Robert E. "Discovery of the Lipoproteins, Their Role in Fat Transport and Their Significance as Risk Factors." *Journal of Nutrition* 128, no. 2 (1998): 439S–43S.

O'Reilly, Jessica, Naomi Oreskes, and Michael Oppenheimer. "The Rapid Disintegration of Projections: The West Antarctic Ice Sheet and the Intergovernmental Panel on Climate Change." *Social Studies of Science* 42, no. 5 (2012): 709–31.

Oreskes, Naomi. "The Scientific Consensus on Climate Change." *Science* 306, no. 5702 (2004): 1686–86.

Oreskes, Naomi, and Erik M. Conway. *Merchants of Doubt: How a Handful of Scientists Obstructed the Truth on Issues from Tobacco Smoke to Global Warming*. New York: Bloomsbury Press, 2010.

Owens, Larry. "Science in the United States." In *Companion Encyclopedia of Science in the Twentieth Century*, edited by John Krige and Dominique Pestre, 821–37. London: Routledge, 2003.

Ozaktas, Haldun M. "Teaching Science, Technology, and Society to Engineering Students: A Sixteen-Year Journey." *Science and Engineering Ethics* 19, no. 4 (2013): 1439–50.

Park, Justin H. "Persistent Misunderstandings of Inclusive Fitness and Kin Selection: Their Ubiquitous Appearance in Social Psychology Textbooks." *Evolutionary Psychology* 5, no. 4 (2007): 860–73.

Partridge, Myra. "Raise the Bar on FXTAS: Recognize It." *Psychiatric Times* 3 (April 1, 2007). https://www.psychiatrictimes.com/articles/raise-bar-fxtas-recognize-it.

Pennock, Robert T. "DNA by Design? Stephen Meyer and the Return of the God Hypothesis." In *Debating Design: From Darwin to DNA*, edited by William A. Dembski and Michael Ruse, 130–48. New York: Cambridge University Press, 2004.

Perrow, Charles. *Normal Accidents: Living with High Risk Technologies*. New York: Basic Books, 1984.

Perry, John. "The Problem of the Essential Indexical." In *Meaning and Truth: Essential Readings in Modern Semantics*, edited by Jay L. Garfield and Murray Kiteley, 613–28. New York: Paragon House, 1991.

Petryna, Adriana. *Life Exposed: Biological Citizens after Chernobyl*. Princeton, NJ: Princeton University Press, 2003.

Phillips, Nelson, and Stelios Zyglidopoulos. "Learning from Foundation: Asimov's Psychohistory and the Limits of Organization Theory." *Organization* 6, no. 4 (1999): 591–608.

Pickering, Andrew. *Constructing Quarks: A Sociological History of Particle Physics*. Edinburgh: Edinburgh University Press, 1984.

Pickering, Andrew. "From Science as Knowledge to Science as Practice." In *Science as Practice and Culture*, edited by Andrew Pickering, 1–26. Chicago: University of Chicago Press, 1992.

Pickering, Andy. "Living in the Material World: On Realism and Experimental Practice." In *The Uses of Experiment: Studies in the Natural Sciences*, edited by David Gooding, Trevor Pinch, and Simon Schaffer, 275–97. Cambridge: Cambridge University Press, 1989.

Pickering, Andrew. *The Mangle of Practice: Time, Agency, and Science*. Chicago: University of Chicago Press, 1995.

Pickering, Andrew, ed. *Science as Practice and Culture.* Chicago: University of Chicago Press, 1992.

Pickering, Andrew, and Adam Stephanides. "Constructing Quaternions: On the Analysis of Conceptual Practice." In *Science as Practice and Culture*, edited by Andrew Pickering, 139–67. Chicago: University of Chicago Press, 1992.

Pigliucci, Massimo, and Gerd B. Müller, eds. *Evolution: The Extended Synthesis*. Cambridge, MA: MIT Press, 2010.

Pinch, Trevor. "Relativism: Is It Worth the Candle?" In *Knowledge as Social Order: Rethinking the Sociology of Barry Barnes*, edited by Massimo Mazzotti. Burlington, VT: Ashgate, 2008.

Pinch, Trevor. "Toward an Analysis of Scientific Observation: The Externality and Evidential Significance of Observational Reports in Physics." *Social Studies of Science* 15, no. 1 (1985): 3–36.

Pinch, Trevor J., and Wiebe E. Bijker. "The Social Construction of Facts and Artifacts: Or How the Sociology of Science and the Sociology of Technology Might Benefit Each Other." In *The Social Construction of Technological Systems: New Directions in the Sociology and History of Technology*, edited by Wiebe E. Bijker, Thomas P. Hughes, and Trevor J. Pinch, 11–44. Cambridge, MA: MIT Press, 2012.

Pinker, Steven. "The False Allure of Group Selection." *Edge*, June 18, 2012. https://www.edge.org/conversation/the-false-allure-of-group-selection#dq.

Pinkus, Lynn B., Larry J. Shuman, Norman P. Hummon, and Harvey Wolfe. *Engineering Ethics: Balancing Cost, Schedule, and Risk—Lessons Learned from the Space Shuttle*. Cambridge: Cambridge University Press, 1997.

Polanyi, Michael. *Personal Knowledge*. Chicago: University of Chicago Press, 1957.

Polanyi, Michael. "The Republic of Science: Its Political and Economic Theory." *Minerva* 1, no. 1 (1962): 54–73.

Popper, Karl R. *Conjectures and Refutations: The Growth of Scientific Knowledge.* New York: Harper and Row, 1965.

Popper, Karl. *Logic of Scientific Discovery*. New York: Basic Books, 1959.

Popper, Karl R. *Objective Knowledge: An Evolutionary Approach*. Oxford: Clarendon Press, 1972.

Popper, Karl. "Three Worlds." In *The Tanner Lecture on Human Values*. University of Michigan, April 7, 1978. http://tannerlectures.utah.edu/_documents/a-to-z/p/popper80.pdf.

Popper, Karl. *Unended Quest: An Intellectual Autobiography*. La Salle, IL: Open Court, 1982.

Porter, Theodore M. *Trust in Numbers: The Pursuit of Objectivity in Science and Public Life*. Princeton, NJ: Princeton University Press, 1996.

Porus, Vladimir N. "What Do the Marxist 'Dialectics of Cognition' and Lakatos's 'Sophisticated Falsificationism' Have in Common?" *Epistemology & Philosophy of Science* 55, no. 3 (2018): 35–40.

Pratt, Mary Louise. "Linguistic Utopias." In *The Linguistics of Writing*, edited by Nigel Fabb, 48–66. New York: Methuen, 1987.

Preston, John. "Science as Supermarket: 'Post-Modern' Themes in Paul Feyerabend's Later Philosophy of Science." *Studies in History and Philosophy of Science Part A* 29, no. 3 (1998): 425–47.

Price, Derek J. de Solla. *Little Science, Big Science*. New York: Columbia University Press, 1963.

Price, Don K. *The Scientific Estate*. Cambridge: Belknap Press of Harvard University Press, 1965.

Price, George R. "Extension of Covariance Selection Mathematics." *Annals of Human Genetics* 35, no. 4 (1972): 485–90.

Price, Weston Andrew. *Nutrition and Physical Degeneration: A Comparison of Primitive and Modern Diets and Their Effects*. Redlands, CA: printed by the author, 1945.

Prigogine, I. *Introduction to Thermodynamics of Irreversible Processes*. New York: Interscience Publishers, 1961.

Prinz, Florian, Thomas Schlange, and Khusru Asadullah. "Believe It or Not: How Much Can We Rely on Published Data on Potential Drug Targets?" *Nature Reviews. Drug Discovery* 10, no. 9 (2011): 712.

Proctor, Robert N. *Cancer Wars: How Politics Shapes What We Know and Don't Know about Cancer*. New York: Basic Books, 1995.

Putnam, Hilary. "Explanation and Reference." In *The Philosophy of Science*, edited by Richard Boyd, Philip Gasper, and J. D. Trout, 171–85. Cambridge, MA: MIT Press, 1991.

Putnam, Hilary. "The Meaning of 'Meaning.'" *Minnesota Studies in the Philosophy of Science* 7 (1975): 131–93.

Putnam, Hilary. *Naturalism, Realism, and Normativity*. Cambridge, MA: Harvard University Press, 2016.

Putnam, Hilary. *Realism with a Human Face*. Cambridge, MA: Harvard University Press, 1990.

Quadrio, Philip A. "Hegel's Relational Organicism." *Critical Horizons: Journal of Social and Critical Theory* 13, no. 3 (2012): 317–36.

Queller, David C. "Kin Selection and Its Discontents." *Philosophy of Science* 83, no. 5 (2016): 861–72.

Radder, Hans. "How Concepts Both Structure the World and Abstract from It." *Review of Metaphysics* 55, no. 3 (2002): 581–613.

Radder, Hans. *In and About the World: Philosophical Studies of Science and Technology*. Albany: State University of New York Press, 1996.

Radder, Hans. "Mertonian Values, Scientific Norms, and the Commodification of Academic Research." In *The Commodification of Academic Research: Science and the Modern University*, edited by Hans Radder, 231–58. Pittsburgh, PA: University of Pittsburgh Press, 2010.

Radder, Hans. "Which Scientific Knowledge Is a Common Good?" *Social Epistemology* 31, no. 5 (2017): 431–50.

Rauchfuss, Horst. *Chemical Evolution and the Origin of Life.* Berlin: Springer, 2008.

Reisch, George A. *How the Cold War Transformed Philosophy of Science: To the Icy Slopes of Logic*. Cambridge: Cambridge University Press, 2005.

Reisch, George A. "Planning Science: Otto Neurath and the 'International Encyclopedia of Unified Science.'" *British Journal for the History of Science* 27, no. 93 (1994): 153–75.

Remmling, Gunter W. *Road to Suspicion: A Study of Modern Mentality and the Sociology of Knowledge.* New York: Appleton-Century-Crofts, 1967.

Rendell, Luke, Laurel Fogarty, and Keven N. Laland. "Runaway Cultural Niche Construction." *Philosophical Transactions: Biological Sciences* 366, no. 1566 (2011): 823–35.

Report of the United Nations Conference on Environment and Development. Rio de Janeiro: United Nations General Assembly, 1992.

Restivo, Sal. "4S, the FBI, and Anarchy." *Science, Technology, & Human Values* 26, no. 1 (2001): 87–90.

Rheinberger, Hans-Jörg. "Gaston Bachelard and the Notion of 'Phenomenotechnique.'" *Perspectives on Science* 13, no. 3 (2005): 313–28.

Richards, Evelleen, and John Schuster. "The Feminine Method as Myth and Accounting Resource: A Challenge to Gender Studies and Social Studies of Science." *Social Studies of Science* 19 (1989): 697–720.

Richerson, Peter J., and Robert Boyd. "Complex Societies: The Evolutionary Origins of a Crude Superorganism." *Human Nature* 10, no. 3 (1999): 253–89.

Richerson, Peter J., and Robert Boyd. "Evolution: The Darwinian Theory of Social Change." In *Paradigms of Social Change: Modernization, Development, Transformation, Evolution*, edited by Waltraud Schelkle, Wolf-Hagen Krauth, Martin Kohli, and Georg Elwert, 257–82. Frankfurt: Campus Verlag, 2000.

Richerson, Peter J., Robert T. Boyd, and Joseph Henrich. "Cultural Evolution of Human Cooperation." In *Genetic and Cultural Evolution of Cooperation*, edited by Peter Hammerstein. Cambridge, MA: MIT Press, 2003.

Richmond, Marsha L. "Women as Public Scientists in the Atomic Age: Rachel Carson, Charlotte Auerbach, and Genetics." *Historical Studies in the Natural Sciences* 47, no. 3 (2017): 349–88.

Rip, Arie. "Constructing Expertise: In a Third Wave of Science Studies?" *Social Studies of Science* 33, no. 3 (2003): 419–34.

Roberts, Jody A. "Creating Green Chemistry: Discursive Strategies of a Scientific Movement." PhD dissertation. Virginia Tech, 2005.

Roberts, Lissa. "Setting the Table: The Disciplinary Development of Eighteenth-Century Chemistry as Read through the Changing Structure of Its Tables." In *The Literary Structure of Scientific Argument: Historical Studies*, edited by Peter Dear, 99–132. Philadelphia: University of Pennsylvania Press, 1991.

Rockmore, Tom. "Is Marx a Materialist?" *Epistemology & Philosophy of Science* 55, no. 3 (2018): 62–75.

Roco, Mihail C., and William Sims Bainbridge, eds. *Nanotechnology: Societal Implications—Maximizing Benefit for Humanity*. Report of the National Nanotechnology Initiative Workshop, December 3–5, 2003. National Science Foundation, 2005.

Rogers, Everett M. *A History of Communication Study: A Biographical Approach*. New York: Free Press, 1994.

Rohde, Robert, et al. "A New Estimate of the Average Earth Surface Land Temperature Spanning 1753 to 2011." *Geoinformatics & Geostatistics: An Overview* 1, no.1 (2012).

Rorty, Richard. *Contingency, Irony, and Solidarity*. Cambridge: Cambridge University Press, 1989.

Rose, Hilary, and Steven Rose, eds. *The Political Economy of Science: Ideology of/in the Natural Sciences*. London: Macmillan, 1976.

Rosen, Zvi. *Bruno Bauer and Karl Marx: The Influence of Bruno Bauer on Marx's Thought*. The Hague: Martinus Nijhoff, 1977.

Rosenbaum, Michael, Ellen M. Murphy, Steven B. Heymsfield, Dwight E. Matthews, and Rudolph L. Leibel. "Low Dose Leptin Administration Reverses Effects of Sustained Weight-Reduction on Energy Expenditure and Circulating Concentrations of Thyroid Hormones." *Journal of Clinical Endocrinology and Metabolism* 87, no. 5 (2002): 2391–94.

Ross, Andrew, ed. *Science Wars*. Durham, NC: Duke University Press, 1996.

Ross, Benjamin, and Steven Amter. *The Polluters: The Making of Our Chemically Altered Environment*. Oxford: Oxford University Press, 2010.

Ross, I. L., J. Jones, and M. Blockman. "We Are Tired of 'Adrenal Fatigue.'" *South African Medical Journal* 108, no. 9 (2018): 724–25.

Rouse, Joseph. *Articulating the World: Conceptual Understanding and the Scientific Image*. Chicago: University of Chicago Press, 2015.

Rouse, Joseph. *Knowledge and Power: Toward a Political Philosophy of Science*. Ithaca, NY: Cornell University Press, 1987.

Rouse, Joseph. "Philosophy of Science and the Persistent Narratives of Modernity." *Studies in History and Philosophy of Science* 22 (1991): 141–62.

Rouse, Joseph. "The Politics of Postmodern Philosophy of Science." *Philosophy of Science* 58 (1991): 607–27.

The Royal Society and the Royal Academy of Engineering. *Nanoscience and Nanotechnologies: Opportunities and Uncertainties*. 2004. http://www.nanotec.org.uk/final Report.htm.

Sahlins, Marshall. *The Use and Abuse of Biology: An Anthropological Critique of Sociobiology*. Ann Arbor, MI: University of Michigan Press, 1976.

Sahlins, Marshall. *What Kinship Is and Is Not*. Chicago: University of Chicago Press, 2013.

Sanders, Lisa. *Every Patient Tells a Story: Medical Mysteries and the Art of Diagnosis*. New York: Broadway Books, 2009.

Sapolsky, Robert M. "Response." *Science* 275, no. 5306 (1997): 1662–63.

Sapolsky, Robert M. *Why Zebras Don't Get Ulcers: An Updated Guide to Stress, Stress-Related Diseases, and Coping*. New York: W. H. Freeman, 1998.

Sayer, Derek. *The Violence of Abstraction? The Analytic Foundations of Historical Materialism*. Oxford: Basil Blackwell, 1987.

Schaffter, Dorothy. *The National Science Foundation*. New York: Frederick A. Praeger, 1969.

Schatzberg, Eric. "Counterfactual History and the History of Technology." *Technology's Stories* 1, no. 3 (2014). https://www.technologystories.org/.

Schatzki, Theodore R. *Social Practices: A Wittgensteinian Approach to Human Activity and the Social*. Cambridge: Cambridge University Press, 1996.

Schatzki, Theodore R., Karin Knorr-Cetina, and Eike von Savigny, eds. *The Practice Turn in Contemporary Theory*. New York: Routledge, 2001.

Schegloff, Emanuel A. "Between Micro and Macro: Contexts and Other Connections." In *The Micro-Macro Link*, edited by Jeffrey C. Alexander, Bernhard Gieson, Richard Munch, and Neil J. Smelser, 207–34. Berkeley: University of California Press, 1987.

Schekman, Randy. "How Journals Like *Nature*, *Cell* and *Science* Are Damaging Science." *Guardian*, December 9, 2013.

Schön, Donald A. *The Reflective Practitioner: How Professionals Think in Action*. New York: Basic Books, 1983.

Schrödinger, Erwin. *What Is Life? With Mind and Matter and Autobiographical Sketches*. Cambridge: Cambridge University Press, 1944.

Schuster, John. *Descartes-Agonistes: Physico-Mathematics, Method and Corpuscular-Mechanism, 1618–33*. Dordrecht, Netherlands: Springer Verlag, 2013.

Schuster, John A. "Methodologies as Mythic Structures: A Preface to the Future Historiography of Method." *Metascience* 1/2 (1984): 15–36.

Schuster, John A. "The Scientific Revolution." In *Companion to the History of Modern Science*, edited by R. C. Olby, G. N. Cantor, J. R. R. Christie, and M. J. S. Hodge, 217–42. London: Routledge, 1990.

Schuster, John A., and Alan B. H. Taylor. "Blind Trust: The Gentlemanly Origins of Experimental Science." *Social Studies of Science* 27, no. 3 (1997): 503–36.

Schuster, John A., and Richard Yeo, eds. *The Politics and Rhetoric of Scientific Method: Historical Studies*. Dordrecht: Reidel, 1986.

Schuurbiers, Daan. "What Happens in the Lab Does Not Stay in the Lab: Applying Midstream Modulation to Enhance Critical Reflection in the Laboratory." *Science and Engineering Ethics* 17, no. 4 (2011): 769–88.

Schwartz, Michael W., Randy J. Seeley, Lori M. Zeltser, Adam Drewnowski, Eric Ravussin, Leanne M. Redman, and Rudolph L. Leibel. "Obesity Pathogenesis: An Endocrine Society Scientific Statement." *Endocrine Reviews* 38, no. 4 (2017): 267–96.

Scimecca, Joseph A. "Paying Homage to the Father: C. Wright Mills and Radical Sociology." *Sociological Quarterly* 17, no. 2 (1976): 180–96.

Sclove, Richard E. *Democracy and Technology*. New York: Guilford, 1995.

Sclove, Richard E. "From Alchemy to Atomic War: Frederick Soddy's 'Technology Assessment' of Atomic Energy, 1900–1915." *Science, Technology, & Human Values* 14, no. 2 (1989): 163–94.

Scott, Janny. "Postmodern Gravity Deconstructed, Slyly." *New York Times*, May 18, 1996.

Scott, Pam, Evelleen Richards, and Brian Martin. "Captives of Controversy: The Myth of the Neutral Social Researcher in Contemporary Scientific Controversies." *Science, Technology, & Human Values* 15, no. 4 (1990): 474–94.

Segerstråle, Ullica. *Defenders of the Truth: The Battle for Science in the Sociobiology Debate and Beyond*. Oxford: Oxford University Press, 2000.

Seid, Heather, and Michael Rosenbaum. "Low Carbohydrate and Low-Fat Diets: What We Don't Know and Why We Should Know It." *Nutrients* 11, no. 11 (2019): 2749.

Semendeferi, Ioanna. "Legitimating a Nuclear Critic: John Gofman, Radiation Safety, and Cancer Risks." *Historical Studies in the Natural Sciences* 38, no. 2 (2008): 259–301.

Shackley, Simon, and Brian Wynne. "Representing Uncertainty in Global Climate Change Science and Policy: Boundary-Ordering Devices and Authority." *Science, Technology, & Human Values* 21, no. 3 (1996): 275–302.

Shadish, William R., and Steve Fuller, eds. *The Social Psychology of Science*. New York: Guilford Press, 1994.

Shapin, Steven. "Discipline and Bounding: The History and Sociology of Science as Seen through the Externalism-Internalism Debate." *History of Science* 30, no. 4 (1992): 333–69.

Shapin, Steven. "History of Science and Its Sociological Reconstructions." *History of Science* 20, no. 3 (1982): 157–211.

Shapin, Steven. *Never Pure: Historical Studies of Science as If It Was Produced by People with Bodies, Situated in Time, Space, Culture, and Society, and Struggling for Credibility and Authority*. Baltimore, MD: Johns Hopkins University Press, 2010.

Shapin, Steven. *The Scientific Life: A Moral History of a Late Modern Vocation*. Chicago: University of Chicago Press, 2008.

Shapin, Steven. *A Social History of Truth: Civility and Science in Seventeenth-Century England*. Chicago: University of Chicago Press, 1994.

Shapin, Steven, and Simon Schaffer. *Leviathan and the Air-Pump: Hobbes, Boyle, and the Experimental Life*. Princeton, NJ: Princeton University Press, 1985.

Shapiro, Warren. "Contesting Marshall Sahlins on Kinship." *Oceania* 84, no. 1 (2014): 19–37.

Shepard, Paul. "The Corvidean Millennium; Or, Letter from an Old Crow." In *Traces of an Omnivore*, by Paul Shepard, 75–86. Washington, DC: Island Press, 1996.

Shipovalova, Lada V. "Contemporary Science Studies with or without Hidden Marxist Roots?" *Epistemology & Philosophy of Science* 55, no. 3 (2018): 47–51.

Shrader-Frechette, Kristin. "Technological Risk and Small Probabilities." *Journal of Business Ethics* 4, no. 6 (1985): 431–45.

Shrum, Wesley. "We Were the Unabomber." *Science, Technology, & Human Values* 26, no. 1 (2001): 90–101.

Silverman, M. P. "Raising Questions: Philosophical Significance of Controversy in Science." *Science and Education* 1 (1992): 163–79.

Singh, Rama S., Costas B. Krimbas, Diane B. Paul, and John Beatty, eds. *Thinking about Evolution: Historical, Philosophical, and Political Perspectives*. Vol. 2. Cambridge: Cambridge University Press, 2001.

Sismondo, Sergio. *Ghost-Managed Medicine*. Manchester, England: Mattering Press, 2018.

Sismondo, Sergio. "Pushing Knowledge in the Drug Industry: Ghost-Managed Science." In *The Pharmaceutical Studies Reader*, edited by Sergio Sismondo and Jeremy A. Greene, 150–64. Chichester, England: Wiley Blackwell, 2015.

Sober, Elliott, and David Sloan Wilson. *Unto Others: The Evolution and Psychology of Unselfish Behavior*. Cambridge, MA: Harvard University Press, 1998.

Soddy, Frederick. "Foreword." In *The Frustration of Science*, edited by Sir Daniel Hall et al., 7–9. New York: Norton, 1935.

Sokal, Alan D. "Transgressing the Boundaries: Toward a Transformative Hermeneutics of Quantum Gravity." *Social Text* 46/47 (1996): 217–52.

Sokal, Alan, and Jean Bricmont. *Fashionable Nonsense: Postmodern Intellectuals' Abuse of Science*. New York: Picador, 1998.

Soler, Léna, Sjoerd Zwart, Michael Lynch, and Vincent Israel-Jost, eds. *Science after the Practice Turn in the Philosophy, History, and Social Studies of Science*. New York: Routledge, 2014.

Sprat, Thomas. *History of the Royal Society*. St. Louis, MO: Washington University Studies, 1958.

Staudenherz, Anton, and Helmut Sinzinger. "Letter to the Editor." *Wiener klinische Wochenschrift* 118, no. 11 (2006): 365–67.

Stefansson, Vilhjalmur. *The Fat of the Land*. New York: Macmillan, 1956.

Stehr, Volker, and Nico Stehr, eds. *Knowledge and Politics: The Sociology of Knowledge Dispute*. London: Routledge, 1990.

Sterelny, Kim. *The Evolved Apprentice: How Evolution Made Humans Unique*. Cambridge, MA: MIT Press, 2012.

Sterne, Jonathan. "'What Do We Want?' 'Materiality!' 'When Do We Want It?' 'Now!'" In *Media Technologies: Essays on Communication, Materiality, and*

Society, edited by Tarleton Gillespie, Pablo J. Boczkowski, and Kirsten A. Foot, 119–28. Cambridge, MA: MIT Press, 2014.

Stich, Stephen P. "A Pragmatic Account of Cognitive Evaluation." In *Naturalizing Epistemology*, 2nd ed., edited by Hilary Kornblith, 393–426. Cambridge, MA: MIT Press, 1994.

Swierstra, Tsjalling, and Jaap Jelsma. "Responsibility without Moralism in Technoscientific Design Practice." *Science, Technology, & Human Values* 31, no. 3 (2006): 309–32.

Szathmáry, Eörs. "Toward Major Evolutionary Transitions Theory 2.0." *Proceedings of the National Academy of Sciences* 112, no. 33 (2015): 10104–11.

Taubes, Gary. "An Exercise in Vitriol Rather Than Sound Journalism." *Reason*, March 2003. http://reason.com/archives/2003/03/01/an-exercise-in-vitriol-rather.

Taubes, Gary. *Good Calories, Bad Calories: Challenging the Conventional Wisdom on Diet, Weight Control, and Disease*. New York: Knopf, 2008.

Taylor, Charles. *Hegel*. Cambridge: Cambridge University Press, 1975.

Taylor, Charles. *Hegel and Modern Society*. Cambridge: Cambridge University Press, 1979.

Teicholz, Nina. *The Big Fat Surprise: Why Butter, Meat, and Cheese Belong in a Healthy Diet*. New York: Simon & Schuster, 2014.

Teicholz, Nina. "Response to Critique of Review of the *Big Fat Surprise*." *Lancet* 393, no. 10186 (May 25, 2019): 2124.

Tetlock, Philip E., Richard Ned Lebow, and Geoffrey Parker, eds. *Unmaking the West: "What-If?" Scenarios That Rewrite World History*. Ann Arbor: University of Michigan Press, 2006.

Tetlock, Philip E., and Geoffrey Parker. "Counterfactual Thought Experiments: Why We Can't Live without Them and How We Must Learn to Live with Them." In *Unmaking the West: "What-If?" Scenarios That Rewrite World History*, edited by Philip E. Tetlock, Richard Ned Lebow, and Geoffrey Parker, 14–44. Ann Arbor: University of Michigan Press, 2006.

Thorpe, Charles. "Community and the Market in Michael Polanyi's Philosophy of Science." *Modern Intellectual History* 6, no. 1 (2009): 59–89.

Tomlinson, Gary. *Culture and the Course of Human Evolution*. Chicago: University of Chicago Press, 2016.

Tomlinson, Gary. *A Million Years of Music: The Emergence of Human Modernity*. New York: Zone Books, 2015.

Tönnies, Ferdinand. *Community and Society*. New York: Harper & Row, 1957.

Tosh, Nick. "Anachronism and Retrospective Explanation: In Defence of a Present-Centred History of Science." *Studies in History and Philosophy of Science* 34 (2003): 647–59.

Trost, Barry M. "The Atom Economy—a Search for Synthetic Efficiency." *Science* 254, no. 5037 (1991): 1471–77.

Trotsky, Leon. *The Permanent Revolution, and Results and Prospects*. New York: Pioneer Publishers, 1978.

Turner, Stephen. "Merton's 'Norms' in Political and Intellectual Context." *Journal of Classical Sociology* 7, no. 2 (2007): 161–78.

Turner, Stephen. “Polanyi Defanged.” *Social Studies of Science* 42, no. 6 (2012): 945–53.

Turner, Stephen P. “Practices as the New Fundamental Social Formation in the Knowledge Society.” *Družboslovne razprave* 24, no. 59 (2008): 49–64.

Turner, Stephen. “Robert Merton and Dorothy Emmet: Deflated Functionalism and Structuralism.” *Philosophy of the Social Sciences* 44, no. 6 (2014): 817–36.

Turner, Stephen. “The Social Study of Science before Kuhn.” In *The Handbook of Science and Technology Studies*, edited by Edward J. Hackett, Olga Amsterdamska, Michael E. Lynch, and Judy Wajcman, 33–62. Cambridge, MA: MIT Press, 2007.

Turner, Stephen. *The Social Theory of Practices: Tradition, Tacit Knowledge, and Presuppositions*. Chicago: University of Chicago Press, 1994.

Uebel, Thomas E. “Logical Empiricism and the Sociology of Knowledge: The Case of Neurath and Frank.” *Philosophy of Science* 67 (2000): S138–S50.

Uebel, Thomas E. *Overcoming Logical Positivism from Within: The Emergence of Neurath's Naturalism in the Vienna Circle's Protocol Sentences Debate*. Amsterdam: Rodopi, 1992.

United States Senate Committee on Finance. “Ghostwriting in Medical Literature: Minority Staff Report.” June 24, 2010.

Urbach, Peter. “Progress and Degeneration in the ‘IQ Debate.’” *British Journal for the Philosophy of Science* 25 (1974): 99–135, 235–59.

Valderrama, Andrés, Juan Camargo, Idelman Mejía, Antonio Mejía, Ernesto Lleras, and Antonio García. “Engineering Education and the Identities of Engineers in Colombia, 1887–1972.” *Technology and Culture* 50, no. 4 (2009): 811–38.

van Veelen, Matthijs. “On the Use of the Price Equation.” *Journal of Theoretical Biology* 237, no. 4 (2005): 412–26.

van Veelen, Matthijs, Julián García, Maurice W. Sabelis, and Martijn Egas. “Group Selection and Inclusive Fitness Are Not Equivalent: The Price Equation vs. Models and Statistics.” *Journal of Theoretical Biology* 299 (2012): 64–80.

Vaughan, Diane. *The Challenger Launch Decision: Risky Technology, Culture, and Deviance at NASA*. Chicago: University of Chicago Press, 1996.

Vinsel, Lee. “The Value of Counterfactual Analysis: Investigating Social and Technological Structure.” *Technology's Stories* 1, no. 3 (2014). https://www.technologystories.org/.

Vishmidt, Marina. “Permanent Reproductive Crisis: An Interview with Silvia Federici.” *Mute*, March 7, 2013. http://www.metamute.org/editorial/articles/permanent-reproductive-crisis-interview-silvia-federici.

Vogel, Sarah. “From ‘the Dose Makes the Poison’ to ‘the Timing Makes the Poison’: Conceptualizing Risk in the Synthetic Age.” *Environmental History* 13, no. 4 (2008): 667–73.

Vogel, Sarah A. *Is It Safe? BPA and the Struggle to Define the Safety of Chemicals*. Berkeley: University of California Press, 2013.

vom Lehn, Dirk, and Robert Dingwall. *Harold Garfinkel: The Creation and Development of Ethnomethodology*. Walnut Creek, CA: Routledge, 2016.

Voutchkova, Adelina M., Jakub Kostal, Justin B. Steinfeld, John W. Emerson, Bryan W. Brooks, Paul Anastas, and Julie B. Zimmerman. "Towards Rational Molecular Design: Derivation of Property Guidelines for Reduced Acute Aquatic Toxicity." *Green Chemistry* 13 (2011): 2373–79.

Walker, Brett L. "Animals and the Intimacy of History." In *The Oxford Handbook of Environmental History*, edited by Andrew C. Isenberg, 52–75. Oxford: Oxford University Press.

Walker, J. Samuel. *Permissible Dose: A History of Radiation Protection in the Twentieth Century*. Berkeley: University of California Press, 2000.

Walker, Kenny, and Lynda Walsh. "'No One yet Knows What the Ultimate Consequences May Be': How Rachel Carson Transformed Scientific Uncertainty into a Site for Public Participation in *Silent Spring*." *Journal of Business and Technical Communication* 26, no. 1 (2012): 3–34.

Wallace, Robert. "Hegel's God." *Philosophy Now: A Magazine of Ideas*, 2011. https://philosophynow.org/issues/86/Hegels_God.

Walton, Steven A., ed. *Instrumental in War: Science, Research, and Instruments between Knowledge and the World*. Leiden: Brill, 2005.

Wartenberg, Thomas E. *The Forms of Power: From Domination to Transformation*. Philadelphia: Temple University Press, 1990.

Watson, James D. *Avoid Boring People: Lessons from a Life in Science*. New York: Knopf, 2007.

Watson, James D. *The Double Helix: A Personal Account of the Discovery of the Structure of DNA*. New York: Simon & Schuster, 2001.

Wax, Paul M. "Elixirs, Diluents, and the Passage of the 1938 Federal Food, Drug and Cosmetic Act." *Annals of Internal Medicine* 122, no. 6 (1995): 456–61.

Weart, Spencer R. "Changing Sun, Changing Climate?" https://history.aip.org/climate/solar.htm; L_0352.

Weart, Spencer R. *The Discovery of Global Warming*. Cambridge, MA: Harvard University Press, 2002.

Weart, Spencer R. "Hockey Stick and Hiatus," S6. https://history.aip.org/history/climate/20ctrend.htm.

Weber, Bruce H., and David J. Depew, eds. *Evolution and Learning: The Baldwin Effect Reconsidered*. Cambridge, MA: MIT Press, 2003.

Weber, Max. *Economy and Society: An Outline of Interpretive Sociology*. 2 vols. Berkeley: University of California Press, 1978.

Weber, Max. "Science as a Vocation." In Max Weber, *From Max Weber: Essays in Sociology*, 129–56. New York: Oxford University Press, 1946.

Webster, Charles. *The Great Instauration: Science, Medicine and Reform, 1626–1660*. London: Duckworth, 1975.

Webster, C. "The Origins of the Royal Society." *History of Science* 6 (1976): 106–28.

Wegner, Daniel M. *The Illusion of Conscious Will*. Cambridge, MA: MIT Press, 2017.

Weinberg, Alvin M. "Impact of Large-Scale Science on the United States." *Science* 134, no. 3473 (1961): 161–64.

Weinstein, Deena. "Fraud in Science." *Social Science Quarterly* 59, no. 4 (1979): 639–52.

Werskey, Gary. *The Visible College: A Collective Biography of British Scientists and Socialists of the 1930s*. London: Free Association Books, 1988.

West, Stuart A., Claire El Mouden, and Andy Gardner. "Sixteen Common Misconceptions about the Evolution of Cooperation in Humans." *Evolution and Human Behavior* 32, no. 4 (2011): 231–62.

Wiggershaus, Rolf. *The Frankfurt School: Its History, Theories, and Political Significance*. Cambridge, MA: MIT Press, 1995.

Williams, George C. *Adaptation and Natural Selection: A Critique of Some Current Evolutionary Thought*. Princeton, NJ: Princeton University Press, 1966.

Williams, L. Pearce. "Normal Science, Scientific Revolutions and the History of Science." In *Criticism and the Growth of Knowledge*, edited by Imre Lakatos and Alan Musgrave, 49–50. Cambridge: Cambridge University Press, 1970.

Wilson, David Sloan. *Darwin's Cathedral: Evolution, Religion, and the Nature of Society*. Chicago: University of Chicago Press, 2002.

Wilson, David Sloan. *Does Altruism Exist? Culture, Genes, and the Welfare of Others*. New Haven, CT: Yale University Press, 2015.

Wilson, David Sloan. "Hunting, Sharing, and Multilevel Selection: The Tolerated-Theft Model Revisited." *Current Anthropology* 39, no. 1 (1998): 73–97.

Wilson, David Sloan. "Species of Thought: A Comment on Evolutionary Epistemology." *Biology and Philosophy* 5 (1990): 37–62.

Wilson, D. S. "A Theory of Group Selection." *Proceedings of the National Academy of Sciences of the United States of America* 72, no. 1 (1975): 143–46.

Wilson, David Sloan. "The Tide of Opinion on Group Selection Has Turned." *This View of Life*, May 26, 2015. https://evolution-institute.org/blog/the-tide-of-opinion-on-group-selection-has-turned/.

Wilson, David Sloan, Yasha Hartberg, Ian MacDonald, Jonathan A. Lanman, and Harvey Whitehouse. "The Nature of Religious Diversity: A Cultural Ecosystem Approach." *Religion, Brain and Behavior* 7, no. 2 (2017): 134–53.

Wilson, Edward O. *Sociobiology: The New Synthesis*. Cambridge, MA: Belknap Press of Harvard University, 1975.

Wilson, Edward O., and Martin A. Nowak. "Natural Selection Drives the Evolution of Ant Life Cycles." *Proceedings of the National Academy of Sciences of the United States of America* 111, no. 35 (2014): 12585–90.

Wilson, James L. *Adrenal Fatigue: The 21st Century Stress Syndrome*. Petaluma, CA: Smart Publications, 2001.

Winch, Peter. *The Idea of a Social Science and Its Relation to Philosophy*. London: Routledge, 1958.

Winner, Langdon. "Do Artifacts Have Politics?" *Daedalus* 109 (1980): 129–36.

Winner, Langdon. "Upon Opening the Black Box and Finding It Empty: Social Constructivism and the Philosophy of Technology." *Science, Technology, & Human Values* 18, no. 3 (1993): 362–78.

Wisnioski, Matthew H. "'Liberal Education Has Failed': Reading Like an Engineer in 1960s America." *Technology and Culture* 50, no. 4 (2009): 753–82.

Wittgenstein, Ludwig. *Philosophical Investigations*. Translated by G. E. M. Anscombe. Oxford: Basil Blackwell, 1958.

Wodak, Jo, and David Oldroyd. "'Vedic Creationism': A Further Twist to the Evolution Debate." Review of *Forbidden Archaeology: The Hidden History of the Human Race*, by Michael A. Cremo and Richard L. Thompson. *Social Studies of Science* 26, no. 1 (1996): 192–213.

Wolf, Michael P. "Reference and Incommensurability: What Rigid Designation Won't Get You." *Acta Analytica* 22 (2007): 207–22.

Wong, Dennis H. "The Oversocialized Conception of Man in Modern Sociology." *American Sociological Review* 26, no. 2 (1961): 183–93.

Wood, P. B. "Methodology and Apologetics: Thomas Sprat's History of the Royal Society." *British Journal for the History of Science* 13 (1980): 1–26.

Woodhouse, Edward J. "Change of State? The Greening of Chemistry." In *Synthetic Planet: Chemical Politics and the Hazards of Modern Life*, edited by Monica J. Caspar. New York: Routledge, 2003.

Woodhouse, E. J., and S. Breyman. "Green Chemistry as Social Movement?" *Science, Technology, & Human Values* 30, no. 2 (2005): 199–222.

Woolgar, Steve. "Interests and Explanation in the Social Study of Science." *Social Studies of Science* 11, no. 3 (1981): 365–94.

Woolgar, Steve. *Science: The Very Idea*. Chichester, England: Ellis Horwood, 1988.

Woolgar, Steve, and Dorothy Pawluch. "Ontological Gerrymandering: The Anatomy of Social Problems Explanations." *Social Problems* 32 (1985): 214–27.

Wrangham, Richard W. *Catching Fire: How Cooking Made Us Human*. New York: Basic Books, 2009.

Wylie, Alison. "The Engendering of Archaeology: Refiguring Feminist Science Studies." In *Science Studies Reader*, edited by Mario Biagioli, 553–68. New York: Routledge, 1999.

Wynne, Brian. "Lab Work Goes Social, and Vice Versa: Strategising Public Engagement Processes." *Science and Engineering Ethics* 17, no. 4 (2011): 791–800.

Wynne, Brian. "Sheep Farming after Chernobyl: A Case Study in Communicating Scientific Information." *Environmental Magazine* 31, no. 2 (1989): 33–39.

Wynne-Edwards, V. C. *Animal Dispersion in Relation to Social Behaviour*. Edinburgh: Oliver and Boyd, 1962.

Yager, Susan. *The Hundred Year Diet: America's Voracious Appetite for Losing Weight*. New York: Rodale, 2010.

Yaworsky, William, Mark Horowitz, and Kenneth Kickham. "Gender and Politics among Anthropologists in the Units of Selection Debate." *Biological Theory* 10, no. 2 (2015): 145–55.

Yehuda, Rachel. "Stress and Glucocorticoid." *Science* 275, no. 5306 (1997): 1662.

Yeo, Giles S. H., and Lora K. Heisler. "Unraveling the Brain Regulation of Appetite: Lessons from Genetics." *Nature Neuroscience* 15, no. 10 (2012): 1343–49.

Yudkin, John. *Pure, White, and Deadly: How Sugar Is Killing Us and What We Can Do to Stop It*. New York: Penguin Books, 2012.

Zahar, Elie. *Einstein's Revolution: A Study in Heuristic*. La Salle, IL: Open Court, 1989.

Zahar, Elie. "Why Did Einstein's Programme Supersede Lorentz's?" *British Journal for the Philosophy of Science* 24 (1973): 95–123, 223–62.

Zander, Joakim. *The Application of the Precautionary Principle in Practice: Comparative Dimensions*. Cambridge: Cambridge University Press, 2010.

Zandvoort, Henk, Tom Børsen, Michael Deneke, and Stephanie J. Bird. "Editors' Overview: Perspectives on Teaching Social Responsibility to Students in Science and Engineering." *Science and Engineering Ethics* 19, no. 4 (2013): 1413–38.

Ziman, John. *Real Science: What It Is, What It Means*. Cambridge: Cambridge University Press, 2000.

Žižek, Slavoj. *The Courage of Hopelessness: A Year of Acting Dangerously*. Brooklyn, NY: Melville House, 2018.

Žižek, Slavoj. *The Sublime Object of Ideology*. London: Verso, 1989.

Zulewski, H., Henryk Zulewski, Beat Müller, Pascale Exer, and André R. Miserez. "Estimation of Tissue Hypothyroidism by a New Clinical Score: Evaluation of Patients with Various Grades of Hypothyroidism and Controls." *Journal of Clinical Endocrinology and Metabolism* 82, no. 3 (1997): 771–76.

Index

www.ingramcontent.com/pod-product-compliance
Ingram Content Group UK Ltd.
Pitfield, Milton Keynes, MK11 3LW, UK
UKHW012035150926
475560UK00001B/14

9 781786 612373